Eliza Ann Youmans

Descriptive Botany

A Practical Guide to the Classification of Plants, with a Popular Flora

Eliza Ann Youmans

Descriptive Botany
A Practical Guide to the Classification of Plants, with a Popular Flora

ISBN/EAN: 9783337060190

Printed in Europe, USA, Canada, Australia, Japan

Cover: Foto ©berggeist007 / pixelio.de

More available books at **www.hansebooks.com**

Appletons' Science Text-Books.

DESCRIPTIVE BOTANY.

A PRACTICAL GUIDE
TO THE CLASSIFICATION OF PLANTS,
WITH A POPULAR FLORA.

BY

ELIZA A. YOUMANS,

AUTHOR OF "THE FIRST BOOK OF BOTANY"; EDITOR OF "HENSLOW'S
BOTANICAL CHARTS."

NEW YORK:
D. APPLETON AND COMPANY,
1, 3, AND 5 BOND STREET.
1889.

CONTENTS.

CHAPTER FOURTH.—COMPARING AND CLASSIFYING PLANTS.

CHAPTER FIFTH.—THE MINUTE STUDY OF THE ESSENTIAL ORGANS OF PLANTS.

CHAPTER SIXTH.—THE PISTIL.

CHAPTER SEVENTH.—FLORAL SYMMETRY, PHYLLO-
TAXY, PREFLORATION, CYMOSE INFLORESCENCE.

CHAPTER EIGHTH.—THE COMPOSITÆ.

CHAPTER NINTH.—THE CRUCIFERÆ.

CHAPTER TENTH.—THE UMBELLIFERÆ.

CHAPTER ELEVENTH.—THE LABIATÆ.

CHAPTER TWELFTH.—THE CONIFERÆ.

INTRODUCTION.

The "First Book of Botany," published in 1870, was prepared as a contribution to better methods in object-teaching. It was not designed as a text-book of Botany; but plants were chosen as objects of study, because they offer special and unequaled advantages for training in observation. It provided that the whole work of the learner should be upon his specimens; that he should find out and record the plant-characters for himself, and thus get important practice in self-education.

But it was soon seen that, in thus cultivating the observing powers, we were laying the true foundation for a real knowledge of Botanical Science; and the desire was often expressed that this method of studying plants should be carried out more fully. Accordingly, the "Second Book of Botany" was prepared upon the same plan. It has, however, been found desirable, for the sake of beginners in the science who are too old for primary lessons, that the abridged contents of the "First Book" should be prefixed to the "Second Book," and also that completeness as a Descriptive Botany should be given to the work, by adding to it a popular Flora. In thus combining the exercises of the former volumes, they have not been materially changed. They provide for the direct study of all those features of plants which are used in classification, and illustrate by practical examples the use to be made of these observations in systematic botany. The ideas given in those works, concerning the value of this study in mental training, are therefore equally applicable here.

By the common practice of the schools, pupils often
"go through" the botanical text-books with only the most
incidental attention to the real objects of study. As there
is no training in observation, there can be no attempt at
the exercise of the reason and judgment of the learner
upon the results of observation. To attain this important
end, botany must be studied in its actual objects. The
characters of plants must become familiarly known by the
detailed and repeated examination and accurate description of large numbers of plants. The pupil must proceed
step by step in this preliminary work—digesting his observations, and making the facts his own. From the beginning he will be engaged in comparing his observations,
and reasoning upon his facts. As he extends his knowledge, the work of comparison and grouping calls for a
higher exercise of thought. In the final classification of
plants, problems of increasing complexity arise. Plants
are to be placed in groups subordinate to each other,
when judged by masses of resemblances, by likenesses,
and differences of unequal values, which involve the exercise of the best powers of the mind.

That the habit of systematic arrangement, in which
the study of botanical classification affords so admirable
a training, is equally valuable in methodizing all the results of thought, is testified to as a result of his own experience by that eminent authority, Mr. John Stuart Mill.
He was a regular field botanist, and cultivated the subject
with a view to its important mental advantages. In the
second volume of his "System of Logic" Mr. Mill says:

"Although the scientific arrangements of organic nature afford as yet the only complete example of the true
principles of rational classification, whether as to the
formation of groups or of series, these principles are applicable to all cases in which mankind are called upon to
bring the various parts of any extensive subject into mental co-ordination. They are as much to the point when

objects are to be classed for purposes of art or business, as for those of science. The proper arrangement, for example, of a code of laws, depends on the same scientific conditions as the classifications in natural history ; nor could there be a better preparatory discipline for that important function than the study of the principles of a natural arrangement, not only in the abstract, but in their actual application to the class of phenomena for which they were first elaborated, and which are still the best school for learning their use.".

But it will be a grave mistake to suppose that these benefits can be secured by the mere use of text-books, however full and valuable the information they contain. Nor are they to be gained by the casual examination of plants, nor by the analyses of a few flowers, with the aid of keys and dictionaries, nor in the limited time usually allotted to the subject. The study must be commenced early, and pursued steadily by direct observation, until its elementary facts and principles are made familiar. It is the claim of this book that, if its method is faithfully followed, it will not only secure an actual acquaintance with an important branch of knowledge, but will enforce a mental discipline of much value in the intellectual work of life, and which is greatly needed in general education.

The exercises of the volume are designed simply as guides to self-education. The pupil is told very little. From the beginning to the end he is sent to the plant to get his knowledge of the plant. The science of botany is especially available for self-culture, because its elementary facts are so simple that their study can be commenced in early childhood, and so numerous as to sustain a prolonged course of observation. From rudimentary and simple facts the pupil may proceed gradually to the more complex; from observation to the truths resting upon observation, through a course of successively higher and more comprehensive exercises. Under the guidance here afforded, the pupil begins his study with leaves, the least complex in structure of the organs of plants, and learns to distinguish all their external characters. At the same time he learns the precise terms by which their parts and features are denoted, and these terms become familiar by use in his written descriptions.

If, in looking over the following pages, objection should be made to so many technical terms, the reply must be that without them it is impossible to gain the mental benefits of this method of study. The learning of words is a large part of education, but learned in the usual loose way they favor lax and careless habits of thought. To counteract this and give clearness to the mental operations requires a discipline adapted to the purpose. Vagueness in the meaning of words necessarily involves vagueness of thought and expression; while to have clear ideas and be able to clothe them in correct language, it is necessary to know precisely what the words represent. This end can only be secured in the best manner by the objective method, in which the mind is directed first to the observed facts, the specific characters, or the definite relations, so that the terms applied to them acquire fixed and accurate meanings. Careful and minute observations recorded in explicit terms make clearness of thinking and precision of language a habit of the mind. To secure this important object, descriptive botany is superior to any other study. Its terms have been slowly perfected, and are much the same in all languages. The vocabulary of botany is more copious, precise, and well-settled than that of any other natural science, and it is therefore unrivaled in the scope it affords for exercise in clear and accurate thinking, and for the best cultivation of the descriptive powers.

The method of instruction developed in these pages was devised and carried into most successful practice by the Rev. J. S. Henslow, Professor of Botany in Cambridge University, England. He had a parish at Hitchin, and resolved to try what might be done in teaching botany to the country children of the village school. His experiments were most interesting, and their results, which are of great value in education, were made public by Dr. J. D. Hooker, Superintendent of the Botanical Gardens at Kew, in evidence that he gave upon the subject before a parliamentary commission.

The following passages from his testimony will give an idea of Prof. Henslow's method:

Question. Have you ever turned your attention to the teaching of botany to boys in classes at school?

Answer. I have thought it might be done very easily. My ideas are drawn from the experience of my father-in-law, the late Prof. Henslow. He introduced the study of plants into the village school of his parish. His system was entirely voluntary. He enrolled the children in a class, and left them to collect plants for themselves; but he visited his parish daily, when the children used to come up and bring the plants they had collected, so that the lessons went on all the week round.

Q. Do you know in what way he taught it? Did he illustrate it?

A. Invariably; he made it practical. He made it an objective study. The children were taught to know the plants, and to pull them to pieces; to give their proper names to the parts; to indicate the relations of the parts to one another; and to find out the relation of one plant to another by the knowledge thus obtained. They learned it readily and voluntarily, and were extremely interested in it and fond of it.

Q. Do you happen to know whether Prof. Henslow thought that the study of botany developed the faculties of the mind—that it taught these children to think? And do you know whether he perceived any improvement in their mental faculties from that?

A. Yes; he used to think it was the most important

agent that could be employed for cultivating their faculties of observation, and for strengthening their reasoning powers.

Q. And Prof. Henslow thought that their minds were more developed; that they were becoming more reasoning beings, from having this study superadded to the others?

A. Most decidedly. It was also the opinion of some of the inspectors of schools, who came to visit him, that such children were in general more intelligent than those of other parishes; and they attribute the difference to their observant and reasoning faculties being thus developed. . . .

Q. So that the intellectual success of this objective study was beyond question?

A. Beyond question. . . . In conducting the examinations of medical men for the army, which I have now conducted for several years, and those for the East India Company's service, which I have conducted for, I think, seven years, the questions which I am in the habit of putting, and which are *not* answered by the majority of the candidates, are what would have been answered by the children in Prof. Henslow's village school. I believe the chief reason to be, that these students' observing faculties, as children, had never been trained—such faculties having lain dormant with those who naturally possessed them in a high degree; and having never been developed, by training, in those who possessed them in a low degree.

It thus appears that Prof. Henslow left his pupils mostly to themselves, meeting them occasionally to consult with them, and advise them when in doubt or difficulty. But he did not rely alone upon the fascination of the subject to secure his purpose.

His profound knowledge of the science and his wisdom as a teacher enabled him to devise and skillfully arrange a series of questions, calling attention to all the points of scientific interest in the structure of flowers, and the answers to which would in each case disclose the important characters of the plant described. The pupils were supplied with copies of these questions—*schedules*, as he called them—and answers were found to them by examining living plants. When a plant had been described in writing by answering these questions, its schedule was pinned fast to it, and it was the examination of the collective work of a scholar, whether by the professor or by a more advanced fellow-learner, that took the place of formal recitation. Left in this way to be his own teacher, and do his own thinking, the method is seen to be chiefly one of self-education.

Prof. Henslow prepared no elementary book upon botany carrying out his method: the printed schedule he used applied only to the flower, the most complex part of the plant, and the attention of children was directed by it chiefly to those features upon which orders depend in classification. But, instead of confining the use of schedules to the study of the flower, I have employed them throughout the work. In the first three chapters, the pupil is provided with leaf, stem, inflorescence, and flower-schedules on which, guided by the questions, he writes down the results of his observations. All the organs of the plant, and all their important modifications, are studied in this way. The presence or absence of botanical features that determine their place and rank among plants is first noted ; and, when found, they are accurately and concisely described.

In Chapter IV the subject of classing plants according to their natural affinities is entered upon. From the beginning of his schedule-work the pupil has really been classing plants in a limited way and without being aware of it. But he is now led to discover that he has been all the while using the principle on which the natural method of classification is based, and that the mastery of Prof. Henslow's flower-schedule has made the grouping of plants by this method both intelligible and easy. When he has answered all its questions concerning any plant, he has possession of the facts upon which its true classification depends.

The next three chapters of the book are devoted to the observation of those minute but especially important characters of plants which require the constant use of magnifying-glasses in their study. Practice with the flower-schedule in describing newly-discovered plants, and in a more searching study of familiar ones, is still continued, and furnishes inexhaustible interest to the learner.

The remaining chapters of this volume are accordingly given to a critical study of six of the most natural orders of plants, specimens of which everywhere abound; and the principles of classification illustrated by these groups will prepare the pupil for a rational use of the Flora, and thereby enable him to dispense with the artificial key that usually accompanies a popular Flora.

I have said that by the common method of studying botany there is no training in observation. The text-book is read and recited in the customary class-room way; and there is only the most incidental attention to the living objects of study, and no attempt to exercise the pupil's own faculties in solving the questions they offer. Accordingly, when classification is attempted, an artificial key has to be resorted to, which takes the place of the actual knowledge which the learner should have. It is at this stage that the contrast in results of the two methods is most apparent. When, by following the key, a pupil seeks for the class, order, genus, etc., to which the plant in hand belongs, he does not use his own knowledge. The structure of the plant is to be compared with an ideal; but he has not the ideal, neither can he interpret structure. So he turns to the key and learns what to look for first. When he has found the part specified, he compares its appearance with the statement of the key. If this seems to agree with the structure under examination, he is directed what to look for next; and if there is no agreement, he is told what to do. The same process is repeated over and over again to the end, with very little mental benefit. The key is simply an elaborate substitution of blind groping for the intelligent action of the pupil's own faculties. The scholar undertakes that for which he has had no preparation and which is beyond his ability; and in most cases he is too worried and confused by this unintelligible process to be able, when he sees another plant of similar structure, to recognize it. The law that time is needed for the accumulation and orderly assimilation of observations and the acquirement of clear ideas has been neglected, and so all his after-work in descriptive botany is wasted. By the present method, however, while the pupil is studying the structure of plants, his reflective faculties are all the while taxed to decide concerning their relationships. And when all those plant-characters upon which science insists have become familiar, so that the eye at once seizes upon them, the exercise of judgment in determining the groups to which a plant belongs is spontaneous and inevitable.

The popular Flora contained in this work will serve as a thorough preparation for the use of complete manuals. It will acquaint the pupil with the leading orders and genera of plants, and with those representative species having the widest range, which are found everywhere, and will most help the learner in mastering the principles of classification. It has been prepared under the immediate supervision of Dr. Byron D. Halsted, Professor of Botany in the Agricultural College of Ames, Iowa, whose extended and thorough knowledge of the science is an assurance that the work is accurate and in accordance with the most advanced views of systematic botany.

While the portion of botany to which this volume is devoted can not be learned from books, there is another

part of this extensive science that may be more success-
fully pursued by ordinary school methods of instruction.
This is physiological botany. By means of diagrams and
the explanations of the text, the scholar is enabled to per-
ceive how and of what the parts of plants are built up,
and what functions these parts perform in its history as a
living being. A valuable manual on this branch of botany
by an eminent authority will shortly appear in this series,
which will complete the exposition of the science here
begun.

DIRECTIONS FOR STUDY.

THE first three chapters of this book were prepared for young children, and are, therefore, very simple and rudimentary. But the course of observations they contain are not to be dispensed with by beginners of any age. The constant temptation of older pupils will be toward haste and inadequate observation. The danger is that plants enough will not be collected, and that the parts of such as are collected will not be studied with sufficient care. The influence of the teacher will therefore be constantly needed to check the too rapid passage of older pupils over that portion of botany included in these chapters.

An excellent way to familiarize pupils with these plant-characters is for them at once to set about preserving and describing specimens of all the varieties they collect.

As good an arrangement as any for pressing plants consists of two stout boards, that will not warp or bend, between which the specimens are placed, with any convenient weight—as stones, or masses of iron, of not less than fifty or sixty pounds—laid on the top. Between the plants you put layers of drying-paper. Newspapers answer very well for this purpose. They should be made into packets of about a dozen thicknesses, stitched together. Lay the specimens smoothly between these packets, having fastened to each of them as full a description as your studies enable you to write. Put unsized paper between the parts of a specimen that overlap each other, to prevent molding and hasten drying. Be careful to dispose the plants so that they will not lie directly above each other; keep the top of the pile as level as possible, to equalize the pressure. The number of packets interposed will depend upon the juiciness of the plants, and must be left to your own judgment. When plants are first put in press, the papers should be changed once a day for three or four days, after which every other day will answer. When the drying packets are changed, they should not be left lying upon the floor, but should be dried upon a line stretched across the room, or in the open air.

At each change of the driers, any further knowledge that has been gained concerning each specimen should be written down, and preserved with it as before. In this way all its features will be observed, and the names denoting them recalled, and by the time they are dried for mounting, it will be possible, by the aid of the last schedule of the chapter, to write, upon the paper holding the specimen, an accurate scientific description of it. Let this be followed by the pressing of entire plants, after comparing their different organs with the examples shown in the book. The attention thus drawn to their characters will be kept alive in changing them and caring for them, and the attempt completely to describe them, when dried and mounted, will go far toward fixing in the mind ideas of the forms and structures of the various organs, and the terms needed in description.

FIG. A.—Collector's Portfolio.

For collecting plants, you will need a small trowel for digging roots, or a large, strong clasp-knife, that will serve both for digging and for cutting branches ; a strong portfolio, from sixteen to twenty inches long, and ten or twelve inches wide, tied with tape or a strong cord. It should be made of two stout sheets of pasteboard, separated at the back (Fig. A), and will be all the better if covered with enameled cloth, to protect it from moisture. This portfolio should contain a stock of thin, unsized paper, such as the poorest printing-paper, or grocer's tea-paper. It is often convenient to have a close tin box, for preserving specimens, to be examined at home while fresh. Such a box, or *vasculum*, is shown strapped upon the collector in Fig. B. It shuts close, and has two compartments: the large one, with a door in the side, nearly as long as the box; and a small one, two or three inches deep, with a door in the end, for receiving small, delicate specimens of any kind.

FIG. B.—A Collector at Work.

If the collector wishes to prepare an herbarium, his specimens must be gathered with great care, and pains must be taken to get average examples of each species. If possible, they should be gathered in dry weather. Herbs should be gathered when in flower and in fruit. They should be taken by the root, and, if it is not too large, this should be pressed, along with the rest, to show whether the plant is annual, biennial, or perennial. Thick roots, bulbs, tubers, and the like, should be thinned with a knife, or cut in slices, lengthwise. Buds and fruit should be obtained, as well as the expanded flower. All three may sometimes be found upon the same plant, but generally they will have to be obtained at different times, unless, indeed, you are able to find buds, flowers, and fruit, all at once, upon plants in different stages of development.

Small herbs may be preserved entire. If the radicle leaves are withered at flowering-time, get a younger specimen in which they are fresh. When herbs are too large for this, they may be cut in sections, or folded, or you must be content with branches and specimen-leaves taken from near the root. In the case of woody plants, one or more shoots should be taken, bearing leaves, flowers, and fruit. Both sterile and fertile flowers should be obtained from monœcious and diœcious plants.

The specimens, when freshly gathered, should be laid between the sheets of the portfolio, the more delicate ones being carefully placed between sheets of drying-paper, so that, on reaching home, they can be transferred to the press without being disturbed. The folds and doublings of leaves and petals of ordinary plants, occasioned by the wind, in the open field, are easily smoothed out when putting the plants in press.

MOUNTING OF SPECIMENS.—When the plants are dry, the next thing is to mount them. For this purpose you will need—1. Strong, heavy, white paper, larger than foolscap; sheets $17\frac{1}{2}$ inches in length by $11\frac{1}{2}$ inches in width is a size, on many accounts, desirable. 2. Corrosive sublimate, for poisoning plants, to keep off insects. 3. Glue, to fasten them upon the paper.

Dissolve about an ounce of sublimate in a quart of alcohol. It should be labeled, and kept with great care, as it is very poisonous. A simple way of applying the solution is to pour a little into a large, flat platter, so as to cover the bottom, and "immerse the whole specimen for a second therein." After poisoning, the specimens are to be laid between driers, and subjected to slight pressure for twenty-four hours, when they are ready to be fastened to the paper. The flowers and tender parts of coarse, tough plants are all that need poisoning.

The specimens are to be fastened to the paper with hot glue, about as thick as cream, laid on to the plants with a camel's-hair pencil. Strips of thin, gummed paper should then be fastened over the thicker parts, to prevent their coming loose in handling. Prepare your glue in an earthen or porcelain-lined vessel, as corrosive sublimate acts on all common metals, and the brush, passing from plant to glue again and again, will be likely to produce stains if there is a trace of metal in the solution.

The labeling and arranging of plants depend upon their classification. When you know the characters upon which classes are founded, have begun to consider the affinities of plants, and have studied a few natural orders, you may intelligently begin to arrange your plants in their proper order. But, before attempting this, you should be so familiar with the assemblages of characters that plants present, and with their relations to each other, that you at once see why a plant is placed *here* and not *there* in your collection. In the Flora you will find a full statement of the characters of each order, followed by those of its leading genera, and of such representative species as will aid in the full comprehension of the principles involved.

THE USE OF CHARTS.—Many of the features of plants are so minute that they are at first difficult to find, and much is gained by consulting beforehand enlarged and colored diagrams showing the botanical characters of the various organs of plants. "Henslow's Botanical Diagrams," published by the Science and Art Department of the English Educational Council, have a high reputation for their scientific accuracy, their completeness of illustration, their judicious selection of typical specimens, and their skillful arrangement for purposes of education. Wishing to furnish pupils with every advantage in this study, the author induced her publishers to incur the very considerable expense of publishing a revised and enlarged American edition of the English Charts. In place of the nine English sheets, this set consists of six large charts in which several American plants have been substituted for species that do not occur in this country, and illustrations of the classes of flowerless plants have been added for which Prof. Henslow did not find room.

In the plan of the charts, the plant is first represented of its natural size and colors ; then a magnified section of one of its flowers is given, showing the relations of the parts to each other. Separate magnified views of the different floral organs, exhibiting all the botanical characters that belong to the group of which it is a type, are also represented. The charts contain nearly five hundred figures colored to the life, and which represent twenty-four orders and more than forty species of plants, showing a great variety of forms and structures of leaf, stem, root, inflorescence, flower, fruit, and seed, with numerous incidental characters peculiar to limited groups. All these are so presented as to be readily compared and contrasted with each other.

The charts are not designed to supersede the study of plants, but only to facilitate it. Their office is the same as the illustrations of the book ; but they are more perfect, and bring the pupil a step nearer to the objects themselves.

Besides this special assistance in object-study, the charts will be of great value in illustrating the Flora. In fact, they are designed to present, fully and clearly, those groupings of characters upon which orders depend in classification ; while in several cases of large and diversified orders the characters of leading genera are also given by typical specimens. The charts will thus be found equally valuable to the beginner, the intermediate pupil, and the advanced student. A Key accompanies the charts, and they can be used with any botanical text-books, and during the season of plants they should be upon the walls of every school-room where botany is studied.

AN EXPLANATION OF THE ABBREVIATIONS USED IN THE BOTANICAL CHARTS.

Seven principal references are made with a Capital Letter, to be looked for below each Illustration ; and the subordinate parts are then noted by small letters. A reference within a ○ implies not magnified; ⊂ on the left indicates a Longitudinal Section, and ⌢ above, a Transverse.

L........Leaf.		Fl......Flower.	
— p......	petiole.	— f. r....	floral receptacle.
— l.......	limb.	— ph....	perianth.
— l. l. ...	leaflet.	— ph. l..	leaves of.
— s	stipule.	— ca....	calyx.
		— ca. s..	sepals.
		— co....	corolla.
I. fl.	Inflorescence (in flower).	— co. p..	petals.
I. fr.	Infructescence (in fruit).	— s.....	stamen.
– p.	peduncle.	— s. f...	filament.
– p. p.	pedicel.	— s. a...	anther.
– b.	bract.	— s. c...	connective.
– b. g.	glume.	— s. p ..	pollen.
– b. p.	pale.	— pi....	pistil.
– g. r.	general receptacle.	— pi. ca.	carpel.
		— o.....	ovary.
		— o. cl..	cell of.
Æ.......Æstivation (diagram).		— o. d..	dissepiment.
green....	sepals.	— o. pl..	placenta.
red......	petals.	— o. f...	funicular cord.
yellow. .	stamens.	— sty...	style.
brown...	carpels.	— sti ...	stigma.
blue.....	ovules.	— oo....	ovule.
shaded...	adhesion of *whorls.*	— oo. rh.	rhaphe.

Fl. oo. ch. chalaze.
— oo. f . . foramen.
— n . . . . . nectary.

Fr . . . . Fruit.
— pe . . . . . . pericarp.
 ep. . . . ⎧ epicarp.
 me. . . ⎨ mesocarp.
 en. . . . ⎩ endocarp.
— ca . . . . . . carpel.
— pe. v . . . . . . valve.
— pe. cl . . . . . cell.
-- pe. d. . . . dissepiment.
— pe. p . . . . . . placenta.
— pe. f . . . . . . funicular cord.
— pe. f. a . . . . arillus.

S . . . . Seed.
— in . . . . . integument.
 ts. . . . ⎰ testa.
 tg . . . ⎱ tegmen.
— h . . . . . . hile.
— m . . . . . micropyle.
— rh . . . . . rhaphe.
— ch . . . . . chalaze.
— ar . . . . . arillode.
— al . . . . . albumen.

E . . . . Embryo.
— ca . . . . . caulicle.
— co . . . . . cotyledon.
— r . . . . . . radicle.
— pl . . . . . . plumule.

CHAPTER FIRST.

THE LEAF.

EXERCISE I.

The Parts of Leaves.

GATHER a variety of leaves, and begin their study by comparing them with the pictures and statements which follow.

A leaf, in its most highly-developed state, consists of three parts (Fig. 1) : The flattened portion is called the

 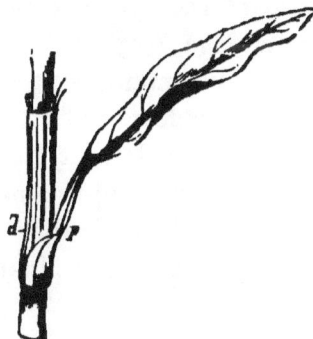

FIG. 1.   FIG. 2.

lamina, or *blade ;* a narrower portion, connecting the blade with the plant, is termed the *petiole,* or *leaf-stalk* (Figs. 2 and 3, *p*) ; and a third portion, at the base of the petiole, which is either in the form of a sheath (Fig. 2, *d*), or consists of two little leaf-like appendages, called *stip-ules*, shown in Fig. 3, *s s*, and still smaller in Fig. 1.

When the petiole is absent, the leaf is said to be *sessile ;* and if stipules are wanting, it is described as *exstipulate*. Fig. 1 represents a petiolate-stipulate leaf — that is, a fully-developed or *complete* leaf.

FIG. 3.

When, as in Fig. 4, the sheath-like leaf-stem, *g*, ends above, at the base of the blade, *l*, in a little membranous appendage, *lig.*, we call this body a *ligule*. It is a very common sort of stipule.

Gather leaves of all kinds, from the grass and herbs underfoot, from bushes, shrubs, and trees, and find and name the parts that compose each one of them. Say whether they are sessile or petiolate, and whether they are stipulate or exstipulate.

FIG. 4.

EXERCISE II.

Venation.

The lines, fine and coarse, that are seen running through the blades of leaves, are called *veins ;* and the various ways in which they are distributed are spoken of generally as the *venation* of the leaf.

When there is but one large central vein, reaching from the base to the apex of the blade, and giving off branches from its sides, it is called a midrib (Fig. 1).

When there are several large veins which thus cross the blade, as seen in Figs. 5 and 6, they are called simply *ribs*. Branches from the ribs are known in botanical description as *veins*, and the smallest of these lines which

FIG. 5.

FIG. 6.

branch off from the veins are known as *veinlets.* Point
out the ribs, veins, and veinlets in all the leaves you have
that exhibit them distinctly.

Now, when you hold a leaf between your eye and the
light, and observe these veinlets uniting with one another
in such a way as to form a kind of *irregular* net-work, you
have in hand a *reticulated* or *net-veined* leaf (Figs. 1 and 6).

If, on the contrary, the leaf you are examining has
veins more or less parallel to one another, or to the edge
of the leaf (Figs. 7 and 8), and if they are connected by
unbranched veinlets, they are termed *parallel-veined* leaves.

There is a further observation to be made
concerning the venation of net-veined leaves.
When, as in Fig. 9 or Fig. 1, the midrib gives

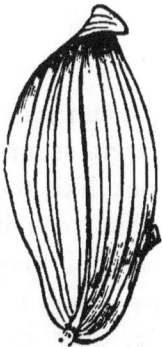

FIG. 7.

FIG. 8.

FIG. 9.

FIG. 10.

off veins right and left from its sides, it is said to be *feather-veined*, or *pinnately* veined. But when the petiole divides, at or near the base of the blade, into several diverging ribs (Fig. 5), the leaf is said to be *radiate*, or *palmate-veined*. If the ribs of a net-veined leaf converge toward the apex, as in Fig. 7, it forms that variety of venation known as *ribbed*.

FIG. 11.

Figs. 10 and 11 represent parallel-veined leaves, in which the veins take the direction seen in feather-veined and palmate-veined leaves. But in this case there is no net-work of veinlets, and so they are not net-veined. You will find many such leaves connected by unbranched veinlets. Remember that it is by the absence of this irregular network that you may know parallel-veined leaves. See Fig. 38.

Determine, in regard to all the leaves you can find, whether they are net-veined or parallel-veined; and whether the net-veined ones are feather-veined or palmate-veined.

EXERCISE III.

Leaf-Margins.

When the edge or margin of a leaf is smooth and even, it is said to be *entire* (Fig. 7). When the margin is uneven, with sharp teeth pointing toward the apex like the teeth of a saw, the leaf is said to be *serrate* (Fig. 12); if the teeth point toward the base, it is *retroserrate*; if they are themselves serrate, as shown in Fig. 13, *b*, they are said to be *biserrate*. When minutely serrate, they are termed *serrulate*.

When the teeth are sharp, without pointing in any particular direction, they form a *dentate* margin (Fig. 5) ; or,

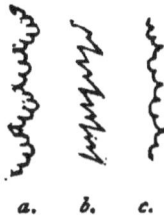

a. *b.* *c.*

FIG. 12.   FIG. 13.   FIG. 14.

when again similarly toothed, the margin is *bidentate* (Fig.
13, *c*).   When the teeth are rounded (Fig. 11), the margin
is *crenate ;* if twice rounded, as in Fig. 13, *a*, it is *bicrenate*.

FIG. 15.   FIG. 16.   FIG. 17.

Margins like the one shown in Fig. 14 are said to be
crisped, or *curled*.   When like Fig 15, they are said to be
wavy, or *undulated*.

When the incisions of a leaf-margin are much deeper than these, reaching half-way to the midrib or petiole, the divisions of the blade so formed are called lobes (Figs. 16, 17), and the spaces between the lobes are called *sinuses,* or *fissures.*

If the blade be divided nearly to the base or midrib (Fig. 18), the partings are termed *partitions,* and the leaf is *partite;* if it is divided quite to the base, or midrib, the

Fig. 20.

Fig. 18.   Fig. 19.

parts are called *segments,* and the leaf is said to be *dissected* (Fig. 19). When the basal lobes, partitions, or segments of a palmate leaf are themselves again divided, so that the whole resembles a bird's foot, the leaf is said to be *pedatifid, pedatipartite,* or *pedatisected,* according to the depth of the divisions. Fig. 20 represents a pedatipartite leaf.

In describing such incised leaves, they are said to be bifid, two-lobed ; trifid, three-lobed, etc. ; or bipartite, tripartite ; bisected, trisected, etc., according to the number of lobes, partitions, or segments. Another way of describing them depends upon the venation. If the leaf is feather-veined, it is said to be *pinnatifid, pinnatipartite,* or *pinnatisected,* etc. When palmate-veined leaves are deeply incised, we describe them similarly as *palmatifid, palma-tipartite, palmatisected.*

When the terminal lobe of a leaf is large and round, with smaller lateral lobes, the leaf is *lyrate* (Fig. 21). When the lateral lobes have their points directed toward the base of the blade, as in Fig. 22, the leaf is said to be *runcinate.*

APICES.—When the apex of a leaf-blade is rounded, as in Fig. 31, it is said to be *obtuse* or *blunt;* when obtuse, with a broad, shallow notch in the middle, it is *retuse.* If this notch is sharp, as in Fig. 23, it is *emarginate.*

When a blade ends abruptly, as if it had been cut

FIG. 23.

FIG. 21.   FIG. 22.

across, it is said to be *truncate;* if the truncated edge is ragged and irregular, as if it had been bitten off, the leaf is said to be *præmorse.* Fig. 24 shows an *acute*, or sharp-pointed apex, while Fig. 27 is *acuminate*, or taper-pointed.

When a blade ends with a rigid point, it is *cuspidate.*

If a blunt apex have a short point standing on it, it is
said to be *mucronate* (Fig. 29).

Before going on to study the figures of leaves, it is
very important that the pupil should be able to answer

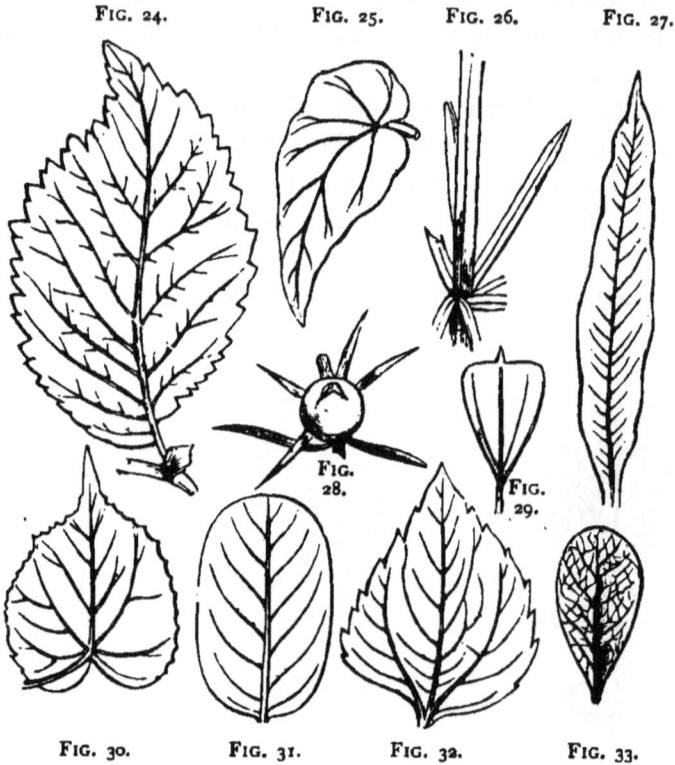

FIG. 24.   FIG. 25.   FIG. 26.   FIG. 27.

FIG.
28.

FIG.
29.

FIG. 30.   FIG. 31.   FIG. 32.   FIG. 33.

accurately the questions, What parts has a leaf? what is its
venation? what its margin? and what its apex? concerning any and every leaf. They should be answered ex-
plicitly in writing. A description of each leaf to this
extent should be written and pinned upon the specimen,
and the collection offered for criticism to the teacher, or,
what is better, to a fellow-pupil.

NOTE.—The schedule-forms for describing leaf, stem, and inflorescence are here omitted because they occupy too much space ; but they
may be found given in full in the " First Book of Botany."

EXERCISE IV.

The Figures of Leaves.

When the blades of feather-veined leaves are unequally developed on the two sides of the midrib, they are said to be *oblique* (Figs. 24, 25). When narrow and of nearly the same breadth at base and apex, with parallel margins, the leaf is *linear* (Fig. 26); and if ending in a sharp, rigid point, it is *acerose*, or *needle-shaped* (Fig. 28). When very narrow, and tapering from the base to a fine point, it is *subulate*, or *awl-shaped*. When broadest at the center, and three or more times as long as broad, tapering both ways, it is *lanceolate* (Fig. 27). When longer than broad, and slightly acute at base and apex, it is *oval*, or *elliptical* (Fig.

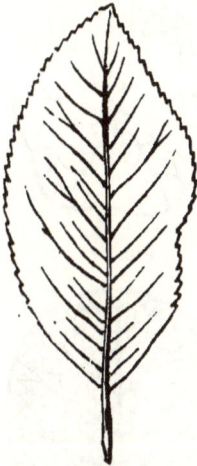

34). If obtuse at base and apex, as in Fig. 31, it is *oblong*. When a leaf is broader at the rounded base than at the apex, as in Fig. 32, it is *ovate*, or *egg-shaped*. If of the same figure, but broader at apex, it is *obovate* (Fig. 33). Fig.

Fig. 34.   Fig. 35.   Fig. 36.

29 shows a *cuneate* or wedge-shaped leaf. It is broad and abrupt-pointed at apex, and tapers toward the base. Fig. 35 shows a *spatulate* leaf, with its broad, rounded apex, and its sudden tapering at the base.

Cordate or *heart-shaped* leaves (Fig. 30) have an acute apex, with their broad, round base hollowed out into two lobes. When this form is reversed, as in Fig. 36, we have

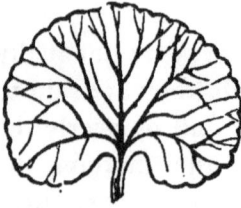

FIG. 37.

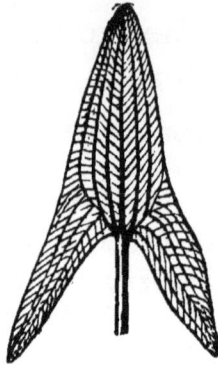

FIG. 38.

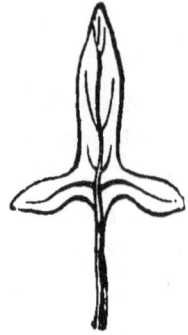

FIG. 39.

an *obcordate* or inversely heart-shaped leaf.   When a cordate base is joined with a rounded apex (Fig. 37), the leaf

FIG. 40.

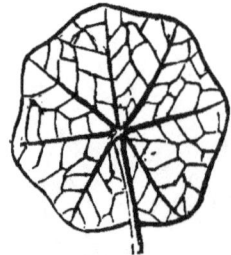

FIG. 41.

FIG. 42.

is *reniform* or *kidney-shaped*.   Fig. 38 shows a *sagittate*
or *arrow-shaped* leaf.   It has an acute apex, and long,
pointed basal lobes.

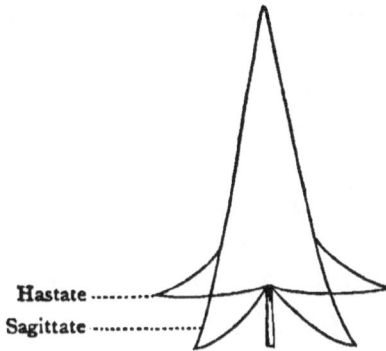

Hastate ....

Sagittate ....

FIG. 43 A.

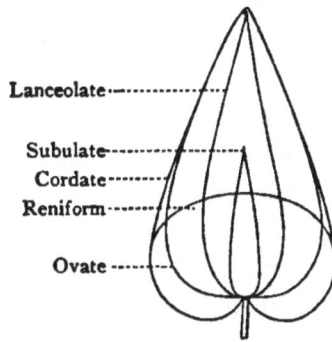

Lanceolate ....

Subulate ....

Cordate ....

Reniform ....

Ovate ....

FIG. 43 B.

The outline shown in Fig. 39 represents a *hastate* or *halberd-shaped* leaf, with its horizontally extending basal lobes. In Fig. 40 these lobes are seen separated from the blade. This is an *auriculate* or *hastate-auricled* leaf. The form shown in Fig. 41 is *orbicular*. It is also spoken of as a *peltate* leaf, because the petiole is inserted on the lower face of the blade, instead of at the base. Fig. 42 represents a *rounded* or *sub-rotund* leaf. A

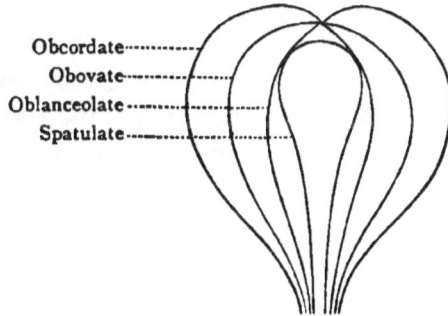

Obcordate ....

Obovate ....

Oblanceolate ....

Spatulate ....

FIG. 43 C.

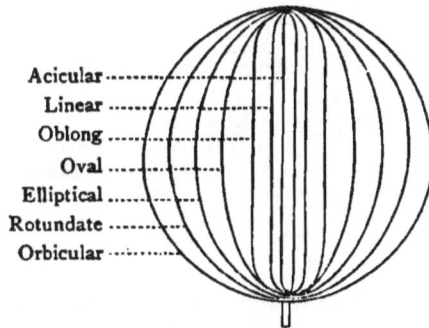

Acicular ....

Linear ....

Oblong ....

Oval ....

Elliptical ....

Rotundate ....

Orbicular ....

FIG. 43 D.

further help in determining the figure of leaves will be found by comparing them with the outlines shown in the above diagrams, Figs. 43 A, 43 B, 43 C, 43 D.

When none of the terms given correctly describe a leaf,

we can often easily and very nearly approach correctness by combining two of them, as *ovate-lanceolate, linear-lanceolate, cordate-ovate, roundish-ovate*, etc.

Facility and correctness of leaf-description depend upon practice. If the scholar will add the question, Figure? to the other questions of Ex. III, and answer them all faithfully in writing, according to his best judgment, concerning every leaf he finds, he will soon have command of this portion of descriptive botany.

EXERCISE V.

Compound Leaves.

The leaves you have been describing have only one blade, and are therefore called *simple* leaves; but there are hosts of leaves, resembling Fig. 44, which, you see,

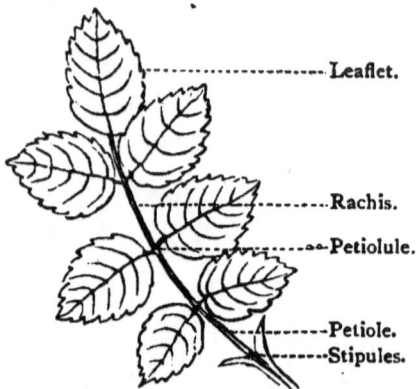

Leaflet.

Rachis.

Petiolule.

Petiole.
Stipules.

FIG. 44.

FIG. 45.

has several blades. A leaf with more than one blade is a *compound* leaf, and each of its blades is called a *leaflet*. Gather all the *compound* leaves you can find and compare them with Fig. 44. Point out and name their parts.

Some compound leaves have no rachis, but the leaflets all grow out from the top of the petiole (Fig. 45). These two kinds of compound leaves correspond to the two

kinds of venation of simple leaves you have been studying.
Pinnately-lobed leaves pass by slight gradations first into
pinnately-sected, and finally into pinnately - compound
leaves.   And, in the same way, palmately-compound leaves
are formed.   It requires a good deal of observation to decide correctly in all cases between deeply-divided leaves
and compound leaves.   If the green matter of the blade
reaches around the framework as far as the midrib, and
is continued along it, however slightly, the leaf is simple ;
or if in palmate-veined leaves the green matter is continued about the summit of the petiole, the leaf is *divided* and
not *compound.* *Leaflets* are often jointed to the rachis or
petiole.   Gather all the compound leaves you can find,
point out and name their parts, and say whether they are
pinnate or palmate.   If pinnate, say how many pairs of
leaflets they have.

The leaflets of compound leaves present the same differences of margin, apex, base, incision, and outline, as the
blades of simple leaves, and the same terms are used in
describing them.   A scientific description of a compound
leaf would require that the kind of leaf should be named,
and its leaflets described as if they were the blades of simple leaves.

EXERCISE VI.

Varieties of Compound Leaves.

Pinnately-compound leaves may have their leaflets in
one, two, three, or many pairs.   They may end with an odd
leaflet, as in Fig. 46, when they are said to be *unequally
pinnate*, or like Fig. 47, which is said to be *abruptly* or
equally pinnate.   If the rachis end in a tendril (Fig. 48) it
is said to be a *cirrous* leaf.   When they resemble Fig. 49,
they are said to be *interruptedly pinnate*, and *lyrately pinnate* when they resemble Fig. 50.   When the leaflets of a
pinnate leaf themselves become pinnate, as seen in Fig. 51,
the leaf is said to be *bipinnate*.   A further continuation of

FIG. 46.

FIG. 47.

FIG. 49.

FIG. 50.

FIG. 48.

FIG. 51.

the process gives the appearance seen in Fig. 52, which
is said to be *tripinnate*.

Palmately compound leaves are said to be *binate, two-
fingered,* or *bifoliate,* when two leaflets spring from a com-

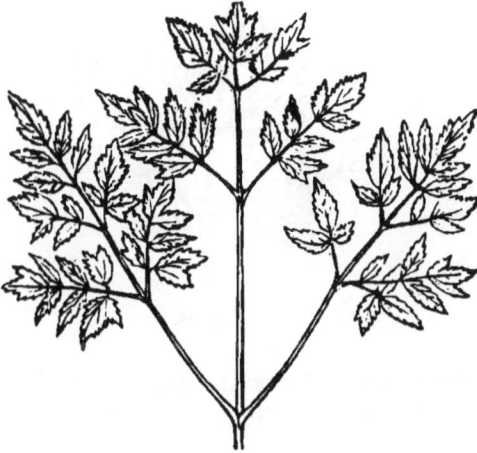

FIG. 53.

FIG. 52.

FIG. 54.

mon point (Fig. 53); *ternate* or *trifoliate* if they have three
leaflets similarly placed (Fig. 54); *quadrinate, four-fin-*

FIG. 55.

FIG. 56.

gered, or *quadrifoliate,* when like Fig. 55 ; *quinate,* or *five-
fingered* (Fig. 45) ; *septenate,* or *seven-fingered* (Fig. 56) ;

and *multifoliate* if there are more than seven leaflets (Fig.
57). When the leaflets of a compound leaf are arranged
in a pedate manner, they are described as pedate leaves.
When the leaflets of palmately-compound leaves become

FIG. 57.   FIG. 58.

themselves compound, the same prefixes are used as in the
case of pinnately-compound leaves. Fig. 58 is a *biternate*
leaf.

When stipules grow to the petiole, as shown in Fig. 44,
they are said to be *adnate ;* when like Fig. 46, they are
described as *thorny ;* when they are large and leaf-like, as
seen in Fig. 48, they are said to be *foliar* stipules. If they
grow around the stem, they are said to be *sheathing;* and
when thin and colorless, they are described as *membranous.*

Observe, also, whether the petiole is long or short, stiff
or limber, round, half-round, channeled, flattened, etc. Describe the color of the two surfaces of the leaves, and state
also whether the surface is smooth, shiny, hairy, woolly, silky,
or the like.

To describe a leaf with scientific precision requires
that you should answer the following questions : Is it simple or compound? petiolate or sessile? stipulate or exstipulate? What venation, margin, and figure has it? If
compound, name the variety. Give the features of both
petiole and stipules when they are present, and mention

also the color and surface aspect. Or these questions may take the form of a schedule, by placing them in a column at the left side of the paper, with space at the right for giving the answers to these questions in regard to any leaf you are describing. Thus:

SCHEDULE FIRST,
FOR SIMPLE LEAVES.

Parts?	
Venation?	
Margin?	
Base?	
Apex?	
Lobes?	
Shape?	
Petiole?	
Color?	
Surface?	

SCHEDULE SECOND,
FOR COMPOUND LEAVES.

Parts?	
No. Leaflets?	
Kind?	
Variety?	

Continue to make written descriptions of all kinds of leaves, until you are so familiar with their features, and the precise words needed to describe them, that you can make an accurate, prompt, and complete oral description of any specimen that comes to hand.

CHAPTER SECOND.

ROOTS AND STEMS.

EXERCISE VII.

Roots.

WHEN you are gathering plants, you will observe their roots. There are two classes of roots that are easily distinguished. Try to decide, in each case, in which one of these classes the root in hand should be placed. Figs. 59 and 60 are examples of these two different classes. In Fig.

FIG. 60.

FIG. 59.

60 a mass of fibers grows downward from the base of the stem. Roots which grow in this fashion are called *fibrous roots*. But when you find a root which seems like a

continuation of the stem, as in Fig. 59, it is a *tap-root*. Tap-roots are often branching, as in Fig. 59, but many

FIG. 63.

FIG. 61.

FIG. 62.

common plants have smooth tap-roots, as shown in Fig. 61, which is known as a *conical root*.

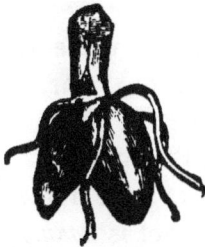

FIG. 64.

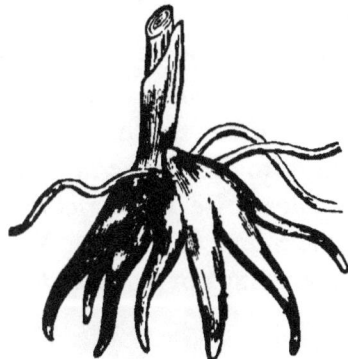

FIG. 65.

Some of the common varieties of tap-root are easily recognized. *Fusiform* or spindle-shaped roots, like Fig. 62, and *napiform* or turnip-shaped roots, like Fig. 63, are familiar to every one.

In the case of fibrous roots, we have seen (Fig. 60) that the stem divides at once at its base into a mass of slender branches or rootlets. These fibers often become enlarged, and, when the swellings take the form seen in

FIG. 66.

FIG. 67.

Fig. 64, the root is said to be *tuberculated*, and each enlargement is called a *tubercule*. Sometimes these tubercules resemble the human hand (see Fig. 65), when they are said to be *palmated tubercules*. When a number of tubercules arise from a common point, the root is said to be *fasciculated* (Fig. 66). When the fibers have numerous small swellings or nodules (Fig. 67), the root is *nodulose*.

EXERCISE VIII.

Stems and their Parts.

Pull up any herb which has a distinct stem and compare the stem with the root. *Herbs* are plants having stems
that die down to the surface of the ground every year. If
the root dies as well as the stem, the plant is called an
annual ; but if it lives and sends up a flowering stem the
second year, and then dies, it is a *biennial ;* while, if the
root lives on from year to year and only the stem dies, the
plant is *perennial.*

Observe the parts growing from the stem. What is at
the top? at the end of each branch? Do you find the same
structures at the tips of the roots? Name all the differences you can find between the stem and root. Compare an herb with Fig. 59, where the stem (*t*) is represented as giving off leaves (*ff*) in a regular manner.
Look at several branching stems, to find if the branches
are put forth regularly. Is there any regularity in the
growth of roots? Observe, in Fig. 59, that the angle
made by the leaf with the stem contains a bud, *b*. What do
you find in this angle in living plants? Botanists call this
angle a *leaf-axil*, and its bud an *axillary bud*. Buds at the
free end of stems and branches are called *terminal buds.*

The points on a stem at which leaves are given off are
called *nodes*, and the spaces between the nodes are *internodes.*

Point out the nodes, internodes, axillary buds, and terminal bud of the main stem (*primary stem*) of as many
plants as you can gather. Point out the same parts upon
the secondary stems or branches.

EXERCISE IX.

Buds.

The time to study winter buds is in early spring.
Choose a swollen bud and observe well its outer covering.

Is it membranous? waxy? gummy? lined with down, wool, or dense hairs? or is it varnished on the outside? Why should these parts of winter buds be called *protective coverings?* Are summer-formed buds *naked* or protected? Can you find the young leaves within these outer bud-scales? The way in which these tiny leaves are folded, rolled, and arranged in the bud is called *vernation.* To study vernation, look for buds that are just opening, where the young leaves still keep the shape they had when packed in the bud. If you have a magnifying-glass, you will find it useful now. The modes of folding and rolling are named as follows : When a leaf is folded so that the apex comes near the base, as shown in the diagram (Fig. 68), it is said to be *reclinate*, or *inflexed;* when it is folded at the midrib, and the margins of the right and left half come together (Fig. 69), the leaf is *conduplicate ;* when the leaf is plaited like a fan (Fig. 70), it is *plicate.* Or a leaf may be rolled from apex to base (Fig. 71), when it is said to be *circinate ;* or, from one margin to the other, in a single

FIG. 68.   FIG. 69.   FIG. 70.   FIG. 71.   FIG. 72.

coil, *convolute*, as Fig. 72 ; or, the two margins may both be rolled inward on the upper surface of the leaf, toward the midrib, *involute* (Fig. 74). When they are rolled similarly on the under surface (Fig. 73), the form is *revolute.*

Leaves are always arranged in the bud either in a *valvate* or *imbricate* manner. The best way to study their arrangement is to cut off the top of the bud with a sharp knife and look down on the cut edges, which will show, not only whether the leaves are imbricate or valvate, but other peculiarities they may exhibit.

The arrangement is *valvate* when the edges of adjacent leaves barely touch each other (Fig. 75).

It is *imbricate* when the edges of the leaves overlap each other (Figs. 76, 77).

When involute leaves are applied together in a circle, without overlapping (Fig. 78), they are said to be *condupli-*

FIG. 73.   FIG. 74.   FIG. 75.   FIG. 76.

FIG. 77.   FIG. 78.   FIG. 79.   FIG. 80.   FIG. 81.

cate. When conduplicate leaves overlap each other at the base (Figs. 79, 80), they are called *equitant.* When a convolute leaf incloses another which is rolled up in a like manner (Fig. 81), the arrangement is *supervolute.*

EXERCISE X.

Stem and Leaves.

The point at which, and the mode by which, a leaf is attached to the stem is called its insertion. The first grown leaves (Fig. 82) are called *cotyledons* (*c c*), and the next, *primordial* leaves (Fig. 82, *d d*). Leaves are called *radical* when they arise at or near the surface of the ground, and *cauline* when they grow from a stem with developed internodes. The small leaves upon flower-stalks are called *bracts.*

When a leaf is enlarged at its base and clasps the stem, it is *amplexicaul,* or *clasping* (Fig. 83). When it forms a complete sheath, as seen in Fig. 84, it is *sheathing.*

3

FIG. 82.

FIG. 83.

FIG. 84.

FIG. 85.

FIG. 86.

FIG. 87.

FIG. 88.

FIG. 89.

A *decurrent* leaf is formed when the blade is prolonged down the sides of the stem (Fig. 85). When the basal lobes of a leaf project beyond the stem and unite, as shown in Fig. 86, it is *perfoliate*. When opposite leaves unite by their bases, as in Fig. 87, they are called *connate*

FIG. 90. FIG. 91.

leaves. When only one leaf arises from a node, and the leaves grow alternately on different sides of the stem (Fig. 88), they are described as *alternate*. If there are two opposite leaves at each node, and the successive pairs are placed at right angles to each other, they are said to *decussate* (Fig. 89). If there are three or more leaves at a node (Fig. 90), they form a *whorl ;* and when all the leaves of a branch grow close together (Fig. 91), they are said to be tufted, or *fascicled*.

EXERCISE XI.

Kinds of Stems.

Stems that have a firm texture can sustain themselves in an upright position, but weak stems must either trail along the ground or attach themselves to other plants or objects for support. If they trail on the ground, they are said to be *prostrate* (Fig. 94). If they lift themselves by tendrils or other means, they are described as *climbing* (Fig. 92) ; and if they grow upward by twisting round other bodies, as shown in Fig. 93, they are said to be *twining*.

The stem of an herb is named a *caulis ;* that of a tree, a *trunk ;* that of grasses, a *culm*, and that of tree-ferns and palms, a *caudex.* Among irregular stems the most common are *runners,* like Fig. 94, which gives off from the main stem a prostrate branch, *a'*, that sends out

FIG. 93

FIG. 92.

FIG. 94.

leaves, *r,* and roots, *f,* so producing a new plant which extends itself in like manner. There is another prostrate stem which creeps along the ground, or partly beneath it, and produces buds from its upper surface and roots from its lower. This form of stem is called a *rhizome* (Fig. 95), where *b* shows the remains of the flowering stem of the present year, *b'* terminal bud, *c c* scars of former flowering stems, *r* roots. Another form of rhizome is

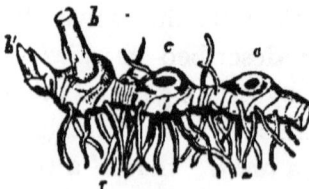

FIG. 95.

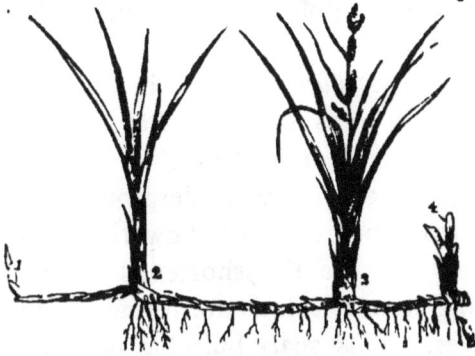

FIG. 96.

FIG. 98.

FIG. 97.

FIG. 99.

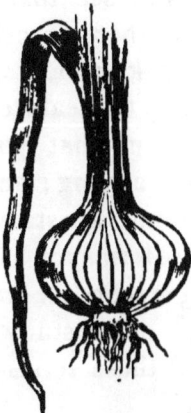

FIG. 100.

FIG. 101.

shown (Fig. 96), which grows wholly underground, and is spoken of as a *creeping root.*

ᐧ Of underground stems the *tuber* (Fig. 97) is a familiar example. The presence of buds, or "eyes," as they are vulgarly called, proves their stem-like nature.

The *bulb* (Figs. 98, 99) is a kind of underground bud which gives off roots from below and a flowering stem above. In both figures you see the shortened stem *a*, roots *b*, scales *c*, flowering stem *d*. Buds are formed in the axils of these scaly leaves. This scaly bulb has no covering, and is called a naked bulb, to distinguish it from the coated or *tunicated* bulb shown in section (Fig. 100), where the scales inclose one another in a concentric manner, and have an outer membranous covering. The *corm* is a solid bulb, which produces one or more buds in the form of young *corms* (Fig. 101, *a'''*).

In answering the questions of Schedule Third, say, as to the *kind of stem*, whether it is annual, biennial, or perennial ; whether it is erect, climbing. twining, or prostrate. If the latter, is it a runner or creeper ? or, if an underground stem, is it a rhizome, tuber, bulb, or corm ? Is the *leaf-insertion* radical or cauline ? Is the *leaf-arrangement* alternate, opposite, or whorled ?

By turning to the FLORA, you will see that all the species there described are chiefly known from each other by the features of leaf and stem that you have been studying ; and these differences are stated in the precise terms you have been using in schedule-work. Do not go on to study flowers till all these terms are familiar.

SCHEDULE THIRD, PERTAINING TO STEMS.

Kind of stem ?	
Leaf-insertion ?	
Leaf-arrangement.	
Vernation ?	
Roots ?	

CHAPTER THIRD.

THE INFLORESCENCE AND FLOWER.

EXERCISE XII.

Kinds of Inflorescence.

THE way flowers are placed upon plants is called their *inflorescence.* When only one flower grows upon a stem, the inflorescence is *solitary ;* but if several flowers grow from the same stem, it is *clustered.* The stem of a solitary flower or of a flower-cluster is called a *peduncle.* The top of the peduncle, from which several flowers start together, is called the *receptacle.* A rounded cluster of flowers, sessile upon the receptacle, is called a head (Fig. 102).

When, instead of a receptacle, the peduncle is prolonged, as shown in Figs. 103, 104, the portion that bears flowers is called the *rachis.*

FIG. 102

Fig. 103 represents a cluster of flowers that are sessile upon the rachis ; *p* is the peduncle ; *b b*, bracts ; *fl,* flowers. Any cluster of flowers sessile upon a *rachis* is described as a *spike.* But if the flowers grow upon short stems

FIG. 105.

FIG. 104.

FIG. 103.

FIG. 107.

FIG. 106.

FIG. 108.

FIG. 109.

of nearly equal length, as in Fig. 104, it is called a
raceme.

The flower-stems that grow from a rachis, or from the
top of a peduncle, are called *pedicels.*

A *spadix* is a spike with a thick rachis covered around
by a large leaf called a *spathe* (Fig. 105).

A spike with sessile bracts among its flowers is called
an *ament* or *catkin.* It grows on trees and shrubs, and
drops off when mature
(Fig. 106).

When you find clus-
ters of nearly sessile
flowers in the axils of opposite leaves, they form
a *glomerule.*

When from the top
of the peduncle there is
given off a number of
pedicels of nearly equal
length, arranged like the
ribs of an umbrella (Fig.
107), the cluster is an
umbel.

FIG. 110.

When you look only
at the top of a corymb (Fig. 108) it resembles an umbel,
but its pedicels are of greatly unequal length (compare
Figs. 108 and 107).

A compound umbel has a small umbel, called an *umbellet,* upon each pedicel (Fig. 109).

In the same way, each of the pedicels of a corymb may
bear a corymb, in which case we have a *compound corymb*
(Fig. 110).

A *compound raceme* is formed of secondary racemes in a
similar manner. When spreading, it is called a *panicle.*

EXERCISE XIII.

The Parts of Flowers.

We now enter upon the study of flowers. There are a great many different kinds of flowers to be examined and compared, even in one small neighborhood. Each one of the specimens you find must be carefully observed and described. An accurate description requires that you should study each part by itself, and note down concerning it all the important particulars you can discover. But before you can do this you must know what parts a flower consists of, and what particulars about these parts are important.

To learn the names of these parts, then, must be your first object. Compare real flowers with Fig. 111, which represents a flower pulled apart so that its main divisions may be fully seen. Begin with the outer leaves of a flower, and compare them with the lower circle of leaves in the diagram, and find the name of this circle. Do the same for the next circle, and so on, to the center of your specimen.

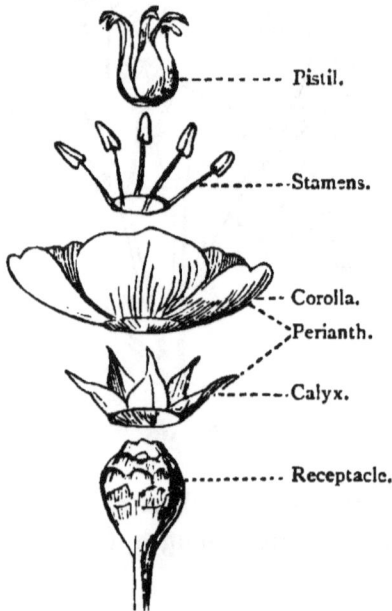

Pistil.

Stamens.

Corolla.
Perianth.

Calyx.

Receptacle.

Fig. 111.

Repeat this process with different flowers till you are able, at once, to point out and name the four divisions of common flowers.

The outer circle of green flower-leaves is named the *calyx.* The inner circle of delicately colored flower-leaves is named the *corolla.* When both circles have the same

color, they take the name of *perianth* (see Fig. 111). Next
inside the corolla come the *stamens*, and within these the
pistil. If there is but one circle of flower-leaves it is called

FIG. 112.

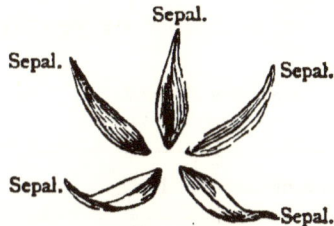

FIG. 113.

a *calyx*, whatever its color.   Point out and name the different circles that compose each of the flowers you have gathered.   But each of these flower-circles is made up of parts
of the greatest importance in the study of plants, and these
parts must be found and named.   Each leaf of a calyx is
called a *sepal* (Figs. 112, 113).   Each leaf of a corolla is
called a *petal*.   When these circles form a perianth, its parts
are called leaves.

Before finding any more parts to name, you may begin to note down and number the parts already found.

FIG. 114.

FIG. 115.

Prepare several flower-schedules like Schedule Fourth.
The first point to be looked for and written down concerning a calyx or a corolla is the number of sepals or

of petals that compose it, as you see has been done for the flowers represented in Figs. 114, 115. When you have done this, pin the schedule to the stem of the flower it describes, so that your observations can be seen and, if need be, corrected by your teacher or a fellow-learner.

SCHEDULE FOURTH,
DESCRIBING FIG. 114.

Names of Parts.	No.
Calyx ?	
Sepals.	5
Corolla ?	
Petals.	5

SCHEDULE FIFTH,
DESCRIBING FIG. 115.

Names of Parts.	No.
Perianth ?	
Leaves.	6

SCHEDULE SIXTH,
DESCRIBING FIG. 115.

Names of Parts.	No.
Perianth ?	
Leaves.	6
Stamens ?	6
Filament.	
Anther.	
Pistil ?	
Carpels.	3
Style.	
Stigma.	

EXERCISE XIV.

Stamens and Pistil.

Prepare new flower-schedules, long enough to make room for the names of the parts of the stamens and pistil of flowers, as shown in Schedule Sixth.

Begin the study of these parts with large, well-developed flowers. Meadow-lilies are good examples. Stamens differ very much in form and proportions in different species of plants, but usually they consist of three parts, shown in Fig. 116. Find the *filament* and *anther* in your living specimens. Observe whether any of the anthers are shedding their *pollen.* As soon as you know just what parts of stamens are meant by these words, write them down in the third place of your schedule under "stamens." Count the stamens (when there are less than twelve) in each of your flowers, and write the number opposite, as you see done in the book. If a flower have more than twelve stamens, make the symbol ∞, which means *many.*

As soon as you can point out and name the parts of stamens, begin the study of the pistil. Its parts are shown

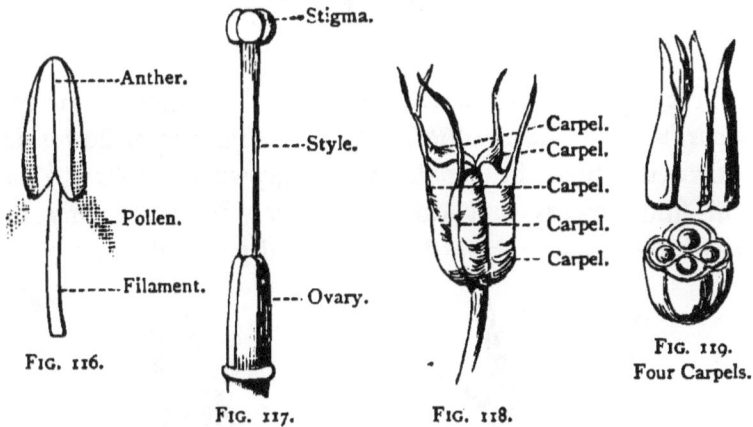

Anther.

Pollen.

Filament.

FIG. 116.

Stigma.

Style.

Ovary.

FIG. 117.

Carpel.
Carpel.
Carpel.
Carpel.
Carpel.

FIG. 118.

FIG. 119.
Four Carpels.

in Figs. 117, 118, 119. The *ovary* is at the bottom; the *stigma* at the top; and the *style* in the middle. If you cut across the swollen ovary, you will find it full of little round bodies that will grow and become seeds. Watch the ovary as flowers fade and disappear; look for old, faded flowers, and in them examine it. But you are not to write "ovary" in the schedule, for the ovary is made up of parts which you are to find, if you can, and count, and put down their number in the schedule. These parts are called *carpels* (Figs. 118, 119). So, after the word "pistil" put *carpels*, then *style*, and then *stigma*. The carpels shown in Figs. 118, 119 are from old ripe ovaries, and may be easily counted. You can count them in Fig. 117, by the grooves down the side, showing where they have grown together. Sometimes they are not grown together at all, and so can be easily counted. If the parts of the ovary are grown smoothly together, count the styles ; and if these, too, are united, count the lobes of the stigma. Or you may cut across the ovary, as shown in Fig. 119, and count the chambers or cells it contains, each of which is a carpel. Count the carpels in the ovary of a lily, and compare the result with the figure giving the number of carpels in Schedule Sixth.

EXERCISE XV.

Kinds of Calyx and Corolla.

Fig. 120 represents a calyx in which the sepals are all separate from each other, while in Fig. 121 a flower is shown in which the sepals are all grown together. You will find flowers that differ in this way, and many in which the sepals are partly joined and partly distinct.

FIG. 120.—Polysepalous Calyx.   FIG. 121.—Gamosepalous Calyx.

When the sepals of a calyx are distinct from each other, so that each one can be pulled off separately, the calyx is said to be *polysepalous*.

A *gamosepalous* calyx has its sepals grown together by

their edges, so that, if you pull one, the whole calyx comes off.

When the petals of a corolla are distinct from each other, so that one can be pulled off without disturbing the rest, it is a *polypetalous* corolla, as shown in Fig. 122, where *p p* are the distinct petals.

When the petals of a corolla are more or less grown together, so that if

FIG. 122.—Polypetalous Corolla.

FIG. 123. Gamopetalous Corolla.

you pull one the whole corolla comes off, it is a *gamopetalous* corolla (Fig. 123).

When the leaves of a perianth are entirely separate from each other, it is described as *polyphyl'lous ;* while, if they are grown together by their edges, however slightly, they are *gamophyl'lous.*

Take time carefully to compare the flower-envelopes of your specimens with these pictures and definitions. You can count the petals of gamopetalous corollas, when other ways fail, by observing their marks of union. Be cautious about calling a corolla *polypetalous* until you have examined several specimens of the same kind of flower. Above all things, *do not guess.* If you can not decide the point, consult with fellow-learners about it.

Another important feature of flowers is their regularity. A *regular calyx, corolla,* or *perianth,* has all its parts of the same size and shape (Figs. 114, 115).

An *irregular calyx, corolla,* or *perianth,* has some of its parts unlike the others in size or form (Figs. 129, 131). The same terms used to describe leaves are applied to the sepals and petals of flowers. Sepals are said to be erect when turned up ; *reflexed,* when turned down ; *connivent,* when turned inward ; and *divergent,* when they spread outwardly. Separate the regular flowers of your collection from those that are irregular. Describe the flower-leaves.

EXERCISE XVI.

Kinds of Corolla.

Gather as many different kinds of flowers as you can find before you begin with this exercise, that you may have living examples of many kinds of corolla. Of course, as your observation extends, you will, all the while, be finding new forms.

A petal is made up of parts, as shown in Figs. 124, 125. The *limb* is the thin, broad, upper part of a petal. The *claw* is the part that is joined to the receptacle. Sometimes it is stem-like. Look over the flowers you have gathered, and put by themselves polypetalous ones, and in another place the gamopetalous ones. Again examine the polypetalous division, and put the regular flowers together, leaving the irregular ones till these are looked over.

Now, there are three kinds of *regular polypetalous* corollas. The first is like Fig. 122. It has four petals growing in the shape of a cross, and so is called a *cruciform*

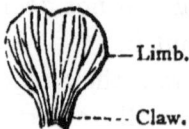

FIG. 124.

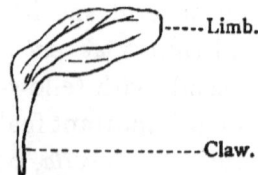

FIG. 125.

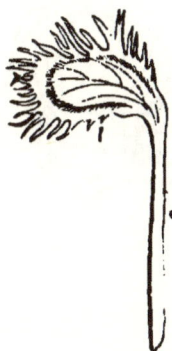

FIG. 126.   FIG. 127.

corolla. When a corolla has five petals, having each a long, slender claw, as shown in Fig. 126, *a*, and a spreading limb (*l*), it is *caryophylla'ceous* (Fig. 127).

A *rosa'ceous* corolla is shown, Fig. 128. Here there are five petals with spreading limb, but the claw is short. A *lilia'ceous* perianth has six leaves, bending away, as seen in Fig. 115.

Among *irregular polypetalous* corollas the most important is the *papiliona'ceous* (Fig. 129), where *c* is calyx; *a*, wings; *car*, keel. The large petal, called the banner (*b*), is the upper one next the stem; the two side ones (*a*) are called wings, and the lower one (*car*) the keel, from its boat-shape. Other forms of irregular polypetalous corollas are said to be *anomalous*. When you have decided to which of these kinds your polypetalous corollas belong, turn to the gamopetalous specimens and separate them, the regular from the irregular.

There are certain parts of a gamopetalous corolla that vary in size and form in different flowers, and that are shown in Figs. 130, 131.

The union of the petals forms the *tube* of a gamopetalous corolla. Any portion beyond this, where the petals are not united, is the *limb* or *border*. The opening into the tube is the *throat*.

FIG. 128.

FIG. 129.

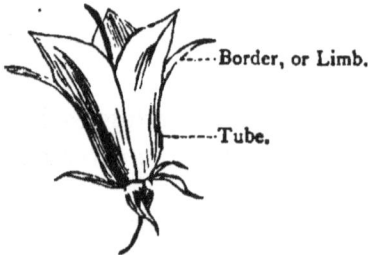

----Border, or Limb.

----Tube.

FIG. 130.

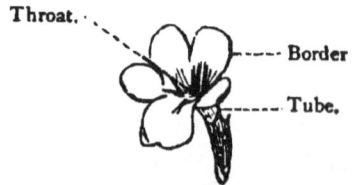

Throat.

----Border

----Tube.

FIG. 131.

FIG. 132.

FIG. 133.

FIG. 134.

FIG. 135.

FIG. 136.

The kinds of regular gamopetalous corollas are *tubular* (Fig. 132), in which the border spreads little or none ;

urceolate (Fig. 133) when the tube is swollen in the middle and has a narrow opening. *Rotate* (Fig. 134) represents a rotate or wheel-shaped corolla, with short tube and flat, spreading border. Fig. 135 represents a bell-shaped, or *campan'ulate* corolla. *Salver-form* corollas (Fig. 136) have a long, narrow tube, with the border at right angles to it. A *funnel-shaped* corolla resembles a funnel (Fig. 123).

In the second column of the flower-schedule you have given the number of sepals, petals, stamens, and carpels of a flower, and now another wide column must be added where further observations may be written, as shown in Schedule Seventh, where the calyx and corolla are described; but the stamens and pistil await more careful study.

SCHEDULE SEVENTH, DESCRIBING FIG. 122.

Names of Parts.	No.	Description.
Calyx ? Sepals.	4	*Polysepalous, regular.* *Oval.*
Corolla? Petals.	4	*Cruciform.* *Claw, long. Limb, spreading.*
Stamens ? Anther. Filament. Pollen.	6	
Pistil ?		

Irregular gamopetalous corollas are *labiate* when the limb divides so as to resemble a pair of lips. They are of two kinds : *personate*, with the throat closed ; and *rin-*

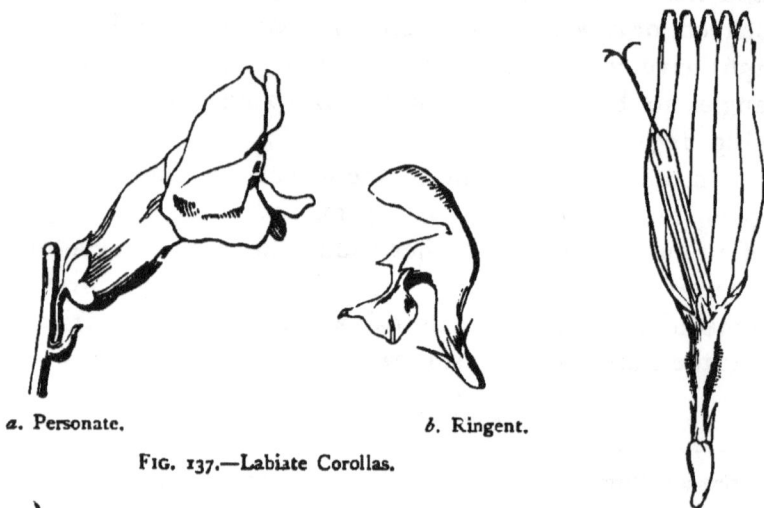

a. Personate.   *b.* Ringent.

FIG. 137.—Labiate Corollas.

FIG. 138.

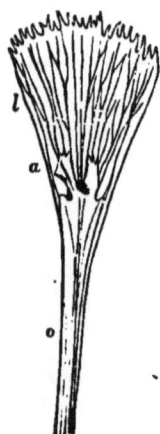

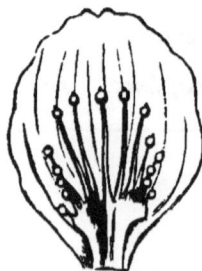

FIG. 140.

FIG. 139.   FIG. 141.

gent, with the throat open (Fig. 137, *a, b*). A *ligulate* or *strap-shaped* corolla is one that seems to be formed by the splitting of the tube on one side (Fig. 138).

Other forms of irregular gamopetalous corollas may be described as anomalous.

A strange aspect is often given to a corolla by a *crown* or *corona*, and by *spurs* and *scales*.

SCALES.—On the inner surface of the petals of many flowers, scales, and hair-like processes of various kinds, are often seen. Figs. 139, 140 are examples.

Sometimes these scales become more or less united, and form a cup-shaped part, as shown in Fig. 141. This

FIG. 142.

FIG. 143.

is called a *corona*, and the corolla is said to be crowned. A petal may grow outward, so as to form a bag or sac (Fig. 143); it is then termed *saccate*. Sometimes this growth is prolonged into a *spur*. All of the petals in Fig. 142 are seen to be spurred.

A *nectary* is a little gland containing sweet liquid, on the claw of a petal.

EXERCISE XVII.

Symmetry of Flowers.

Look carefully at the pictures and explanations of this exercise. Count the sepals in Fig. 144. Count the petals and stamens. Observe the two-lobed stigma of the pistil, which shows the number of carpels. Has each of the floral circles the same number of parts?. Then it is a symmetrical flower. Any flower that has the same number of parts in each of its circles is symmetrical; and even if some of the circles have just twice, or three or four times, as many as others, it is still symmetrical. Count the parts

FIG. 144.—Binary Symmetry.

FIG. 145.—Ternary Symmetry.

FIG. 146.—Quaternary Symmetry.

FIG. 147.—Quinary Symmetry.

in the floral circles of Fig. 145. Is this flower symmetrical? Does Fig. 146 represent a symmetrical flower? 147?

These kinds of symmetry are described as *binary, ternary, quaternary*, and *quinary*. Examine the flowers you have collected and discover, if you can, the symmetrical ones, naming the symmetry they show.

EXERCISE XVIII.

Complete and Incomplete Flowers.

A *complete* flower consists of calyx, corolla, stamens, and pistil. If any one or more of these flower-circles is absent, the flower is *incomplete*.

If you have the botanical charts, look at the magnified flowers represented on them, and point out the symmetrical ones. Find also examples of *complete* and *incomplete* flowers.

The stamens and pistil of flowers have been called *essential* organs, because seeds can not be formed without their presence. As the calyx and corolla cover and nourish these, they have taken the name of *protecting organs.*

When the protecting organs are both present in a flower, it is said to be *dichlamyd'eous.*

When there is only a calyx, it is *monochlamyd'eous.*

If both calyx and corolla are absent, it is *achlamyd'eous,* or *naked.* A *perfect* flower (Fig. 148) has both the *essential* organs ; while, if one of these be absent, it is *imperfect* (Figs. 149, 150) ; and, if both are wanting, it is said to be

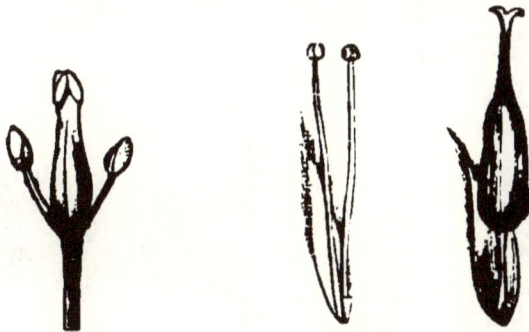

FIG. 148.
A Perfect Flower.

FIG. 149.   FIG. 150.
Imperfect Flowers.

neutral. A *staminate* flower has no pistil. A *pistillate* flower has no stamens. *Staminate* flowers (Fig. 149) are said to be *sterile,* because they do not produce seed. They are also spoken of as *male* flowers. Pistillate flowers are said to be fertile, because they may bear seed. They are also called *female* flowers (Fig. 150).

A perfect flower is indicated thus, ☿ .

A staminate, sterile, or male flower, thus, ♂ .

A pistillate, fertile, or female flower, thus, ♀ .

When both staminate and pistillate flowers grow upon the same plant (Fig. 151), it is said to be *monœcious.*

When staminate and pistillate flowers grow upon separate plants (Figs. 153 and 154), such plants are said to be

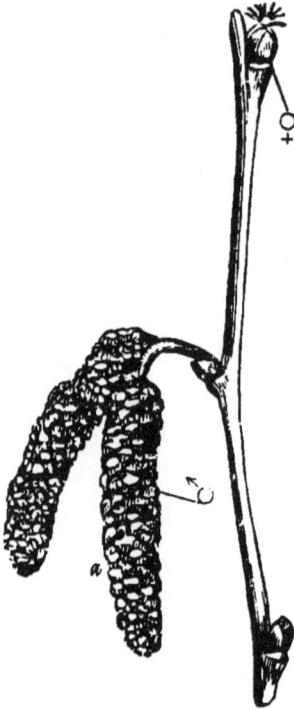

FIG. 152.—Pistillate Flower, from Catkin (Fig. 153).

FIG. 151.—A Monœcious Plant.

FIG. 153.—Female Catkin of a Diœcious Plant.

diœcious. Fig. 152 represents a pistillate flower from the female catkin (Fig. 153). Fig. 155 represents a staminate flower from the male catkin (Fig. 154). These catkins grow upon different trees; so the willow from which they were taken is *diœcious.*

When staminate, pistillate, and perfect flowers are all found upon the same plant, it is *polygamous.*

When you have filled out a schedule with the description of a flower, ask yourself the following questions about it, and answer them, if you can, in writing, at the back of the schedule :

Is this flower symmetrical or unsymmetrical? Is it complete or incomplete? Is it dichlamydeous, monochla-

FIG. 155.
Staminate Flower, from
Catkin (Fig. 154).

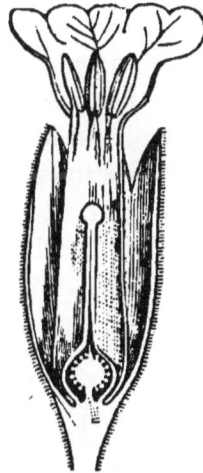

FIG. 154.
Male Catkin of a Diœcious Plant.

FIG. 156.

mydeous, or achlamydeous? Is it perfect or imperfect? Did it grow upon a monœcious, diœcious, or polygamous plant?

EXERCISE XIX.

Form of the Receptacle and Insertion of Floral Organs.

INSERTION.—In botanical language, organs are said to be *inserted* at the place from which they seem to grow. For instance, in Fig. 156 it will be seen that the pistil is inserted upon, or seems to grow from, the receptacle ; the stamens are inserted upon the corolla ; the corolla is in- .

4

serted upon the receptacle, and the calyx also is inserted
upon the receptacle.

Look at the magnified flowers shown in section on
Chart 1, and point out the receptacle in each case. Are
all these receptacles alike in form? State, in regard to
each flower, where the pistil is inserted; where the stamens; where the corolla; and where the calyx. Which
floral whorl in each flower occupies most space upon the
receptacle? Are these flowers perfect? Are they complete? Are they symmetrical?

Repeat these observations upon the magnified flowers
shown in section in Chart 2; in Charts 3, 4, 5, 6.

Make a longitudinal section of each of your living
flowers, and look for the insertion of the floral organs. If
you sometimes fail to discover it, do not be discouraged.
It will not, of course, be as clearly visible as it is shown
to be on the chart. Try again. Make frequent attempts,
as failure is often due to lack of experience.

EXERCISE XX.

Polyandrous Stamens.

We now return to the study of the flower at the point
where it was left in Schedule Seventh.

The third column of this schedule, you remember, is
the place where you wrote whether the parts of floral
whorls are grown together or not. You have studied the
calyx and corolla to learn whether their parts are grown
together. If the sepals are not grown together, you say
the calyx is *polysepalous ;* and, if they are grown together,
you say it is *gamosepalous.* So, also, when the petals of the
corolla are distinct, you say the corolla is *polypetalous ;* and,
when grown together, *gamopetalous.*

Gather all the flowers you can find, and observe the
stamens to see if they are grown together. Put aside all
that are in the least grown together.

Now look at the flowers with distinct stamens, and put by themselves all that have more than twelve.

A flower with more than twelve distinct stamens is said to have its stamens *indefinite*.

They are *definite* when there is a fixed number not above twelve.

Separate those with indefinite stamens, and label them *polyandrous* (from *poly*, many, and *andria*, stamens), which means many distinct stamens.

Now examine the flowers with definite stamens, and label each one with the name that, in the following table,

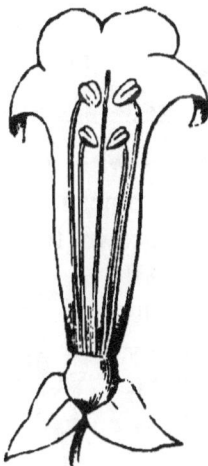

FIG. 157.
Didynamous Stamens.

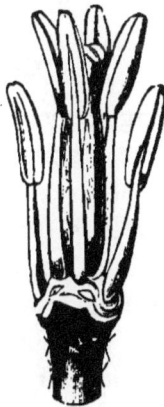

FIG. 158.
Tetradynamous Stamens.

is placed opposite its number of stamens. The Greek numeral prefix denotes the number of distinct stamens :

Mon-androus—one stamen.
Di-androus—two stamens.
Tri-androus—three stamens.
Tetr-androus—four stamens.
Pent-androus—five stamens.
Hex-androus—six stamens.

Hept-androus—seven stamens.
Oct-androus—eight stamens.
Enne-androus—nine stamens.
Dec-androus—ten stamens.
Dodec-androus — twelve sta-
mens.

Poly-androus—more than twelve.

Like the word polyandrous, these terms apply only to distinct stamens ; at the same time they have the important advantage of giving the precise number.

But, if a tetrandrous flower has two stamens long and two short (Fig. 157), it is said to be *didynamous ;* and, if an hexandrous flower has four stamens long, and two short (Fig. 158), it is said to be *tetradynamous.*

These words, applied to the stamens of a flower, give at the same time their number, the fact that they are distinct, and the proportion of long to short ones.

Can you find upon the charts any flowers with tetradynamous stamens? Have any of them didynamous stamens?

EXERCISE XXI.

The Growing together of Stamens.

Having disposed of all your flowers with distinct stamens, next examine those with united stamens.

First observe whether they have grown together by their filaments, or by their anthers. All those having their anthers united, whether into a tube, around the pistil, or in any other way, may be put together and labeled *syngenesious* (Figs. 160 and 161).

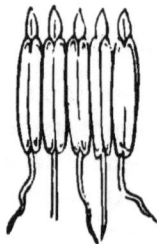

FIG. 159.
Syngenesious Stamens.

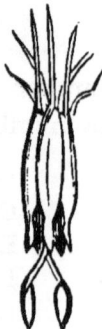

FIG. 160.
Syngenesious Stamens.

FIG. 161.
Syngenesious Stamens.

Fig. 159 shows this tube laid open. Those that have grown together by their filaments have to be further studied. Are all the filaments grown together in one bundle? If so, the stamens are *monadelphous* (Fig. 162).

FIG. 162.
Monadelphous Stamens.

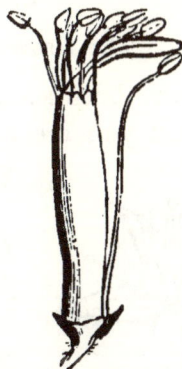

FIG. 163.
Diadelphous Stamens.

FIG. 164.
Tri- or Polyadelphous Stamens.

FIG. 165.—Polyadelphous

Are they grown together in two bundles? Then they are *diadelphous* (Fig. 163).

Are they in three or more bundles? Then we say they are *polyadelphous* (Figs. 164 and 165). Fig. 164 has one bundle cut away.

The number and length of the hard words in this exercise may discourage pupils, but by use they will become familiar, and they will then greatly help the process of description.

Collect all the plants in the neighborhood, from garden, road-side, fields, and woods, and, in describing their stamens, you will become well acquainted with all the necessary terms.

EXERCISE XXII.

The Growing together of Carpels.

You have been accustomed to counting the carpels of flowers, and you are now to find whether or not they are grown together.

Fig. 166.
Apocarpous Pistil.

All such as are not grown together at all you may label *apocarpous* (Fig. 166).

Those that are grown together, whether slightly at the base of the ovary or through the whole length of the pistil, you label *syncarpous* (Figs. 167, 168).

Find all the apocarpous ovaries pictured upon the charts. All the syncarpous ones.

Find also the apocarpous ovaries in your collection of flowers. The syncarpous ones.

For this exercise, faded flowers, and even those that have lost their floral leaves, will serve better than such as are fresh.

COHESION.—In botany this word is used for the growing together of parts with their fellows, as of petals with petals, carpels with carpels. Figs. 173 and 177 illustrate this.

Professor Henslow, the author of the flower-schedule we are using, places the word *cohesion* above the third column, and devotes it to observations upon the cohesion of parts in flowers.

Fig. 169 represents half a buttercup. It has been sliced down through the middle, making what is called a vertical section of the flower, that you may see the structure of the stamens and pistil. This flower is used here because of its simplicity, its parts being all quite distinct from each other. It is without cohesion, and, in describ-

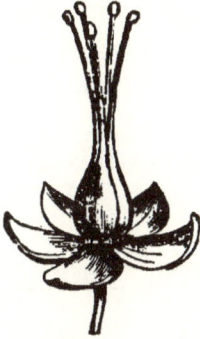

FIG. 167.
Syncarpous Pistil.

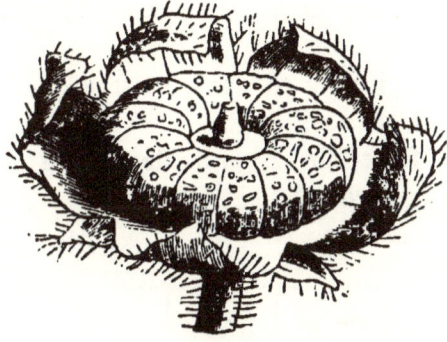

FIG. 168.
Syncarpous Pistil.

ing it, you have to use terms which apply to distinct stamens and carpels.

The learner will, of course, provide himself with a real flower, and fill out a schedule from his own examination of it. The buttercup is easily found, for it grows almost everywhere, and blossoms throughout the summer. I must

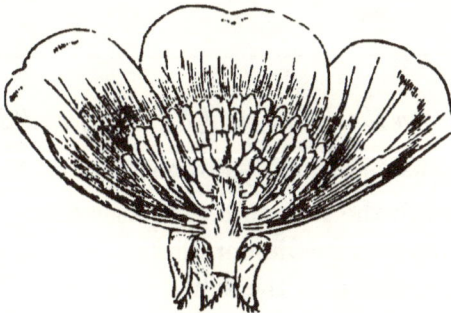

FIG. 169.

insist that the pupil be not content with simply looking over the description in the book. The example is given, not as a substitute for the pupil's own effort, but as a means of testing his observations ; of letting him know whether his own way of carrying out the schedule description is the correct one. Any lack of confidence he may feel in beginning a new process will disappear when he sees that his observations and statements agree with the printed ones. A schedule or two thus employed, when he is beginning to use new terms, will assist him in gaining self-reliance.

Schedule Eighth, describing Fig. 169, gives this arrangement :

SCHEDULE EIGHTH.

Organs.	No.	Cohesion.
Calyx ? *Sepals.*	5	Polysepalous.
Corolla ? *Petals.*	5	Polypetalous.
Stamens ?	∞	Polyandrous.
Pistil ? *Carpels.*	∞	Apocarpous.

Questions upon the Buttercup (*Fig.* 169) *and Schedule.*

Is there cohesion in the calyx ?
What word in the schedule expresses this ?
Is there cohesion in the corolla ?
How is this stated in the schedule ?
Are the stamens definite or indefinite ?
Are they grown to each other ?
What word in the schedule answers this question ?
Do the carpels cohere ?
How is this expressed ?

Questions reviewing the subject of Cohesion in the Parts of a Flower.

What is meant by cohesion in botany?

How do you describe a calyx with no cohesion (Fig.

FIG. 170.
Polysepalous, no cohesion.

FIG. 171.
Gamosepalous, coherent.

170)? A corolla (Fig. 172)? Stamens (Exercise XX)? Pistil (Fig. 176)?

When the sepals are coherent, how do you describe the calyx (Fig. 171)? The corolla (Fig. 173)?

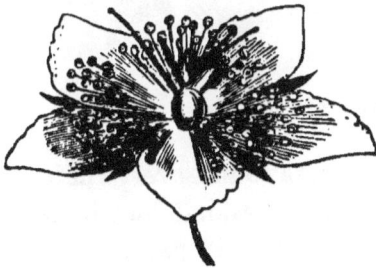

FIG. 172.
Polypetalous, no cohesion.

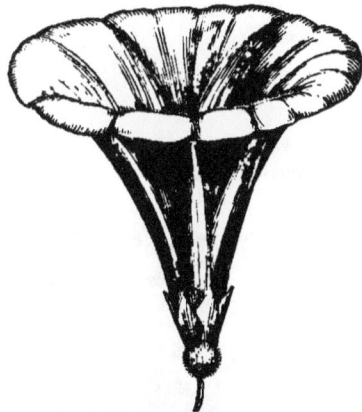

FIG. 173.—Gamopetalous, coherent.

When stamens cohere by their anthers, what word do you use in describing them (Figs. 159, 160, 161)?

When, by their filaments in one bundle, what word is used (Fig. 162)?

In two bundles (Fig. 163)?
In three or more bundles (Figs. 164 and 165)?
How do you describe a coherent pistil (Fig. 177)?

FIG. 174.—Polyandrous,
Stamens not coherent.

FIG. 175.
Triadelphous, Stamens coherent.

FIG. 176.
Apocarpous, no cohesion.

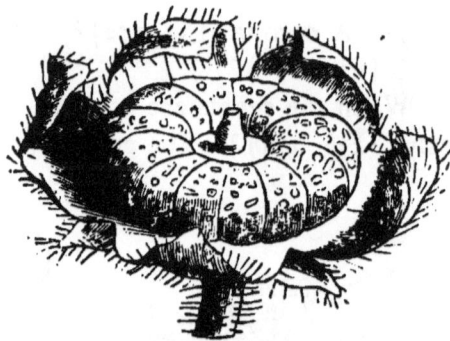

FIG. 177.—Syncarpous, coherent.

There are a few common flowers found everywhere in
the country, in which there is no cohesion ; but, in most
flowers, the parts of some
of the floral circles will be
found more or less united.

Figs. 178, 179, and 180
represent the flower of the
Saint - John's - wort.   Fig.
179 is a vertical section of
the flower, and Fig. 180 one
of the bundles of stamens.

FIG. 178.

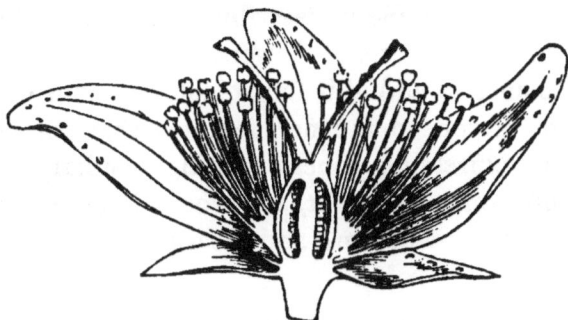

FIG. 179.

Schedule Ninth, describing Fig. 178, is an example where cohesion of stamens and pistil is described.

SCHEDULE NINTH.

Organs.	No.	Cohesion.
Calyx ? *Sepals.*	5	Polysepalous.
Corolla ? *Petals.*	5	Polypetalous.
Stamens ?	∞	Tri- or Polyadelphous.
Pistil ? *Carpels.*	3	Syncarpous.

By turning to page 63 you will see that another column is there added to the schedule. After three more exercises, which introduce new observations and new terms, this addition becomes necessary. Your attention is called to it now, to give urgency to the advice that you make diligent use

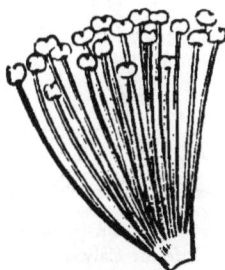

FIG. 180.

of the present schedule in describing all kinds and degrees of cohesion in all sorts of flowers. If you do this, when the time comes to add this fourth column, your mind will be free to attend to the new features that belong to it. The terms expressing cohesion being familiar, there will be no confusion of thought, and you will enter upon the new observations with ease and pleasure.

EXERCISE XXIII.

Union of Floral Whorls with each other—Calyx and Pistil.

In your study of pistils, did you always find the calyx at the base of the ovary?

Have you ever seen upon the apex of ripened fruit the withered calyx, or the scar left by its fall?

Point out upon the charts all the cases where the calyx is below the ovary.

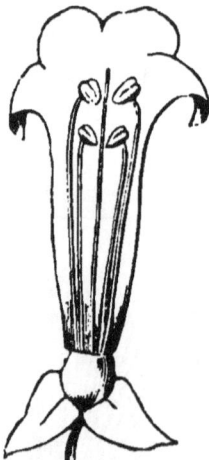

FIG. 181.
Inferior Calyx.
Superior Ovary.

FIG. 182.   FIG. 183.
Superior Calyx.—Inferior Ovary.

Point to those where the calyx is above it.

Is the calyx in all the pictures upon the chart either at the base or at the apex of the ovary?

For this exercise select flowers that have their parts so well developed that you can see distinctly where each organ is inserted. Take, for example, the morning-glory, and observe whether the calyx arises below the ovary or not. If you find it is inserted below the ovary, label it calyx below, or *inferior* (Fig. 181), and lay it aside. If the calyx is inserted above the ovary, label it calyx above, or *superior* (Fig. 182). Of course, if the calyx is below the ovary, or *inferior*, the ovary will be above the calyx, or *superior;* and, when the calyx is *superior*, the ovary will be *inferior.*

Examine all your flowers in the same way, giving each its proper label. If some specimens have the calyx inserted neither at the bottom nor at the top of the ovary, but somewhere along its side (Fig. 184), you describe these as having the calyx half inferior, and the ovary half superior. These words, inferior and superior, came into use before the facts about this matter were understood. We now know that when the calyx seems to be inserted at the top of the ovary, it is *really* inserted on the receptacle, and has its tube grown to the ovary. The true expression is "calyx adherent to ovary," in place of calyx superior; and "calyx free from ovary," in place of calyx inferior. But the words superior and inferior are in general use, and so are retained in schedule description.

FIG. 184.
Calyx, half inferior.
Ovary, half superior.

EXERCISE XXIV.

The Union of Floral Whorls with each other.

There is, perhaps, no part of the study of plant-forms that will tax your patience as much as the subject of this exercise.

Try first to determine the insertion of the corolla.

Compare the arrangement of parts in each of your flowers with that shown in Fig. 185, and, when you find the corolla inserted below the ovary, and free from the calyx, label the specimen corolla, *hypogynous*.

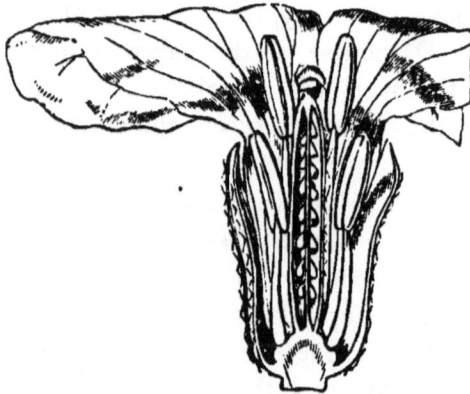

FIG. 185.—Corolla, hypogynous.

Examine the remainder of your flowers, and, when you find one with the corolla inserted, as shown in Fig. 186, say corolla upon the calyx, or *perigynous*.

How is the corolla inserted in Fig. 187? Point out upon the charts instances where the corolla has a similar insertion.

Look at the flowers not yet described, and, if you find cases where the corolla is inserted upon the ovary, describe them as *epigynous*, from *epi*, upon, and *gynia*, pistil (Fig. 187).

If not quite certain about these characters in your specimens, write your label with a mark of interrogation, to show doubt. Do not be discouraged if these points of structure remain for some time troublesome ones to discover. Try to find them out, and, if you succeed, it is well ; but if you fail, your labor will not be lost.

As some flowers upon the same plant are more perfectly developed than others, you should gather several of
each kind, and examine them all, to find the best examples of the structure you are studying.

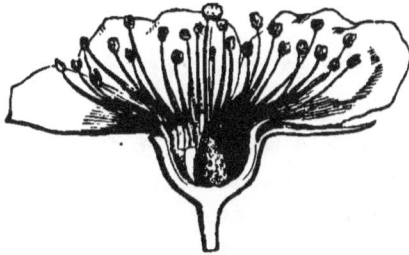

FIG. 186.—Corolla, perigynous.

FIG. 187.—Corolla, epigynous.

Look at the flowers in Chart 1, and observe in each
case whether the corolla arises from the receptacle, and
whether the calyx is free from the corolla.

Find upon the other charts all the cases where the corolla is inserted under the ovary, and is free from the calyx.

Observe the flowers on Chart 2. Where is the corolla
inserted in these figures? Can you find upon the other
charts any pictures of flowers where the corolla has a similar insertion?

EXERCISE XXV.

Union of Floral Whorls with each other—Stamens.

If the stamens have the same insertion as the corolla,
use the same words to describe them. For instance, in
Fig. 188 the stamens are *hypogynous ;* in Fig. 189, *perigynous ;* in Fig. 190, *epigynous.*

When you find them arising from the corolla, as seen
in Fig. 191, they are said to be *epipetalous.*

Sometimes they are consolidated with the pistil, as shown
in Fig. 192 ; then they are *gynandrous,* or upon the pistil.

Observe the flowers upon the chart in this respect.

Examine all the flowers you can find, and label them by the insertion of the stamens ; as, stamens under the ovary, or *hypogynous ;* stamens upon the calyx, or *perigynous ;* stamens upon the ovary, or *epigynous ;* stamens upon the corolla, or *epipetalous ;* stamens consolidated with the pistil, or *gynandrous.*

Adhesion in botany means the growing together of different floral whorls, while *cohesion*, as you have seen, means the growing together of the parts of the same whorl.

The word *free* is used to express absence of adhesion, and the word *distinct*, absence of cohesion.

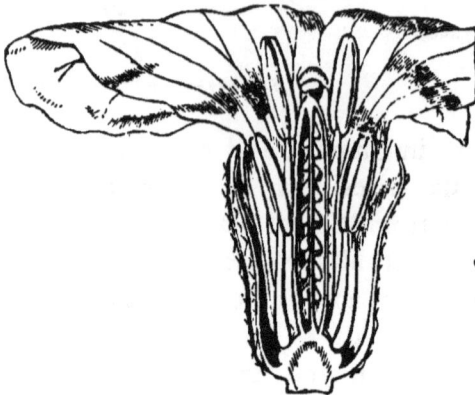

FIG. 188.—Stamens, hypogynous.   FIG. 189.—Stamens, perigynous

FIG. 191.
Epipetalous Stamens.

FIG. 192.
Gynandrous Pistil.

FIG. 190.—Stamens, epignyous.

In Fig. 193 there is neither cohesion nor adhesion. Not only are the sepals and petals *distinct* from each other, not only is each stamen and each carpel *distinct*, but the whorl of sepals is inserted upon the receptacle, and is *free*

FIG. 193.
Parts, distinct.
Organs, free.

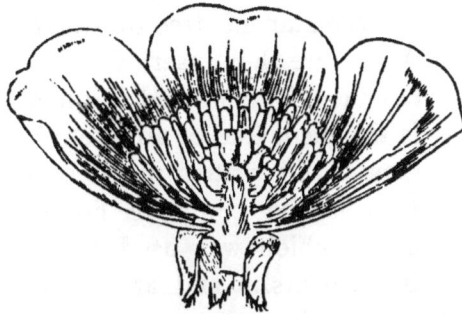

FIG. 194.

from the whorls within it. The corolla is inserted upon the receptacle, and is also *free*. The stamens and pistil are also inserted upon the receptacle, and are likewise *free*.

The last column of Schedule Tenth, describing Fig. 194, is for the record of observations on adhesion.

SCHEDULE TENTH.

Organs.	No.	Cohesion.	Adhesion.
Calyx ? *Sepals.*	5	Polysepalous.	Inferior.
Corolla ? *Petals.*	5	Polypetalous.	Hypogynous.
Stamens ?	∞	Polyadelphous.	Hypogynous.
Pistil ? *Carpels.*	∞	Apocarpous.	Superior.

Questions upon the Buttercup (*Fig.* 194) *and its Schedule.*

Is the calyx free or adherent?
How is this expressed in the schedule?
Where is the corolla inserted?
How is this stated in the schedule?
Are the stamens free or adherent?
Where are they inserted?
How is this expressed in the schedule?
Is the pistil free or adherent?
How is this written in the schedule?

This is the complete flower-schedule of Professor Henslow, which was used by his classes both at Cambridge University and at his parish school at Hitcham. Complaints have been made that it is difficult. Pupils who commence its use before they fully understand the features of plants to which it calls attention will, no doubt, get confused when they attempt to fill up the blanks one after another, but those who have examined a variety of flowers, in connection with the foregoing pages, will have no such trouble.

FIG. 195.

The presence or absence of cohesion and adhesion in flowers is of great importance in determining the relationships of plants, and scholars can not do better than continue the use of this schedule throughout the summer season, along with the making of an herbarium. Always write from your own observation. Never give a word of description unless it be of something your own eyes have seen, and that you could point out to any one who might contradict you. We give some further examples of the use of the schedule in flowers of very unlike structure.

Find living flowers like those here described, and fill out schedules of them yourself. Be careful not to copy statements from the book. Depend upon your own judgment.

Fig. 195 represents a flower of cow-parsnip. That of the carrot, or any umbelliferous plant, will do as well.

SCHEDULE ELEVENTH, DESCRIBING FIG. 195.

Organs.	No.	Cohesion.	Adhesion.
Calyx ? *Sepals.*	5	Gamosepalous.	Superior.
Corolla ? *Petals.*	5	Polypetalous.	Epigynous.
Stamens ?	5	Pentandrous.	Epigynous.
Pistil ? *Carpels.*	2	Syncarpous.	Inferior.

FIG. 196.

Fig. 196 shows a vertical section of the flower of daffodil. It is common enough in gardens ; but, if there are
pupils who can get neither this flower, nor the jonquil, nor
the snow-drop, they can certainly find a lily of some kind,
wild or cultivated, and observe the features in which it is
unlike this picture.

SCHEDULE TWELFTH, FOR FIG 196.

Organs.	No.	Cohesion.	Adhesion.
Perianth ? *Leaves.*	6	Gamophyllous. Crowned.	Superior.
Stamens?	6	Hexandrous.	Perigynous.
Pistil ? *Carpels.*	3	Syncarpous.	Inferior.

Fig. 197 is a blossom of wild geranium. Fig. 198 shows
its stamens and pistil. The flower of the garden geranium
will serve in its place, if it can be more easily obtained.

SCHEDULE THIRTEENTH, FOR FIG. 197.

Organs.	No.	Cohesion.	Adhesion.
Calyx ? *Sepals.*	5	Polysepalous.	Inferior.
Corolla ? *Petals.*	5	Polypetalous.	Hypogynous.
Stamens?	10	Decandrous.	Hypogynous.
Pistil ? *Carpels.*	5	Syncarpous.	Superior.

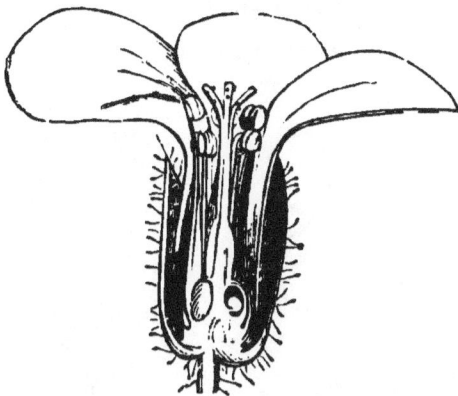

FIG 197.

FIG. 198.

EXERCISE XXVI.

The Receptacle.

The peculiarities of plants pointed out in this exercise are not very common. But pupils who are using the flower-schedule, and collecting all the plants they can find, will be sure to meet with examples of them sooner or later. This exercise should, therefore, be carefully

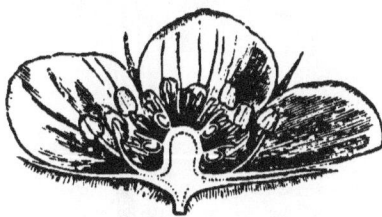

FIG. 199.—Convex Receptacle.

FIG. 200.—Receptacle, greatly enlarged.

read and borne in mind, so that when the things it describes are met with they will be recognized.

You have seen the receptacle forming a swelling like that of Fig. 199, and gradually expanding into a structure like Fig. 200. Sometimes the receptacle is prolonged between the carpels, and coheres with their styles, which separate from it at maturity, as seen in Figs. 201 and 202.

FIG. 201.

FIG. 202.

FIG. 203.—Cup-shaped Receptacle.

It sometimes appears as a cup-shaped depression (Fig. 203), in which the pistil is almost concealed, and again as shown in Fig. 204.

FIG. 204.—Elevated Fleshy Receptacle.

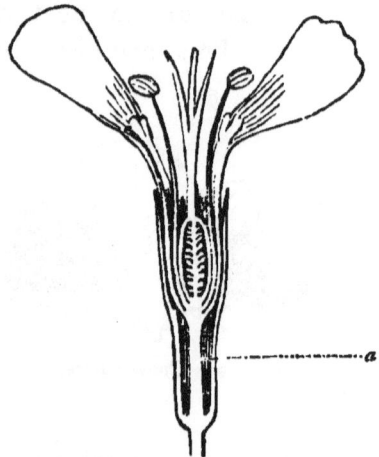

FIG. 205.—*a.* Anthophore.

When the receptacle becomes elongated, so that one circle of floral organs is separated from another by a stalk-like internode; the circle thus raised is said to be *stipitate*,

and the stalk supporting it is
called a *stipe*. In Figs. 205
and 206 the stamens, pistil,
and corolla are *stipitate*, and
the stalk which bears them
is the *stipe*.

When the stipe supports
corolla, stamens, and pistil, it
is called an *anthophore* (Fig.
205). When it supports only
stamens and pistil, it is known
as the *gonophore* (Fig. 206, *a*).

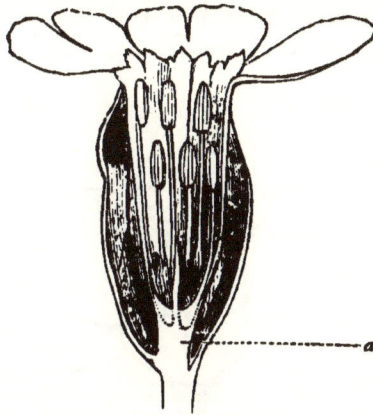

FIG. 206.—*a*. Gonophore.

EXERCISE XXVII.

Appendages of the Receptacle.

Examine the receptacle in the magnified flowers upon
Charts 1, 2, 3, and 4.

Carefully observe the space between the calyx and
ovary in the following figures. You see a sort of fleshy

FIG. 207.
Hypogynous
Disk.

FIG. 208.
Hypogynous
Disk.

FIG. 209.—Hypogynous Disk.

rim around the pistil, which is called a *disk*. It takes on
very different shapes in different plants. In Figs. 207 and
208 it is merely a raised cushion ; in Fig. 209 it is seen

BOTANY.

partly inclosing the ovary.   In Figs. 210 and 211 the
disk is seen surrounding the ovary ; while in Figs. 212

FIG. 210.—Perigynous Disk.

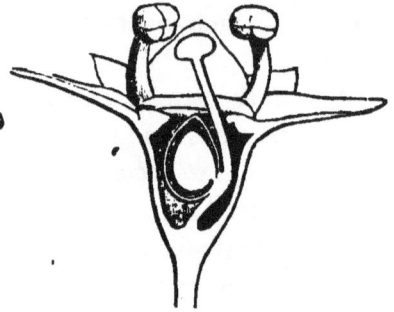

FIG. 211.—Perigynous Disk.

FIG. 212.
Epigynous Disk.

FIG. 213.
Epigynous Disk.

FIG. 214.

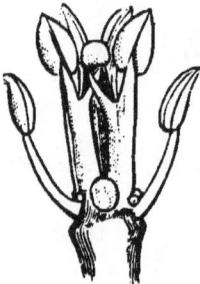

FIG. 218.

FIG. 215.

FIG. 216.

FIG. 217.

and 213 it is shown above the ovary, and at the base of the style.

The little glands upon the receptacle are known as *nectaries.* They contain sweet fluids, and are found among the stamens (Figs. 214, 215) or at the base of the pistil, forming a part of the disk (Figs. 216, 217, 218).

Turn to pages 186 to 191, and observe that the orders of flowering plants are divided into the following groups, according to the characters of cohesion and adhesion they exhibit :

1. Inferior Polypetalous Exogens.*
2. Discifloral Polypetalous Exogens.
3. Superior Polypetalous Exogens.
4. Superior Monopetalous Exogens.
5. Inferior Monopetalous Exogens with Regular Flowers.
6. Inferior Monopetalous Exogens with Irregular Flowers.
7. Apetalous Exogens with Perfect Flowers.
8. Apetalous Exogens with Imperfect Flowers.
9. Superior Endogens.†
10. Inferior Endogens.
11. Gymnosperms.‡

---

* *Outside-growers.* Stems that grow by an annual addition of a ring of wood outside the previous wood, and hence they are called Exogens (from two Greek words signifying *outside growers*). All the trees and large shrubs of temperate and cold climates are exogenous in their growth. Multitudes of herbaceous plants are also classed as exogens. They may be known by their venation. All plants with net-veined leaves are exogens. The seeds of exogenous plants contain a two-leaved embryo (see Ex. X), and are hence called *dicotyledonous* plants.

† *Inside-growers.* Stems that grow by the addition of new wood directed toward their interior. All plants with parallel-veined leaves are endogens. Their seeds also contain a one-leaved embryo, whence the name *monocotyledonous* plants.

‡ See page 161.

5

CHAPTER FOURTH.

COMPARING AND CLASSIFYING PLANTS.

EXERCISE XXVIII.

Plant Characters and Affinities.

You are now to take a step forward in the study of plants. Having acquired considerable knowledge of their parts by direct observation, you will begin to compare them—to note the resemblances and differences of whole plants, and, by these resemblances, to arrange or group them in a systematic way. This is classification

You have been classifying the *parts* of plants ever since you commenced observing them. For instance, those with parallel-veined leaves have been classed by themselves, and those with flowers in umbels have been classed together, and kept distinct from such as blossom in heads and panicles; but your groupings have thus far been made upon single features of plants. Now, however, you know their parts so well that you can begin to compare whole plants with each other.

If, for example, you have put into one group all square-stemmed plants, simply because they have square stems, it is time to consider whether these plants are alike in other respects. "Oh, yes," some of you will say; "they have opposite leaves." Well, look at their inflorescence; do they all agree in that? Is it always axillary? Are the flowers similar in all the square-stemmed plants you know? When you have answered these questions, you will understand what I mean by studying plants as wholes.

Provide yourself with the following plants: The buttercup (which is found almost everywhere), the wild columbine, and the poppy. If the columbine is not to be found, get monk's-hood, or larkspur, or anemone, and proceed with them in the way pointed out for the columbine. If the poppy can not be found, you might substitute bloodroot or celandine. Having got the plants, proceed according to the plan laid down, and do not accept the statements or conclusions of the book, unless, on comparing them with your own plants, you see that they are true.

There are two botanical expressions of which, at the outset, you should learn the meaning. One of these is *the characters of plants*, and the other *the affinities of plants*. And, first, what is meant by *plant-characters?*

If you will describe a buttercup, I think we can easily find just what is meant.

You say, "CALYX, *sepals*, 5, polysepalous, inferior; COROLLA, *petals*, 5, polypetalous, hypogynous; STAMENS, many, hypogynous; PISTIL, *carpels,* many, apocarpous, superior. It has simple, exstipulate, alternate divided leaves; petiole spreading at base; stem, erect; flowers, in a loose cluster; juice, watery, acrid."

Now, this is a description of a particular buttercup, and yet it applies to all buttercups. Are all buttercups, therefore, exactly alike? By no means. They differ in size, shape, thriftiness, number of blossoms, etc.; but, in our botanical description, we do not record these individual peculiarities.

Well, the points of form and structure in which all buttercups agree—that is, their *permanent features*—are called by botanists the *characters* of the buttercup. All such unchanging features of plants are *plant-characters.* A plant is an assemblage of characters, and the description of a plant is but a list of its characters.

Now, it is by comparing groups of characters that we reach the idea of *affinities.* If, as we have seen, each plant bears a fixed group of characters, the resemblance of one plant to another is only the resemblance of one group of characters to another. Let us make such a comparison between the buttercup and the columbine.

Do not rely upon the descriptions in the book, but make similar tables yourself.

BUTTERCUP.—*Flower.*	COLUMBINE.—*Flower.*
Calyx.—Sepals, 5, polysepalous, inferior.	*Calyx.*—Sepals, 5, polysepalous, inferior, colored like the petals.
Corolla.—Petals, 5, polypetalous, hypogynous, obcordate, yellow.	*Corolla.*—Petals, 5, polypetalous, hypogynous, spurred, red.
Stamens.— ∞, hypogynous.	*Stamens.*— ∞, hypogynous.
Pistil.—Carpels, ∞, apocarpous, superior.	*Pistil.*—Carpels, 5, apocarpous, superior.

Comparing the above lists, you see agreements and differences. The calyx and corolla of one plant agree with those of the other in number of parts and in the position of parts. They differ only in color and outline. The stamens of one are like those of the other in being numerous and hypogynous. The pistils agree in structure, but differ in the number of carpels. If you compare the leaves, stems, inflorescence, etc., you also get a list of their resemblances and differences. This is comparing plants by the groups of characters they present.

These resemblances of character among plants are called their affinities.

The *degree* of affinity between plants depends upon two circumstances : First, upon the *kind* of characters in which they agree ; and, second, upon the *number* of characters in which they agree.

The characters of plants differ in importance. Color, size, and odor, being usually more variable than position and number, they are said to be less important than these. The characters of the leaf, for the same reason, are not usually as important as those of the flower. In the beginning of study, you may assume that those plants have the strongest *affinities* that resemble each other most in the characters recorded in the cohesion and adhesion columns of the schedule.

To make this plainer, compare the poppy and buttercup, as, before, you compared the columbine and buttercup.

BUTTERCUP.	POPPY.
Calyx.—Sepals, 5, polysepalous, inferior.	*Calyx.*—Sepals, 2, polysepalous, inferior.
Corolla.—Petals, 5, polypetalous, hypogynous.	*Corolla.*—Petals, 4, polypetalous, hypogynous.
Stamens.—Polyandrous, hypogynous.	*Stamens.*—Polyandrous, hypogynous.
Pistil.—Carpels, many, apocarpous, superior.	*Pistil.*—Carpels, many, syncarpous, superior.
Leaves.—Net-veined, divided.	*Leaves.*—Net-veined, divided.
Juice.—Watery.	*Juice.*—Milky.

To find which has the strongest affinity for the butter-cup, the columbine or the poppy, all that is necessary, at present, is, to ascertain which of them is nearest like the buttercup in respect to cohesion and adhesion of the parts of the flower.

On examination, you see that the columbine, like the buttercup, is perfectly destitute of cohesion, while in the poppy you have a coherent, or syncarpous, pistil. This settles the question. The affinity of the columbine for the buttercup is greater than the affinity of the poppy for the buttercup.

If you compare their leaves, you will find those of the poppy more like buttercup-leaves than are those of the columbine, but differences in leaf-structure do not usu-ally signify as much in classification as differences in the pistil.

Compare, in the same way, the hollyhock and the Saint-John's-wort with mallows, and decide which has the strongest affinity for the mallows.

Compare the flower of the locust and of the geranium with that of the pea or bean.

I mention these plants, not because they are useful above all others for your purpose, but to start you in the work. It really matters little what plants you take, if you only carefully compare the group of characters of each one with that of the others, and endeavor to discover the affinities they present.

EXERCISE XXIX.

How to begin Classification.

If you have made the comparisons pointed out in Ex. XXVIII, you are prepared for an explanation of the plan by which you are to begin to classify plants. As we made use of the buttercup and columbine to learn the meaning of *affinity* in botany, a little further statement about them will, perhaps, be helpful before we pass to the regular work of the exercise.

The buttercup thrives best in low, damp places. It is like frogs in this respect ; and, because of this, it is named after them. Its botanical name is *Ranunculus*, from *Rana*, a frog. The Ranunculus has certain characters with which you are familiar. Now, when you find other plants which are very much like it, that is, which present nearly the same group of characters, particularly those of cohesion and adhesion, you class them with it, you say they belong with the buttercup ; or, in more botanical language, they belong to the Ranunculaceæ. In some regions this plant, from the form of its leaf, is called the Crowfoot, and plants closely resembling it are said, therefore, to belong to the Crowfoot family. Now, the resemblance of the columbine to the buttercup entitles it to belong to the Ranunculaceæ. The monk's-hood and larkspur also belong to the same family, and this will give you some idea of the degree of similarity that should exist between members of one family.

Our object in the present exercise is, to fix upon a method by which to begin the work of classifying plants, by comparing the groups of characters they present, and putting together those that are most alike.

Get a pocket note-book. Write in it, boldly and plainly, the flower-schedules of the following plants : Buttercup, shepherd's - purse, mustard or radish, catchfly, mallows, Saint-John's-wort, clover, pea or bean, wild rose, strawberry, geranium, violet, morning-glory.

Now, why have we put these particular schedules into the note-book ? Compare them with each other. Do you not see that the statements in the cohesion and adhesion columns are widely unlike ? This is why we have chosen them. They are so many different examples of the make-up of flowers, and you have simply to compare each flower you describe with one and another of these examples, to

see which it most resembles. If unlike them all, then set
up your new acquaintance as another example, and see
if you can find any similar plants in the course of the
summer. So, do not confine yourself to comparisons between your specimens and the patterns in your note-book.
Compare them freely with each other, and you will soon
have many little collections of plants bearing very strong
resemblances to each other.

Your thought will be something like this : While you
are observing and describing a plant, you will ask yourself, " Have I ever before described one like this in the
matters of cohesion and adhesion ? " If you can think of
none, you will try to recall those nearest like it. By pur-
suing this plan, you will be surprised to find how quickly
many of the plants of a region, that were never before
thought of as at all alike, fall into company on the ground
of these deeper resemblances which your studies have led
you to discover.

The reason why you are set systematically to classifying plants now, and have not been asked to do it before,
is, that among the characters of plants that belong to roots,
leaves, stems, etc., there are none that are so uniform
throughout large numbers of different plants as these features of cohesion and adhesion in flowers. Since you began to observe plants, you have not been taught to notice
any points of structure that would serve so well for uniting plants into groups, the members of which are truly and
somewhat nearly related to each other.

But the grounds on which you are to begin to classify
plants, although important, and, in many cases, quite sufficient, are not the only ones on which classification is based.
Though they may sometimes be found too narrow, yet you
must begin somewhere, and to make your beginning as
free as possible from complexities, you start with the features named in the flower-schedule. In working with this,
much of your experience will be clear and satisfactory,

but you may meet with difficulties. By-and-by, however, the subject will be resumed, and, if you have sometimes been confused and puzzled in classifying by the flower-schedule alone, new ideas will be all the more welcome.

Students who have the botanical charts will find them very helpful in the work of classification. Upon these charts there are pictured in the colors of Nature some forty pattern-plants, magnified, and shown in section, so that their structure is easily seen. These plants have been selected because the differences they present are just those broad contrasts that separate groups of plants in Nature. At this stage of your study, while your thoughts are confined to the features of the flower-schedule, the first, second, third, and fifth charts present plants of all varieties in these respects. Their great value lies in the distinctness of the idea they give as to how pattern-plants are constructed.

The work of classification being now entered upon, it will be resumed, from time to time, with further explanations as we proceed, particularly when we come to study such groups of plants as the grains and grasses, the cone-bearing plants, the Compositæ, familiarly known as compound flowers, the Umbelliferæ, etc. These striking natural orders will introduce us to new principles in judging of affinities, and pupils who are specially fond of this part of the study, and are apt in tracing resemblances, will do well to look over the chapters upon these plants without waiting to reach them in the course of regular study.

There is often, among both teachers and pupils, an aversion to skipping about. The idea of thoroughness with them seems to imply moving steadily on from page to page of a book, without ever deviating from its order. But in such a science as botany it is not necessary to proceed in this way. The subject can not be marked off sharply into parts that must be learned in a certain order. Of course, plant-characters must be known before they

can be used in classification ; but, when a few are known.
they may be at once put to service. A pupil can not do
better than to acquaint himself with the group of cruciferous plants as soon as the special characters that belong to
this group are familiar. Any group of plants may be classified as soon as the characters upon which it is founded
are fairly known. To get a knowledge of classification
requires much time, and its study should, therefore, be
commenced at the earliest possible moment.

There is another reason for skipping about, which wili
be at once appreciated. It is this : Plants have their time
to flower, and their flowers must be studied at that time.
For example : the Coniferæ blossom in spring, and spring
is the time to study them. Stamens may be found throughout the entire season, and so may be studied at any time.
It would be folly, therefore, to let the period pass in which
the Coniferæ might be studied, because you "hadn't come
to them" in the book, and pursue the study of stamens
because they are next in order. Again, the characters of
orchids are illustrated by a plant which has its season, and
the time to study orchids is when this plant makes its appearance.

CHAPTER FIFTH.

THE MINUTE STUDY OF THE ESSENTIAL ORGANS OF PLANTS.

EXERCISE XXX.

Parts of Stamens.

In Fig. 219 you see the parts of a well-formed stamen. The ANTHER-LOBE is the cell which holds the pollen (Fig. 219). CONNECTIVE, a continuation of the filament which unites the two lobes of the an-ther. It is often inconspicuous or absent, but is sometimes easi-ly seen (Fig. 219). VALVES, the sides of an anther-lobe.

LINE OR POINT OF DEHIS-CENCE.—The opening through which the pollen escapes.

Do you see in your specimen a groove down the middle of the anther on one of its sides ? Is there anything like a ridge on the other side of the anther, opposite the groove ? Can you divide the anther at this place without coming upon the pollen ? What name is given to this part of the anther in Fig. 219 ? What are the two halves it connects called ?

Look at your living anther for the line along each lobe, called the line of dehiscence.

It may help the learner in forming a distinct idea of these different parts of the anther, to know that the sta-

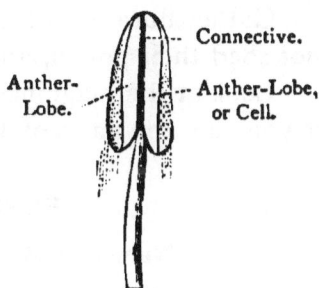

men is looked upon by botanists as a sort of leaf, the filament answering to the petiole, and the anther to the blade
The connective corresponds to the midrib of a leaf, and
the line of dehiscence to its margin, each lobe being half
of a leaf-blade, and the valves of an anther corresponding
to the upper and under sides of a leaf.

Examine the anthers of as many different flowers as
possible, and try to find the cells, connective, line of dehiscence, valves.   Do not be disappointed or discouraged if,
in many cases, you fail to distinguish some of the parts.

Gather flowers with large, perfect stamens, which have
not shed their pollen, and compare them with Fig. 219.

Look at the magnified stamens on the charts, and find,
if you can, the parts of the anther named in this exercise.

EXERCISE XXXI.

Number and Shape of Anther-Lobes.

An anther-lobe is said to be EMARGINATE when the
summit, or base, of the anther-cell extends upward or
downward, a little beyond the connective (Fig. 226).

NUMBER OF ANTHER-LOBES.

FIG. 220.
One-celled Anther.

FIG. 221.
Two-celled Anther.

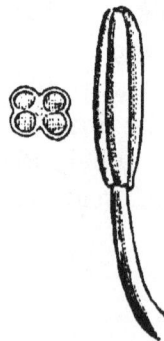

FIG. 222.
Four-celled Anther.

Label each flower of your collection with the number
and shape of the anther-cells of its stamens.

SHAPE OF ANTHER-LOBES.

FIG. 225.
Kidney-shaped
Anther.

FIG. 227.
Sinuous Anthers.

FIG. 224.
Oblong Anthers.

FIG. 223.
Arrow-shaped
Anther.

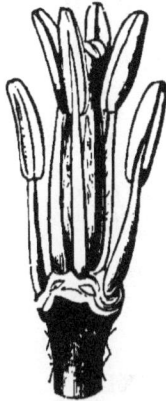

FIG. 226.
Emarginate Anthers.

Find upon the charts one-celled anthers, two-celled anthers, four-celled anthers.

EXERCISE XXXII.

Dehiscence of the Anther.

VERTICAL OR LONGITUDINAL DEHISCENCE. — When the anther opens by a slit along its length to emit the pollen (Fig. 228).

TRANSVERSE.—When the line of dehiscence is across the anther (Fig. 229).

POROUS.—When the anthers emit the pollen through little pores (Fig. 230).

VALVULAR.—When a portion of the anther is lifted up to emit the pollen (Figs. 231 and 232).

What modes of dehiscence of anther-cells are shown upon the charts? In describing the stamens, name the kind of dehiscence the anther exhibits.

FIG. 228.
Vertical,
or Longi-
tudinal.

FIG. 229.
Transverse.

FIG. 231.
Valvular.

FIG. 230.
Porous.

FIG. 232.
Valvular.

EXERCISE XXXIII.

Introrse and Extrorse Anthers.

The projecting side of the anther-cell is called its *face*, and the opposite side its *back*, whether the valves are unequal or not.

When the valves of the anther are of equal size, the dehiscence will occur laterally (Fig. 235) ; but, if one valve be wider than the other, it will throw the line of dehiscence

FIG. 233.—Face.　　　FIG. 234.—Back. 　　　FIG. 235.—Lateral
Dehiscence.

nearer to the connective on one side than on the other. These narrowed valves are usually on the face or projecting side of the anther-cell (Fig. 233).

It is on the other side that the connective is usually visible, if seen at all, and that the filament is in most cases attached (Fig. 234).

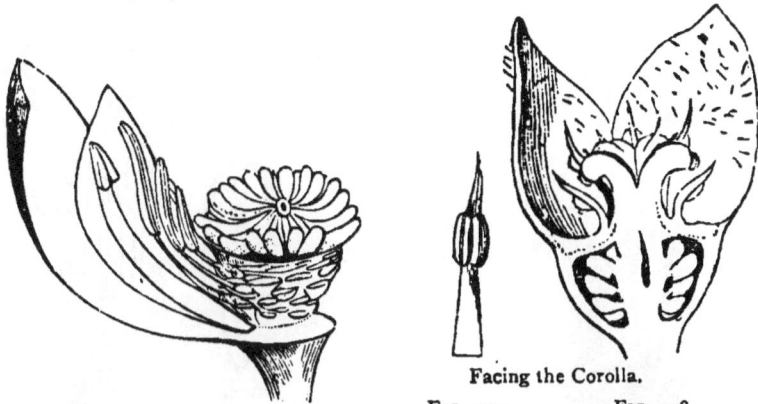

FIG. 236.—Introrse Anthers.

Facing the Corolla.

FIG. 237.   FIG. 238.
Extrorse Anthers.   Extrorse Anthers.

Anthers are INTRORSE when the line of dehiscence, or face of the anther, is toward the pistil (Fig. 236).

Anthers are EXTRORSE when the line of dehiscence, or face of the anther, is turned toward the corolla (Figs. 237 and 238).

Look over the charts for examples of extrorse and introrse anthers. Observe this feature when you study flowers.

EXERCISE XXXIV.

Attachment of Filament to Anther.

INNATE.—Anthers are *innate*, or *basifixed*, when the fila-ent runs directly into the base of the connective (Figs. 239, 240, and 244).

ADNATE.—Anthers are *adnate*, or *dorsifixed*, when the filament runs up the back of the anther, joining the connective in such a way that the anther is hung in front of it (Figs. 241 and 242).

VERSATILE.—If the filament is attached by a slender apex to the middle of the anther, the ends of which swing freely up and down, the attachment is said to be *versatile* (Fig. 243).

The modes of attachment, pictured and named above,

FIG. 240.
Innate.

FIG. 241.
Adnate.

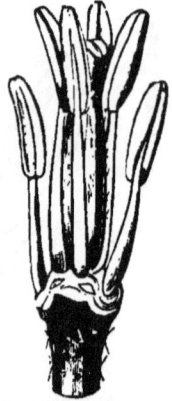

FIG. 239.—Innate.

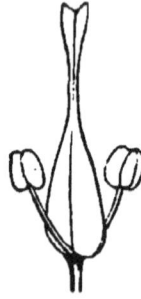

FIG. 242.
Adnate.

shade into each other, so that, in practice, it is often difficult to determine them. The versatile passes into the ad-

FIG. 244.
Basifixed.

FIG. 245.
Dorsifixed.

FIG. 246.
Apsifixed.

FIG. 243.—Versatile.

nate, and the adnate into the innate, and a nice exercise of judgment is sometimes needed in describing this feature of flowers.

Find these several modes of attachment on the charts. Determine and describe the mode of attachment in each of your living specimens.

EXERCISE XXXV.

Forms of Filaments.

FILIFORM filaments are thread-like, as the name denotes, but strong enough to support the anther (Fig. 239).

SUBULATE filaments taper like an awl (Fig. 247).

CAPILLARY filaments are hair-like, and too slender to support the anther (Fig. 248).

DILATED filaments are flattened out like Fig. 249.

PETALOID filaments resemble petals in form, and bear the anther at the summit, as seen in Figs. 250 and 251.

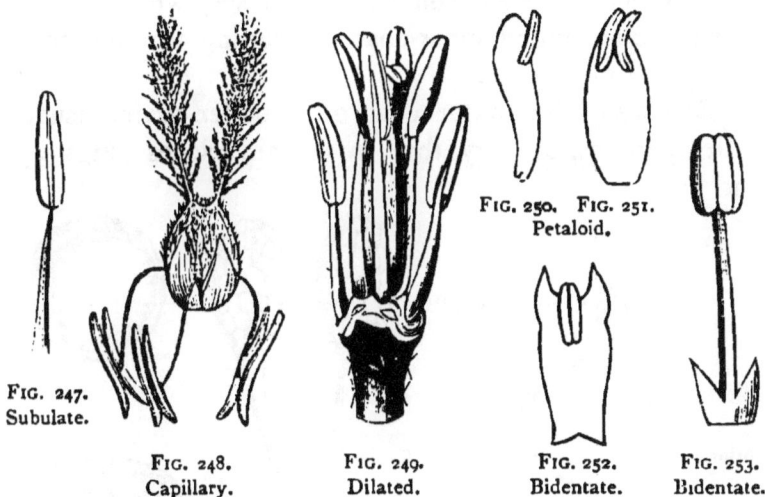

FIG. 250. FIG. 251.
Petaloid.

FIG. 247.
Subulate.

FIG. 248.   FIG. 249.   FIG. 252.   FIG. 253.
Capillary.   Dilated.   Bidentate.   Bidentate.

BIDENTATE, or BICUSPID, filaments are toothed at the summit or at the base, as seen in Figs. 252 and 253.

Find examples of the several kinds of filaments upon the charts. Describe the different forms of filaments in your collection of plants.

EXERCISE XXXVI.

Structure and Forms of Pollen.

The pollen-grain is generally composed of two membranes, or coats, filled with a thick liquid substance containing minute grains, which is its essential portion. The outer coat is frequently marked with bands, lines, and

FIG. 254.　　　FIG. 255.　　　FIG. 256.

grooves, or covered with bristling points (Fig. 254). The inner coat is very thin, and swells when wet. If you moisten pollen-grains, you may often see, with a microscope, the expanded inner coat protruding through openings in the outer coat.

EXTINE.—The outer coat of a pollen-grain, usually with openings, or very thin in certain places (Figs. 254 and 255).

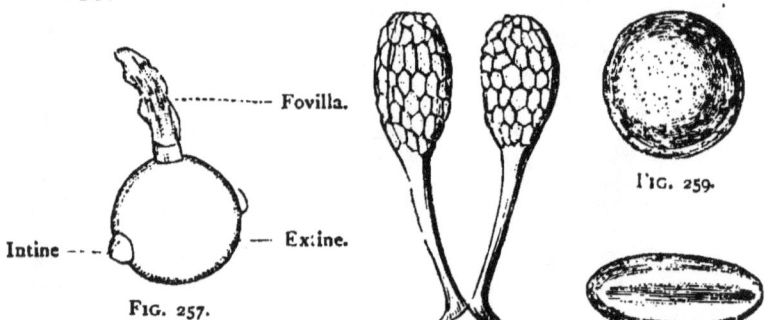

Fovilla.

Intine

Extine.

FIG. 257.

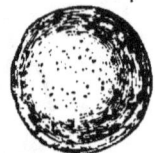

FIG. 259.

FIG. 258.—Pollinia.

FIG. 260.

INTINE.—The inner coat of a pollen-grain, very thin, tough, and elastic, often seen protruding through holes in the extine (Figs. 255 and 256).

FOVILLA.—The rich protoplasmic liquid contained within the intine (Fig. 257).

POLLINIA.—Pollen-grains cohering in masses. In Fig. 258 they are in pairs, and are furnished with stalk-like processes; but in some plants they are single, and without a stalk.

Pollen-grains display a great variety of shapes. Besides the round and oblong (Figs. 259 and 260), you will find them angular, lobed, and joined together in various ways (compound pollen) by threes, fours, and even larger numbers.

Look at the various forms of pollen pictured upon the charts.

Examine the pollen of flowers with your magnifying-glass, and note the shape of the grains, and the kind of surface they present. Observe the moistened pollen of various plants under the microscope.

EXERCISE XXXVII.

Forms of Connective.

APPENDICULAR.—When the connective, extending above or below the anther, takes the form of a feather, or a lengthened point, or

FIG. 261. FIG. 262. FIG. 263.
Appendicular.

FIG. 264. FIG. 265.
Connective, widened.

a fleshy mass, or spur-like appendages, or stipules (Figs. 261, 262, and 263), it is said to be appendicular.

When one lobe of an anther is abortive, or suppressed, the anther is said to be *dimidiate.* Fig. 266 represents a dimidiate anther and a connective developed into arms, so that the lobes are entirely disconnected.

Observe the abortive anther-lobe of Fig. 266. The entire stamen, as well as each of its parts, is liable to suppression, abortion, or imperfect development. The symmetry of flowers is often destroyed in this way. In some plants the non-development of organs that exist in the rudimentary state is a constant character, and should be regarded in describing them.

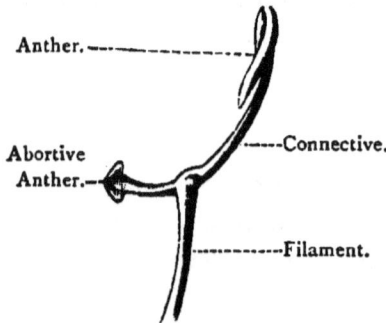

Anther.-------------

Abortive Anther.-------

-----Connective.

-----------Filament.

FIG. 266.—Dimidiate.

Observe the figures on the chart which illustrate these forms of connective. Look over the flowers of your collections, and in future describe the form of connective when you can distinguish it.

EXERCISE XXXVIII.

General Features of Stamens.

EXSERTED.—Stamens are said to be *exserted* when they extend beyond the corolla (Fig. 267).

INCLUDED.—When the stamens are not as long as the corolla, they are said to be *included* (Fig. 268).

The entire whorl of stamens is called the *androecium.*

When the filament is wanting, the anther is described as *sessile.*

When the anther is wanting, the stamen is said to be *sterile.*

FIG. 268.

FIG. 267.

Converging stamens are said to be *connivent*.

In observing and describing stamens, the following questions will be found useful by calling attention to the several characters pointed out in the present chapter:

Parts? Number of anther-lobes? Shape of anther-lobes? Attachment of filament and anther? Facing? Form of filament? Form of pollen? Form of connective? General features.

CHAPTER SIXTH.

THE PISTIL.

EXERCISE XXXIX.
Kinds of Style and Stigma.

FIG. 270.
Sessile and Lateral.　FIG. 271.—Bifid.　FIG. 272.—Trifid.

FIG. 273.—Trifid.

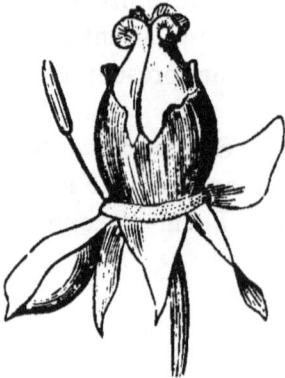

FIG. 274.—Scrolled

FIG. 276.
Lobed.

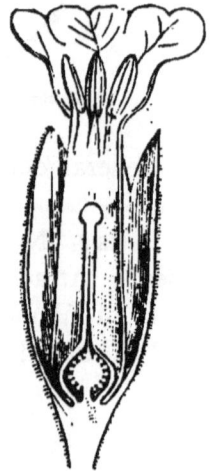

FIG. 275.—Globose

NAME the kinds of stigma shown on the chart.

EXERCISE XL.

Form and Position of Styles.

| FIG. 277. | FIG. 278. | FIG. 279. | FIG. 280. |
| Sigmoid. | Lateral. | Basal. | Terminal. |

The shapes of styles may be named by the same words as the shapes of filaments.

Observe, in faded flowers and young fruit, whether the styles are persistent or deciduous.

EXERCISE XLI.

Pistil, Ovary, Fruit.

It will be convenient to apply the following names to certain distinctions among pistils with which pupils are now familiar :

A COMPOUND PISTIL (Fig. 281) consists of several united carpels—is *syncarpous.*

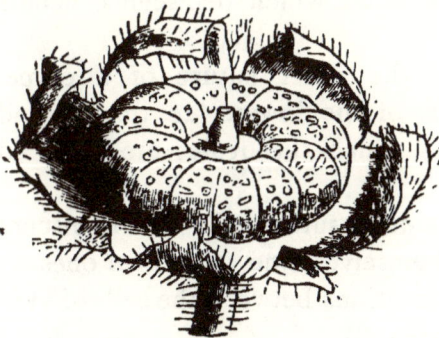

FIG. 281.
A Compound Pistil.

FIG. 282.
A Simple Pistil.

A SIMPLE PISTIL (Fig. 282) consists of only a single carpel, and is, of course, *apocarpous.*

A MULTIPLE PISTIL (Figs. 283 and 284) consists of several *distinct* carpels— is also *apocarpous.*

Pluck from the pea or bean vine pods of different ages and com-pare them. The soft, small bodies in the young pods are called *ovules.* The ripe, full-grown con-tents of the mature pod are *seeds.* Pod and con-tents form the *fruit.* The fruit of a plant is its rip-ened ovary. Find the ovules of unripe apples, tomatoes, cucumbers, etc. Count the carpels in all the ovaries you examine.

FIG. 283.
Multiple Pistil.

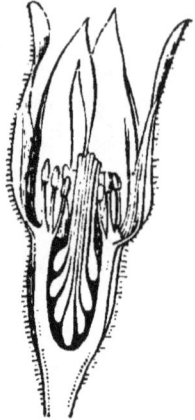

FIG. 284.
Multiple Pistil.

Look among dry pea or bean pods for those that have begun to open. Examine the edges of the separate parts.

DEHIS'CENCE is the self-opening of an ovary at ma-turity.

A SUTURE is the line along which dehiscence occurs (Figs. 317, 319).

VENTRAL SUTURE — the inner suture of a carpel looking toward the center of the flower. In the pea and bean it is the suture along which the ovules are attached (Fig. 314).

DORSAL SUTURE—the outer suture. Besides dehiscent ovaries, which open of themselves, find *indehiscent* ones.

DISSEP'IMENTS—the partitions between the cells of syn-carpous ovaries (Figs. 291–318).

PARIETES—the walls of the ovary.

AXIS—the central part of an ovary. In compound ovaries it is where the ventral sutures join together. Find the axis in Figs. 291–318.

VALVES—the parts into which carpels separate by dehiscence (Fig. 319).

PLACENTA—the cord along the ventral suture to which ovules are attached.

Point out and *name* the various kinds of pistil shown upon the charts.

EXERCISE XLII.

The Structure of Ovaries.

Whether a pistil is simple, multiple, or compound, each carpel may be looked upon as a single leaf. The simple pistil of the pea, for instance, may be regarded as the blade of a leaf folded at the midrib, so that its inner portion answers to the upper face of a leaf, and its outer portion to the under face. Its dorsal suture will correspond to the midrib, and its ventral suture to the margin of the leaf.

To make this plainer, take any strong oblong leaf (Fig. 285), and fashion it into a carpel, like the pea-pod, taking the upper part of the leaf for the inner part of the carpel. Fold in the margins slightly to represent the placentæ (Fig. 286). If the fold will not stay in place, take a stitch or two along it with a needle and thread. Now fold it together at the midrib (Fig. 287), and compare it with a pea-pod. Find the valves; the dorsal and ventral portions; the stigma; the base.

Gather old, faded pea-blossoms, in which the ovary is somewhat enlarged, and observe that the ventral suture is turned inward; that is, it lies along the central line, or axis, of the flower. It is along this axis, then, that the double placentæ are formed. Observe the position of the dorsal suture, or back of the pod. It is important to bear in mind that, in the case of the simple pistil, the ovules are attached centrally along the axis of the flower.

Roughly to imitate a multiple pistil, you have only to bind together, by their petioles, several leaf-blades that

6

have been converted into carpels, as above. Observe the placentation of any multiple pistil, and you will invariably find that the placenta of each carpel is central in the same way that, in the artificial one, you have made the margins of your carpellary leaves turn inward, and the midribs outward.

After thus preparing simple and multiple pistils from foliage leaves, let us try to construct a compound pistil from leaf-blades. If we can do this, it will give us a clear understanding of the structure of syncarpous ovaries.

Form, from foliage leaves, an artificial ovary of three coherent carpels. A three-celled compound pistil consists

Fig. 285.   Fig. 286.   Fig. 287.

of three carpellary leaves grown together. It is as if, by pressing together the carpels of your multiple pistil, they should unite by their sides. To make an artificial compound pistil, then, you have only to select three large, symmetrical foliage leaves, and pin or stitch them together in such a way that their margins will meet in the center, and their under surfaces will form its outer wall. If you can

not get leaves of firm texture that will hold a pin or a
stitch without tearing, try lining them with some thin cloth
or paper.   Fold each of the leaves at the midrib, with the
upper surface inward, as seen in Fig. 288.   Fasten the left
half of one leaf-blade to the right half of another, so that

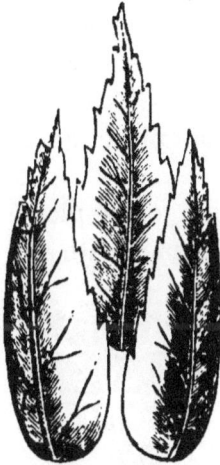

FIG. 288.

FIG. 289.

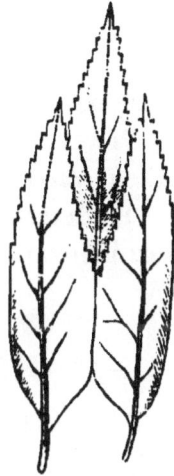

FIG. 290.

the united portions will form a double wall between the
cells, and the six edges will meet together at the center, as
represented in Fig. 289.

Your aim being simply to understand how, and from
what, each part of a compound pistil is formed, you need
not care for the clumsiness or shapelessness of your manu-
factured ovary.

Point out its cells. Its dissepiments. Explain why
they are double. Point out the dorsal and ventral suture
of each carpel of your syncarpous structure. Where
should you look for ovules in this pistil?

Prepare a compound ovary by joining three leaves at
their margins, as seen in Fig. 290. In what part of an
ovary so formed are the leaf-margins? In what part of
the ovary would you look for the ovules? The theory
that the pistils are made from leaves is important, because

it gives clear ideas of the varied and complex characters of ovaries; and these characters are of the greatest value in classification.

EXERCISE XLIII.

Placentation.

After studying the structure of ovaries, as explained in Ex. XLII, the following definitions will be easily understood :

PLACENTATION. — The arrangement of placentæ is called *placentation.*

To determine the mode of placentation of a plant, slice its ovary across, and compare its appearance with the

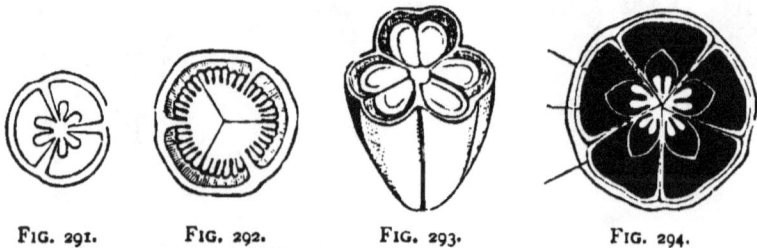

FIG. 291.   FIG. 292.   FIG. 293.   FIG. 294.

following figures. The formation and arrangement of placentæ are so various, that we have given an unusual number of drawings to illustrate the definitions.

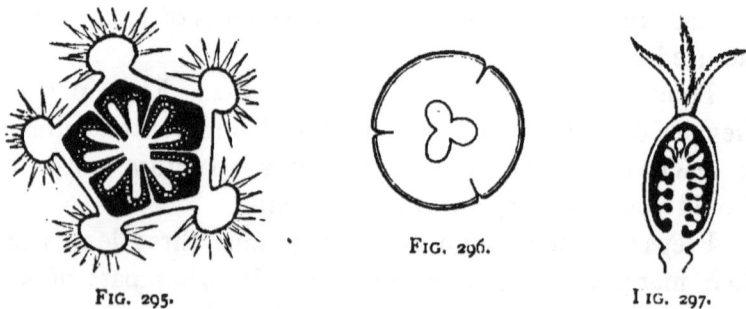

FIG. 295.   FIG. 296.   I IG. 297.

AXILLARY PLACENTATION. — When the ovules are found along the central line, or axis of the pistil, the placentation is called *axillary*, or *axile* (Figs. 291, 292, 293, 294, 295).

FREE-CENTRAL PLACENTATION.—When the dissepiments, or double partitions between the cells, are absent, leaving the placentæ and ovules at the center, and all the

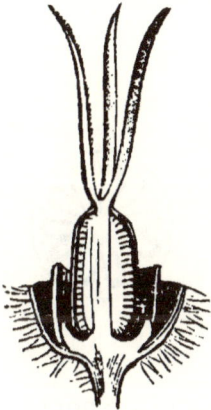

FIG. 298.

FIG. 299.

FIG. 300.

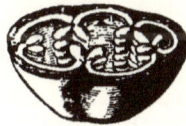

FIG. 302.

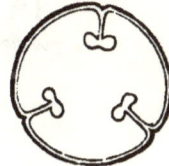

FIG. 301.

cells opening into one chamber, the placentation is said to be *free-central* (Figs. 296, 297, 298, 299).

PARIETAL PLACENTATION is seen when the placentæ are attached to the walls, or projections from the walls, of the ovary, as is illustrated in Figs. 300–307.

FIG. 303.

FIG. 304.

FIG. 305.

FIG. 306.

FALSE DISSEPIMENTS.—It will be well to know that, in many ovaries, there are partitions not formed in the way described in Ex. XLII. The following are instances of what are known as false dissepiments :

Observe in Fig. 308 a partition going inward from the

dorsal suture, and nearly reaching the center of the seed-vessel.

Fig. 309 shows a similar false partition not quite so much extended.

Fig. 310 is a section across the middle of an ovary, and Fig. 311 is a section across the upper part of the same

FIG. 307.   FIG. 308.   FIG. 309.

ovary. The partitions that appear in one and are not seen in the other must be false—they can not be formed by the sides of adjacent carpels.

In Fig. 312 the placentæ are parietal, but a membrane is formed, reaching across the ovary, and forming a false

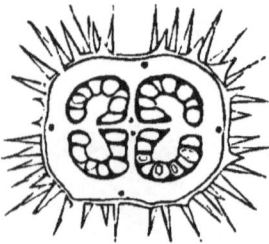
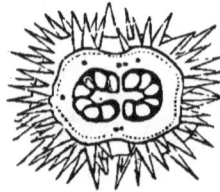
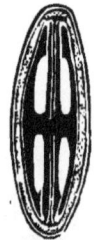

FIG. 310.   FIG. 311.   FIG. 312.

dissepiment. These false dissepiments, you see, are developed, in some cases, from the dorsal suture; in others, from the placentæ.

It may sometimes be difficult to decide between true and false dissepiments; but, as your knowledge of plants increases, the different members of the same group will often be found to afford transitional characters that make evident what otherwise would be uncertain.

Observe and name all the forms of placentæ seen upon the charts.

EXERCISE XLIV.

Modes of Dehiscence.

To understand the modes of dehiscence pictured in this exercise, you have only to prepare a three-celled compound ovary, as directed in Ex. XLII, observing the place of the dorsal and ventral sutures, the relations of the valves, and that the partitions are double.

REGULAR OR VALVULAR DEHISCENCE occurs when the ovary separates into the regular pieces called valves.

Dehiscence is SEPTICIDAL when the ovary splits through the partitions, each dissepiment separating into its two layers, one belonging to each carpel (Figs. 313, 314, and 315),

Dehiscence is LOCULICIDAL when the splitting opens into the cells by the dorsal suture, as seen in the dia-

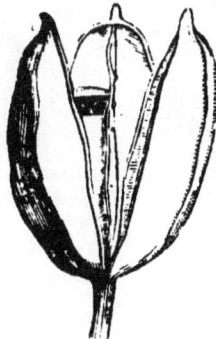
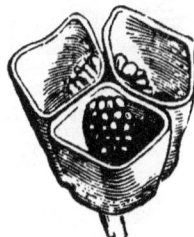

FIG. 313.   FIG. 314.   FIG. 315.

gram 316 and in Fig. 317, which represents the ovary of a violet, where the carpels flatten out as soon as they are released from each other.

Dehiscence is SEPTIFRAGAL where the valves fall
away, leaving the dissepiments behind attached to the
axis (Figs. 318 and 319).

FIG. 316.   FIG. 317.   FIG. 318.

IRREGULAR DEHISCENCE.—Seeds are sometimes dis-
charged through chinks, or pores (porous dehiscence)
(Fig. 320), or the ovary may burst in some part irregularly.
Name the modes of dehiscence given on the charts.

Now compare the
capsules in your collec-
tion with the figures
and definitions given in
this exercise, and deter-

FIG. 319.   FIG. 320.

mine, if you can, the mode of dehiscence of each of them.

How would you produce loculicidal dehiscence in the
compound ovary you have made with leaves, as directed
in the opening of this exercise?

How septicidal? How septifragal?

EXERCISE XLV.

Direction of Ovules and Seeds.

Ovules have a *horizontal* direction when they are neither turned upward nor downward, as in Figs. 321 and 322.

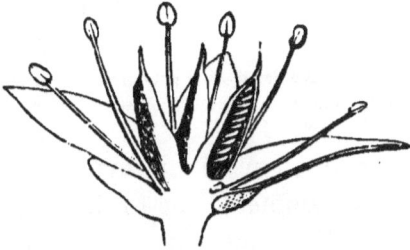

They are *ascending* when
rising obliquely upward,
as in Fig. 323.

Ovules are said to be

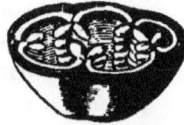

FIG. 321.

FIG. 322.

erect when rising upright from the base of the cell (Fig.
324). They are *suspended* when hanging perpendicularly

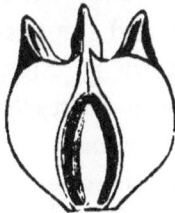

FIG. 323.

FIG. 324.

FIG. 325.

FIG. 326.

from the summit of the cell (Fig. 325). They are *pendulous* when hanging from near the top (Fig. 326).

Find examples of ovules having different directions in
the magnified ovaries upon the charts.

EXERCISE XLVI.

Parts of the Ovule.

FIG. 327.—Growth of Ovule of Celandine.—*a.* Nucleus. *b.* First-formed covering. *c.* Second covering. *d.* Funiculus, very greatly enlarged. *e.* Base of Ovule.

BASE OF OVULE.—The little stem of an ovule—the funiculus—has two points of attachment, one to the ovule and the other to the placentæ. Now, the base of the ovule is at the point where it is attached to the funiculus, and not at the point where the funiculus is attached to the placentæ.

APEX.—The apex of the ovule is opposite to the base.

PRIMINE.—The outer covering of the ovule—seen at *b*, Fig. 327.

SECUNDINE.—The inner covering of the ovule—*c*, Fig. 327.

NUCLEUS.—The substance within the coverings—*a*, Fig. 327.

RHAPHE.—The connection between the base of the nucleus and the base of the ovule. This is shown in Fig. 328 by the fine, irregular lines representing tissue and connecting the base of the nucleus with the base of the ovule.

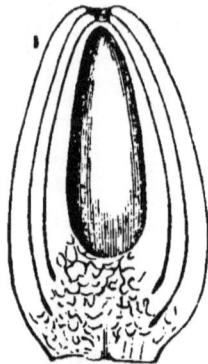

FIG. 328.—Section of the Ovule of Pelargonium before fertilization. (Magnified.)

MICROPYLE.—The opening in the coats of an ovule or seed. In Fig. 328 the micropyle is shown at the top of the ovule.

CHALAZA.—The place where the coats and nucleus grow together. In Fig. 328 it can not be distinguished from the *rhaphe.*

·   HILUM.—The scar left by the separation of a seed
from its placenta.

It is not supposed that pupils will find all these parts
of the ovule in plants.   Some of them are usually discernible, and they may all be understood in their proper relations by studying the diagrams.

EXERCISE XLVII.

Kinds of Ovule.

The STRAIGHT, or ORTHOTROPOUS OVULE, has the
base of the nucleus and the base of the ovule in the same
position, while the micropyle is at the apex (Fig. 329).

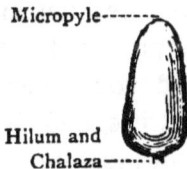

FIG. 329.
Straight, or Orthotropous.

FIG. 330.
Curved, or Campylotropous.

In the CURVED, or CAMPYLOTROPOUS OVULE, the micropyle, or apex, is bent over close to the base (Fig. 330).

In the INVERTED, or ANATROPOUS OVULE, the funiculus lengthens, and bends round, growing fast to the coat,

FIG. 331.
Inverted, or Anatropous.

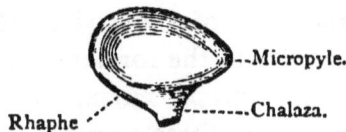

FIG. 332.
Half-inverted, or Amphitropous.

until the base of the nucleus is at the apex of the ovule
(Fig. 331).

In the HALF-INVERTED, or AMPHITROPOUS OVULE, the
funiculus only lengthens till the ovule turns a quarter of
the way over, as in Fig. 332.

In describing the pistil of flowers, answer the following questions: What is the form and position of the stigma? Of the style? Is the pistil simple or compound? What is its placentation? Its dehiscence? What is the direction of the ovules? Can you determine the kind of ovule?

EXERCISE XLVIII.

The Composition of Fruit.

FRUIT.—The ripened ovary, with its contents, is the fruit of plants. Whatever adheres to the ovary also becomes part of the fruit.

In studying fruit, observe with care what parts, besides the pistil, have been concerned in its formation. In describing flowers, you note whether the pistil is inferior or superior; is there any reason to suppose that inferior fruit would be most likely to have other parts of the flower besides the pistil united with it? Did you observe the flowers of the cherry, plum, or peach trees, and those of apple and pear trees when they were in blossom? and if so, will you compare your recollection of them with the appearances presented by their fruit? If you have forgotten their structure, perhaps you have kept a description of them, and can refresh your memory.

Observe the ripe fruit of the cherry. Look at the top of the peduncle for scars left by the parts of the fallen flower. Look for a dot at the top of the fruit, showing the place of the style. Has anything but the pistil entered into the formation of this fruit? Observe the plum, peach, grape, currant, etc., and see if they are like the cherry in these respects.

Now examine an apple or pear. What do you find at the top of the fruit, opposite the peduncle? It must be the remains of the calyx-limb, the tube of which you saw united to the pistil when you studied it in flowering-time. Of what, then, does the fruit consist? Divide an apple or pear, as shown in Fig. 333. Find the parts shown in this diagram. The remains of the flower are seen at C. The calyx-tube, grown fleshy and succulent, is marked T. The outer border of the ovary is seen at E. From what part of the flower is the eatable portion of a pear or apple developed? To repeat our former question, would the fruit of a superior pistil be more likely than that of an inferior pistil to consist of the ovary alone?

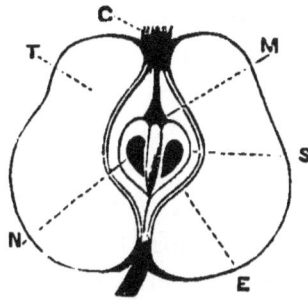

Fig. 333.

I have illustrated the composition of fruit with apples and cherries because they are so common; but these observations may, and should be, repeated upon every variety of fruit that can be found.

Trace the formation of each of the fruits pictured upon the charts, and point out those that consist of the pistil alone, and those which do not. In the latter case, name the parts that are consolidated with the pistil in the fruit.

When fruit is formed from the pistil alone, the wall of the ovary is called a *pericarp* (from *peri*, around).

Gather specimens of every kind of fruit that grows within reach. In late summer or early autumn, the fruit of garden, field, and forest, if carefully collected, will give you a large and various assortment. For example: you may have at the same time cucumbers, melons, beans, peas, grapes, apples, pears, elder and pokeweed berries, chestnuts, walnuts, pumpkins, etc., and the less conspicuous seed-vessels of mullein, Saint-John's-wort, lettuce, radish, cabbage, etc., etc. Earlier in the season the list will be different, and it will vary somewhat with the locality, but, wherever collected, and whatever its components, be sure to gather every kind that can be had.

Look over your collection, and separate the superior

from the inferior fruits. Observe the structure of those formed from inferior pistils, and point out the pericarp in those formed from superior pistils.

Preserve, for further study, the specimens you have gathered.

Parts of the Pericarp.

EPICARP.—When the walls of a pericarp are formed of two or more layers of different texture, as in the peach, plum, or cherry, the outer one (the skin, in the case of these fruits) is called the *epicarp*.

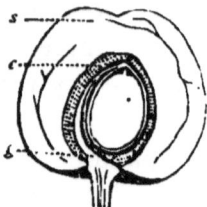

FIG. 334.

ENDOCARP.—The stony case around the seed of the peach, plum, or cherry, is called the *endocarp*. But the endocarp of fruits is not always stony. Whatever its texture, the inner layer of a pericarp is named the endocarp.

MESOCARP. — Sometimes, between the outer and inner parts of a pericarp, there is found a third layer of different aspect, like the pulp of a peach. This third layer is called the *mesocarp*. The distinction between the epicarp and mesocarp is often very slight, and then both together are called the *epicarp*.

In Fig. 334, *e* is the endocarp, *s* the mesocarp, and *g* the epicarp.

In Fig. 333, E is the epicarp, N the endocarp, and S the seeds. At N is shown a slight development of the mesocarp. Point out these parts in an apple and a peach. Point out the parts of the pericarp in the different fruits pictured upon the charts.

Classify your collection of fruits by the structure of the pericarp. Put by themselves all those that have but one layer in the pericarp. Put those with two layers—an epicarp and endocarp—by themselves, leaving those with three layers—epicarp, mesocarp, and endocarp. Describe the layers that make up the fruit ; that is, say 'whether, in each case, the layer is pulpy, woody, stony, membranous, leathery, etc.

Preserve your collection for further study, and add to it all you can get.

EXERCISE L.

The Classification of Fruit.

Look over your collection and separate the dehiscent from the indehiscent fruits. The indehiscent group may now be further separated into juicy fruits and dry fruits. Compare your specimens of juicy fruit, one by one, with the following pictures and definitions of fruits. The first picture is that of a berry ; so you may first find the berries of your collection. To determine whether a particular fruit is a berry or not, cut it across, and see if it agrees in structure with Fig. 335, and the requirements of the definition. Never mind whether your conclusion accords with common speech or not ; whether a strawberry turns out to be a berry or not; but follow the definition wherever it leads.

Indehiscent Juicy Fruits.

BERRY.—A thin-skinned, indehiscent, fleshy fruit, having the seeds imbedded in the pulpy mass (Figs. 335, 336).

FIG. 335.　　　　　FIG. 336.

HESPERIDIUM.—A kind of berry with a leathery rind (Fig. 337). (Example, lemon and orange.)

PEPO.—The pepo is an indehiscent, fleshy fruit, with seeds borne on parietal placentæ, and with the epicarp more or less thickened and hardened. (Example, squash.)

FIG. 337.

POME is the term applied to a fleshy, indehiscent, several-celled fruit, with a leathery, or cartilaginous, endocarp, inclosed by the calyx-tube. Figs. 338 and 339 are transverse and vertical sections of a pome. (Example, apple and pear.)

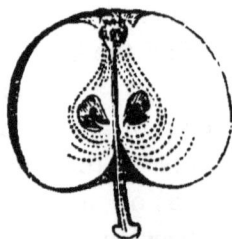

FIG. 338.

FIG. 339.

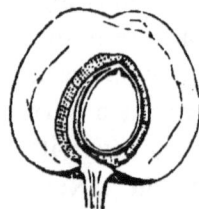

FIG. 340.

DRUPE (example, peach or cherry) is a pulpy, indehiscent, one-celled, one or two seeded fruit, with a succulent or fibrous epicarp, and hard, stony, distinct endocarp (Figs. 340 and 341).

If you have blackberries, raspberries, and the like, among your fruits, compare one of the little cells of which they are formed with this definition of a drupe. To one or other of these classes you should be able to refer any form of indehiscent juicy fruit.

FIG. 341.

Indehiscent Dry Fruits.

Select from among your dry, indehiscent fruits all those that resemble Figs. 342, 343, 344, and 345, and that are usually miscalled seeds. You will find upon many of them such appendages as hairs, teeth, plumes, bristles, etc.

An ACHENIUM is a dry, indehiscent, one-seeded fruit,

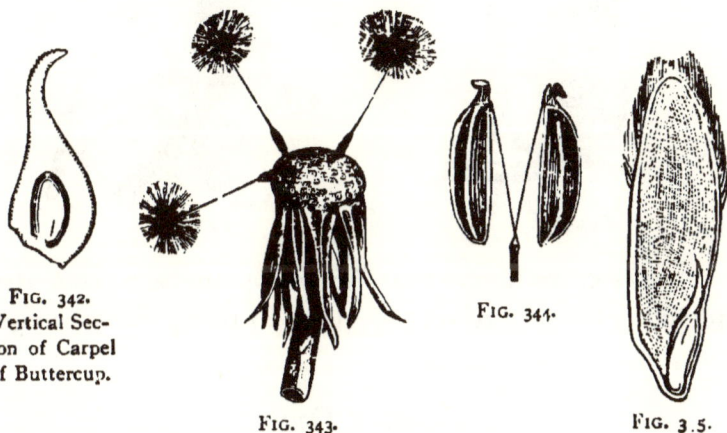

FIG. 342.
Vertical Section of Carpel of Buttercup.

FIG. 344.

FIG. 343.

FIG. 3.5.

with a separable pericarp, tipped with the remains of the style (Fig. 342). (The dark-colored, seed-like bodies on the outside of a strawberry are achenia.)

UTRICLE.—By this term is understood a kind of achenium, with a thin, bladdery pericarp which is sometimes dehiscent.

CARYOPSIS.—A dry, indehiscent, one-celled, one-seeded fruit, with the pericarp adherent to the seed, as seen in wheat, barley, oats, maize, etc. (Fig. 345).

CREMOCARP.—Pendent achenia (Fig. 344). (See Ex. LXVI.)

CYPSELA.—Still another variety of achenium, with an adherent calyx-tube, as in compositæ (Fig. 343).

NUT.—A hard, one-celled, one-seeded, indehiscent fruit, produced from a several-celled ovary, in which the cells have been obliterated, and all but one of the ovules have disappeared during growth. It is often inclosed in an involucre, called a *cupule* (Fig. 346), or it has bracts at the base.

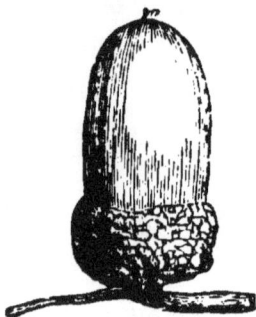

FIG. 346.   FIG. 347.

SAMARA, or KEY-FRUIT (example, the elm).—A dry, indehiscent fruit, growing single or in pairs, with a winged apex, or margin (Fig. 347).

Dehiscent Fruits.

Any dry, dehiscent fruit, whether simple or compound, may properly be called a pod.

FOLLICLE.—A pod of a single carpel, with no apparent dorsal suture, and dehiscing by the ventral suture (Fig. 283). You will seldom find an

FIG. 349.   FIG. 348.   FIG. 350.   FIG. 351.

ovary consisting of but one follicle ; but it is a common kind of carpel in multiple pistils. Observe the ripe ovary

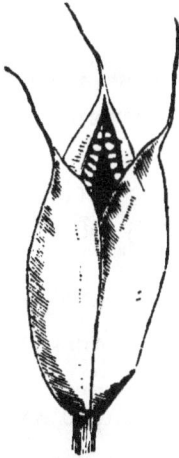

FIG. 353.

FIG. 352.

of columbine or pæonia. Each carpel is a follicle, and you may find them slightly coherent at the base, as if forming a transition between the apocarpous and syncarpous pistil.

LEGUME.—A pod of a single carpel, with dorsal and ventral sutures, and dehiscing by both or either, as the pea and bean pod. It assumes many different forms.

One of these, the LOMENT, is a sort of legume with transverse joints between the seeds, and falling to pieces at these joints (Fig. 348).

Another variety, the SILIQUE, is a two-valved, slender pod, with a false dissepiment, from which the valves separate in dehiscence. It has two parietal placentæ (Fig. 349).

SILICLE.—A short, broad silique (Fig. 350).

PYXIS.—A pod which dehisces by the falling off of a sort of lid (Fig. 351).

CAPSULE.—The pod of a compound pistil; the dry, dehiscent fruit of syncarpous pistils (Figs. 352 and 353). The pieces into which a capsule falls at dehiscence are called valves, the same as in one-carpeled fruit.

Those fruits that consist of achenia on a dry receptacle, as the sunflower, or on an enlarged, pulpy receptacle, as the strawberry, or those which consist of small drupes on a dry, spongy receptacle, crowded almost into one mass, as the blackberry, are *aggregate fruits.* They are sometimes called *etærio.*

Accessory, or anthocarpous fruits, are such as consist of other parts of the flower only apparently joined with the ovary.

MULTIPLE, COLLECTIVE, or CONFLUENT FRUITS, are formed by the union of many separate flowers into one mass (Figs. 354 and 355).

The *sorosis* is a kind of multiple fruit, to which the

FIG. 354.   FIG. 355.

pineapple (Fig. 354) belongs. The fig is a multiple fruit of the kind known as *syconus,* while *strobilus* is the name given to the multiple fruit of trees of the pine family (Fig. 355).

EXERCISE LI.

The Seed.—Its Form and Surface.

The forms of seeds vary very much. They may be globular, ovoid, reniform, oblong, cylindrical, top-shaped, angular, etc. Some seeds are small and fine, like sawdust; others are flattened and bordered, as seen in Fig. 356.

The surfaces of seeds may be smooth, striated, ribbed, furrowed, netted, and tubercular, as shown in the following figures :

Seeds are said to be *definite* when few and constant in number ; *indefinite* when numerous and variable.

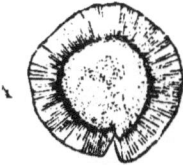

FIG. 356.

FIG. 357.
Smooth.

FIG. 358.
Striated.

FIG. 359.
Ribbed.

FIG. 360.
Netted.

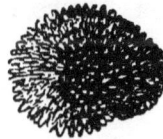

FIG. 361.
Tuberculous.

FIG. 362.
Furrowed.

Seeds are *solitary* when single in the ovary, or in a cell of the ovary.

The albumen of seeds is the mass of tissue in which the embryo is imbedded. It is said to be *mealy* when it may be readily broken down into a starchy powder ; *oily*, when loaded with oil; *mucilaginous*, when tough, swelling up readily in water ; and *horny*, when hard, and more or less elastic.

EXERCISE LII.

Parts of the Seed.

Prepare for the study of the parts of seeds by planting all the kinds of seeds that you can get that are large enough for easy examination.

The seeds of the pumpkin, squash, four-o'clock, bean, pea, apple, Indian corn, oats, and barley, are good examples for the purpose. Plant two or three dozens of each sort, one inch deep, in a box of soil or sawdust, which must be kept warm and moist. Put the different kinds in rows by themselves, and mark each row, so that, when you want any particular one, you can get it without mistake.

When your seeds have soaked for a day or two in the wet earth, take a bean from the box and compare it with one that has not been planted.

How has it changed in appearance?

Cut it in two and see whether, like a piece of chalk, it looks alike outside and inside, or whether the parts are unlike.

Has it a skin or shell that you can loosen?

Take a second bean from the box, cut carefully around it, and try to peel off the outer part.

SEED-COAT, OR INTEG'UMENT. — The skin or shell around the outside of a seed.

BODY, KERNEL, OR NU'CLEUS.—The substance within the seed-coat.

Compare your specimen with Fig. 363.

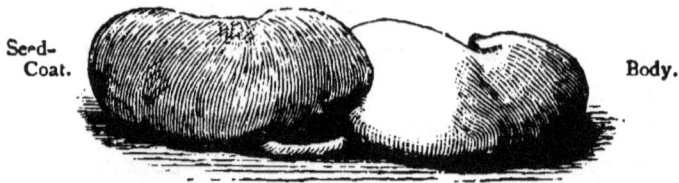

FIG. 363.

Can you separate the seed-coat from the body of the bean as it is seen to be separated in the picture?

Now take a pea from your box and see if it is made up of parts.

Has it a seed-coat? Is there a kernel or body within the seed-coat?

Try a pumpkin-seed. Compare the coat of a pumpkin-seed with that of the pea or bean.

Are they alike in thickness? in hardness? in color? in transparency? Name all the differences you see between them.

In the same way, take up and examine, one after another, seeds from each of the rows. Find their parts, and compare the parts of one kind of seed with those of another kind.

If you are not able at first readily to separate a seed into distinct portions, do not hastily conclude that it is without them. Let it lie in its warm, wet bed a while longer, and then try again.

EXERCISE LIII.

Parts of the Body, or Kernel.

When you have carefully examined all the kinds of seeds you planted to find the parts that make them up, you will be ready to study one of these parts by itself. After

FIG. 364.

FIG. 365.
Albumen. Embryo.

FIG. 366.
Embryo.

taking off the skin or coat of a seed, look closely at the body of it. Begin with a well-soaked seed of Indian corn. Compare it with Fig. 364.

Is your seed narrower at one end than the other? Are the two sides of it alike? Is there a little pointed or rounded figure to be seen on one side?

Remove the skin and look carefully at the figured side of your specimen. Can you see a thick, lumpy body like the one marked *a* in the picture?

Try, with a dull knife or the finger-nail, to pry this lump out of its bed. If the seed is soaked to its center,

you can easily do this. Look carefully at the hole it leaves.
Is not its surface smooth? Do you see any spot where
the lump seems to have been grown to the other part, and
to have broken away when you took it out?

Compare the parts you have got with Figs. 365, 366.

Em'bryo.—The young plant contained in a seed.

Albu'men, En'dosperm.—The material in which the
embryo is imbedded.

What names are given to the two parts of the body of
a seed of Indian corn?

Which is the embryo in your specimen? Which is the
albumen?

Now examine the kernel of a pea or bean. Can you
separate this into two parts without breaking it somewhere?

Compare it with the parts of Indian corn.

What name is given to the entire kernel? What part,
found in the Indian corn, is missing here?

            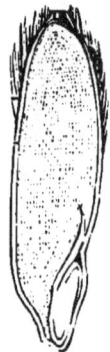

Fig. 367.   Fig. 368.   Fig. 369.   Fig. 370.

Look at the body of a seed of four-o'clock
and see how many and what parts it has.
Look also at the body of a pumpkin-seed.

Fig. 371.

Examine the kernel of each of the kinds of
seed you have planted, and observe which consist of embryo alone, and which are part embryo and part albumen.

Albuminous Seeds are those which have albumen.

Exalbuminous Seeds are those in which the body
consists of the embryo alone.

The relations of embryo to albumen in various seeds

are here shown. But they may be better seen upon the charts. Your own observation, however, will supply you with much information upon this subject.

EXERCISE LIV.

Parts of the Embryo.

Take out of the soil a bean which has begun to sprout. Remove the seed-coat, and let the parts of the embryo separate, as seen in Figs. 371 and 373.

COTYLE'DON.—The bulky first leaf or leaves of the embryo—more or less formed before the growth of the seed begins.

RAD'ICLE.—The lower, or root end, of the embryo.

PLU'MULE.—The first—the terminal bud—the upper end of the embryo.

GERMINATION.—The beginning of growth in a seed.

Read the names of the parts of the embryo given in

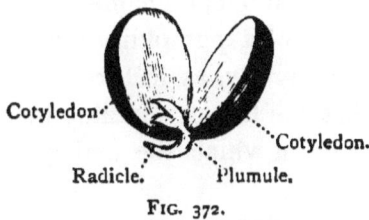

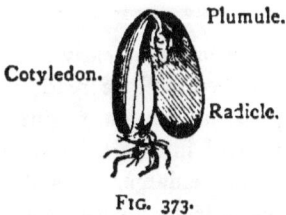

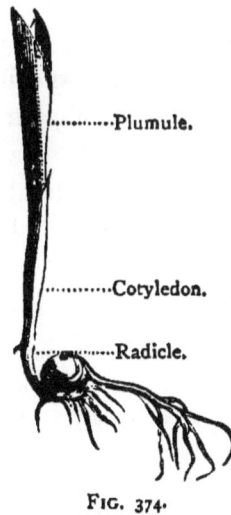

Cotyledon·

Cotyledon.

Radicle.   Plumule.

Fig. 372.

·······Plumule.

·······Cotyledon.

Plumule.

Cotyledon.

·······Radicle.

Radicle.

Fig. 373.   Fig. 374.

Figs. 372 and 373. Look at the definitions of these words. Compare your specimen with the figures, and point out its cotyledons; its radicle; its plumule. Handle your em-

7

bryo with care, for it breaks easily. Has its radicle begun to put forth roots?

Take from your box a vigorous seed of Indian corn in which the roots have begun to grow, and compare it with Fig. 374.

Separate the embryo and albumen, and, if it has grown as much as the one pictured above, you may easily find the cotyledon, the plumule, and the radicle.

When you are sure that you have found the radicle or root-end of your embryo, that you know which part is cotyledon, and which plumule, take another seed of the same kind, but less grown—one where the root-end of the embryo has scarcely begun to swell —and see if you can find the parts.

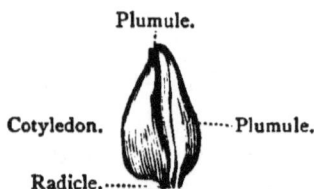

Plumule.

Cotyledon.   ······Plumule.

Radicle.········

FIG. 375.

Fig. 375 represents such an embryo with the parts shown.

Point out and name the parts of the embryo of an apple-seed; of a pumpkin-seed; and of each of your specimens successively, as in former exercises. Which of your seeds has the largest plumule before growth begins? Have you any in which the embryo has at first no plumule at all?

Have you failed to find cotyledons in any embryo looked at?*

---

* If these experiments with seeds are made as early as April, in this climate, the children who have made them will be ready for more extended observations when planting in the garden begins. Most garden-seeds are too small to be separated into parts by young children. But, when growth begins, their parts enlarge, and a child, who has before studied larger seeds, will be able to identify the radicle, cotyledons, and plumule, without difficulty. In the kitchen-garden, a universal appendage of country-houses, the sprouting of the radish, onion, beet, parsnip, lettuce, tomato, carrot, cabbage, cucumber, etc., will furnish an excellent continuation of the study of seeds.

EXERCISE LV.

Monocotyledons and Dicotyledons.

A MONOCOTYLED′ONOUS embryo has one cotyledon or seed-leaf (Fig. 376).

A DICOTYLED′ONOUS embryo has two cotyledons or seed-leaves (Fig. 377).

These are long, hard words, hard to pronounce and hard to spell. But they are very necessary words in describing seeds.

Go over the seeds you have planted, and point out the dicotyledons. Find the two thick leaves that were packed within the seed-coat when the seed ripened.

Are any of your seeds monocotyledonous? If so, which?

Figs. 376 and 377 were drawn from plants that had grown a little. When your seeds have also grown a little,

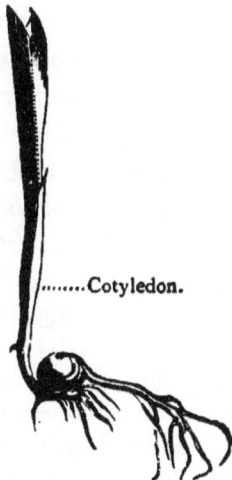

......Cotyledon.

FIG. 376.

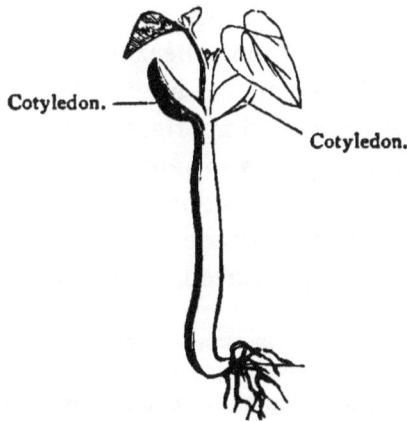

Cotyledon. ——

Cotyledon.

FIG. 377.

compare them one after another with these pictures. Look at your young bean-plant. Find the first node above the cotyledons. How many leaves are growing there? how many at the first node of the corn-stem? how many in each of your growing seeds?

Observe whether the cotyledons in all cases rise into the light and air. Observe whether all cotyledons are shaped alike, and also whether they resemble the true leaves of the plant.*

EXERCISE LVI.

Position of the Embryo in Seeds.

You are now familiar with the different aspects of the embryo in many different seeds. You have seen it large and small, straight and curved, outside the albumen and imbedded within it; sometimes with flat cotyledons, and sometimes with cotyledons folded or coiled in various ways and degrees. We are now to observe its relation to the parts of the seed.

In studying ovules, you found the hilum and the micropyle, and you may find the same parts in the seeds that were once ovules. The hilum of seeds is usually obvious enough, and the micropyle may be easily found. You have only to soak the seed till its coats are distended with water, and, on squeezing, the micropyle is made apparent by the escape of water at that point. The place of the micropyle is important, because the radicle of the embryo always points toward it, and, in sprouting, issues through it, and the relation of the micropyle to the hilum determines the attitude of the embryo. Seeds are straight, half inverted, inverted, and curved, the same as ovules, and

---

* A word of caution may not here be amiss. There is danger that the sympathy of teachers with bright and interested pupils will lead them to tell in advance what children can find out for themselves by continued observation. The relation between number of cotyledons and venation is an instance of such temptation. By-and-by, when the leaves of his growing plants are well developed, the pupil might be put in the way of discovery, by asking him to make a list of his monocotyledons, and to give their venation in each case. Let him do the same with his dicotyledons. He will now see a perfect uniformity of relation in a few cases, and will be curious to know if it is everywhere constant. He will thus arrive at the induction by his own observation.

the same terms are used to express these facts in regard to them. In a straight or orthotropous seed, the micropyle being at the apex, you find an inverted embryo, like Fig. 378. In this case the embryo is said to be *antitropal,* or reversed.

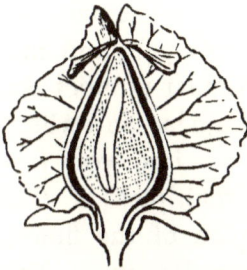

Fig. 378.

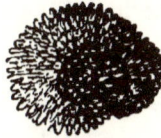

Fig. 379.

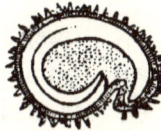

Fig. 382.

Fig. 380.

Fig. 381.

If the micropyle be turned to one side, the embryo will be oblique, as seen in Fig. 380. In this case the embryo is said to be *heterotropal.* Fig. 379 represents the seed which is shown in section in Fig. 380.

If the seed be inverted, or antitropous, the embryo will be erect, as shown in Fig. 381. Here the embryo is said to be *orthotropal.*

When a seed is curved upon itself so as to bring the orifice next the hilum, or point of attachment (campylotropous seed), you may find the embryo presenting the appearance shown in Fig. 382.

When the embryo is in the center of the albumen (Fig. 381), it is said to be *axial ;* and when not in the center, it is said to be *excentric.*

There are two modes of folding to which the embryo is subject, which occur uniformly in certain groups of plants. They are *cotyledons accumbent*—that is, with the radicle folded against their edges ; and *cotyledons incumbent,* having the radicle folded against the back of one of them.

FLORAL SYMMETRY, PHYLLOTAXY, PREFOLIATION,
CYMOSE INFLORESCENCE, ETC.

EXERCISE LVII.

Numerical Plan of Flowers.

FIG. 383.

WHEN, in examining a flower, you count the parts of its calyx and corolla, the stamens and the carpels, and find that some particular number occurs again and again; and when, in case of deviation, you frequently find multiples of this number, the plan of the flower is said to be based upon it. For instance, the plan of the flower represented in Fig. 383 is based on the number three. The plan of the flower represented in Fig. 384 is based on the number four, and that of Fig. 385 upon the number five. In other words, in Fig. 383, three, or its multiple, six, is the con-

FIG. 384.

FIG. 385.

stant number ; in Fig. 384, four is the prevailing number ;
while in Fig. 385 it is five.

What numbers have occurred oftenest in your written
descriptions of flowers ? When you describe a flower,
observe always what figures you use in numbering its
parts, and decide what number the plan of the flower is
based upon.

EXERCISE LVIII.

Alternation of Parts in Flowers.

Figs. 387 and 388 represent the stamens and pistil of
the flower shown in Fig. 386. Does this picture represent
a perfect flower ? Does it repre-
sent a complete flower? a regular flower? a symmetrical flower?
Fig. 389 is a cross-section of this
flower, given to illustrate the relation of the parts to each other.
Observe that the petals alternate
with the sepals; that is, they stand
opposite to the openings between
the sepals. In the same way the
stamens alternate with the petals,

FIG. 386.   FIG. 387.   FIG. 388.

and the carpels with the stamens. This regular alternation of parts is spoken of as a symmetrical arrangement

of the flower. Fig. 392 is the cross-section of Fig. 390, and Fig. 391 gives a vertical section of the same flower.

FIG. 389.

FIG. 390.

Are its parts arranged symmetrically? that is, is the alternation perfect?

You see that flowers present symmetry of arrangement as well as symmetry of numbers, and it is important that you should observe them in this respect. Determine what

FIG. 391.

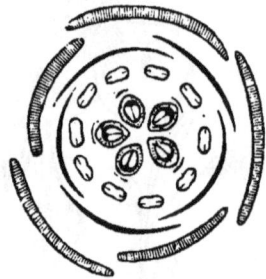

FIG. 392.

parts of the flower you are studying alternate symmetrically, and where the symmetry fails. You will often find these observations valuable in classification.

EXERCISE LIX.

Leaf-Arrangement.—Phyllotaxis.

To study leaf-arrangement, get straight leafy stems, or shoots, a foot or more in length, such as are shown in Figs. 393 and 394, from any vigorous tree, shrub, or herb.

First separate the specimens having opposite and verticillate leaves from those with alternate leaves.

Observe that the successive pairs of leaves in opposite-leaved plants are placed at right angles to each other,

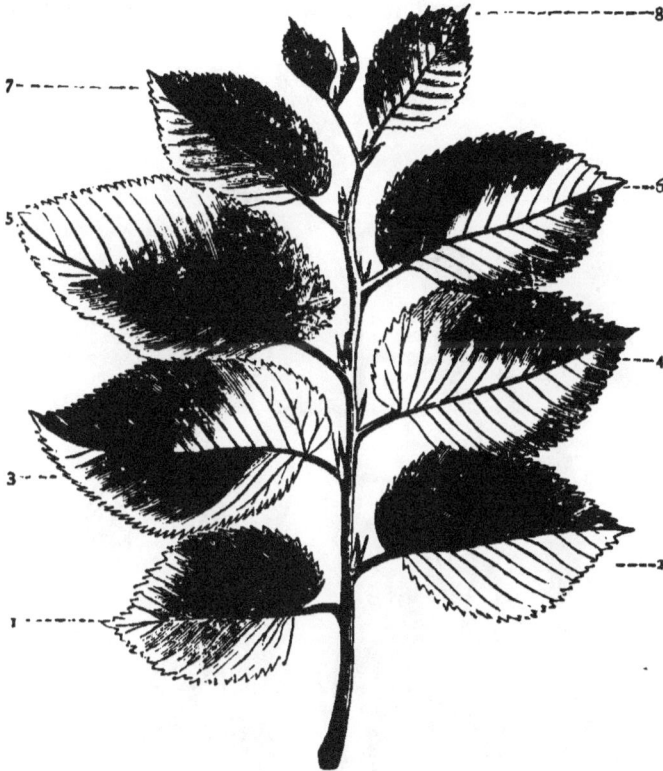

FIG. 393.

each leaf of the upper pair being placed over a space left by the lower pair. They are hence called *decussate* leaves. In the same way the whorls of leaves in verticillate-leaved stems are so placed that they alternate with each other.

Observe the arrangement of leaves in the stems of grasses, and in stems with equitant leaves.

Put by themselves all the stems in which the leaves are neither decussate nor whorled.

Examine them, one after the other, thus : Take a

small string, and, holding one end of it just below one of
the lower leaves of your specimen, carry it up and around
the stem (Fig. 395), so that it shall pass just under each
successive leaf.   Proceed in this way till you reach a leaf
standing directly over the one you started with.   Your
string now includes what is called a leaf-cycle ; that is,

FIG. 394.

the distance in a spiral line around the stem, from one
leaf to another placed exactly above it.

Holding the string in place, observe, first, how many
times it has wound around the stem ; and, second, how

many leaves it passes on its way. If, in passing from the first leaf to the one directly over it, the string makes but one circuit around the stem, and the third leaf is over the first, so that the cycle includes but two leaves, the fourth leaf being over the second, and so on, you have an arrangement like that seen in Fig. 393. The leaves in this example are seen to form two rows along the side of the stem, which are separated by half its diameter.

This is the distichous, two-ranked, or ½ arrangement.

If, in passing from one leaf to another directly above it, the string goes but once round the stem, and the fourth leaf is over the first, giving a cycle of three leaves, the arrangement is like that shown in Figs. 394 and 395. There are three perpendicular rows of leaves along the stem, separated from each other by ⅓ its circumference.

This is the tristichous, three-ranked, or ⅓ arrangement.

Again, the string may pass twice around the stem before it reaches the leaf placed just over the first, which, on counting, proves to be the sixth (Fig. 397). There are five longitudinal rows along the stem, separated from each other by ⅖ its circumference.

This is the pentastichous, quincuncial, or ⅖ arrangement.

Fig. 395.   Fig. 396.

Observe that the numerator in the foregoing fractions gives the number of times the string winds around the stem in completing a cycle, while the denominator gives the number of leaves in the cycle.

This fraction is sometimes called the angle of diver-

gence of the leaves. In Fig. 393 the angle of divergence is ½ the circumference of the stem ; in Fig. 394 it is ⅓, and in Fig. 396 it is ⅖ its circumference.

In studying some of your specimens, the string may pass three times round the stem in its spiral course before you come to a leaf placed over the first, and this leaf may be the ninth in the upward succession, eight leaves being required to complete the cycle. Here you have eight per-

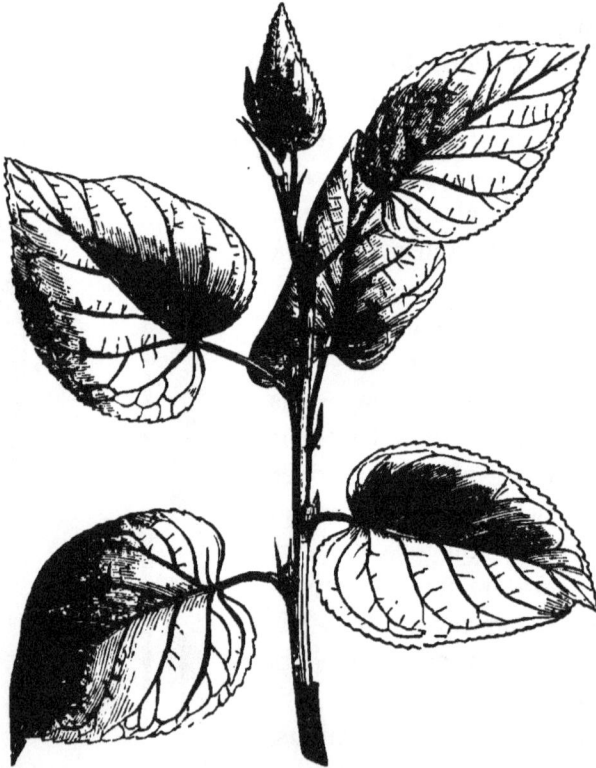

FIG. 397.

pendicular rows of leaves, with an angular divergence of ⅜ the circumference of the stem ; it is, therefore, called the ⅜ arrangement.

In some plants the leaf-cycle includes five turns of the spiral and thirteen leaves, so that the fourteenth is placed over the first. This is the $\frac{6}{13}$ arrangement. There are also the $\frac{8}{21}$, the $\frac{13}{34}$ arrangements, and so on. But these more complex modes are only found where leaves grow in rosettes, as the house-leek, or in the case of crowded radical leaves, or in the scales of cones. In these cases the vertical rows are not distinguishable, and the order has to be made out by processes of reasoning rather than by simple observation.

There is a curious feature of the fractions expressing the angular divergence of leaves. Observe that any one of the fractions of the series is the sum of the two preceding simpler ones. For example, the angles of divergence in Figs. 393 and 394 are $\frac{1}{2}$ and $\frac{1}{3}$. Adding these numerators and these denominators, we have $\frac{2}{5}$, the pentastichous, or next more complex arrangement. By adding in the same way $\frac{1}{3}$ and $\frac{2}{5}$, we get $\frac{3}{8}$, while $\frac{2}{5}$ and $\frac{3}{8}$ give $\frac{5}{13}$, and so on.

The $\frac{1}{2}$, $\frac{1}{3}$, and $\frac{2}{5}$ modes of arrangement are so definite and simple as to be easily discovered ; but it is not worth while, ordinarily, to continue the study of a specimen if it does not belong to one of these modes. A slight twisting of the stem, a considerable lengthening of internodes, or their absence altogether, renders observation difficult, and the decision uncertain. So, when commencing the study of leaf-arrangement, take care to select the straightest and thriftiest stems for the purpose.

Examine the arrangement of bracts, and see if they follow the same order as leaves.

Observe whether the spirals take the same direction in branches as in the parent stem. When they do, they are called *homodromous ;* but when they turn in opposite directions, they are said to be *heterodromous.*

Give the numbers of the leaves in each perpendicular series in your specimen showing the $\frac{1}{2}$ arrangement (Fig. 393).

In the $\frac{1}{3}$ arrangement, what leaf stands over the first ?

over the second? the third? the fourth? the fifth? Give the series of numbers that belong to the leaves of each row.

The name applied by botanists to these modes of leaf-arrangement is *phyllotaxis.*

EXERCISE LX.

Arrangement of Floral Leaves in the Bud.—Æstivation, or Prefloration.

In most common flowers, the floral circles, calyx, corolla, etc., appear quite distinct; but have you never observed cases in which it was doubtful where the calyx ended and the corolla began? or, where the corolla ended and the calyx began? or, even, where the bracts ended and the calyx began? Have you never seen sepals with the color and delicacy of petals, and in the same flower some sepals that were green, and some more or less like petals? or, the same sepal green without and petal-like within? Have you not seen the involucre made up of colored bracts, which looked like a corolla? Have you not sometimes met with flowers in which you could see the gradual transition from petals to stamens? or in which some of the stamens or carpels were changed to green foliage-leaves? Have you ever known of single flowers becoming double by cultivation, and of stamens and carpels replaced by petals? Did you ever see a leafy shoot growing out from the center of a flower, or of a flower-bud? These appearances are not uncommon, and may be easily observed if you are watchful.

It is from these singular aspects of plants, joined with the study of their development, that botanists have come to regard flowers as altered branches, and floral leaves as changed foliage-leaves. They speak of carpels as carpellary leaves, stamens as staminal leaves, petals as corolla-leaves, and the sepals as calyx-leaves.

If this be so, the laws of arrangement of floral leaves ought to agree with the phyllotaxy of foliage-leaves. Botanists say that they do so agree, and the place where this agreement is best seen is in the flower-bud. The arrangement of floral leaves is an important help in determining the affinities of plants.

To observe this arrangement, make a horizontal section of a bud just before it opens. Be careful to make the section in the upper part of the bud, where the petals

Fig. 398.   Fig. 399.   Fig. 400.

and sepals are most easily seen. Observe with a magnifying-glass the disposition of parts, and compare your examples with the modes of arrangement here pictured and named.

In VALVULAR præfloration there is no overlapping of parts. The edges of the sepals and petals just meet, and the flower is almost always regular (Fig. 398).

INDUPLICATE is a form of valvate æstivation, in which the edges are turned slightly inward, or touch by their external face (Fig. 399).

REDUPLICATE is a form of valvate æstivation, in which the edges turn slightly outward, or touch by their internal face (Fig. 400).

In the CONTORTED arrangement, each leaf overlaps its neighbor, and the parts seem twisted together (Fig. 401). It becomes CONVOLUTE when each sepal or petal wholly covers those within it.

In IMBRICATE æstivation, the parts of a floral circle, usually five, are placed as seen in Fig. 402. The first leaf is external, the fifth internal, and the intermediate ones successively overlap each other.

FIG. 401.

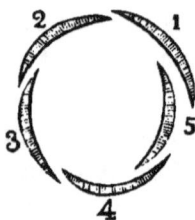

FIG. 402.

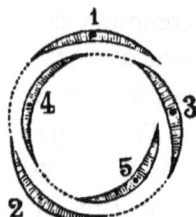

FIG. 403.

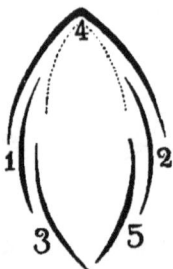

FIG. 404.

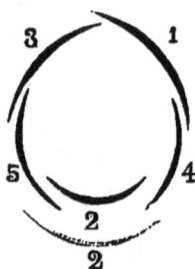

FIG. 405.

FIG. 406.

The QUINCUNCIAL arrangement is seen in Fig. 403. There are two exterior leaves, two interior, and one intermediate.

FIG. 407.

The VEXILLARY arrangement (Fig. 404) is a form of the quincuncial, where one of the petals, that ought to be internal, has, by rapid growth, become larger than the others, and external to them, so as to cover them in.

FIG. 408.

In the COCHLEAR arrangement, inequality of development has produced the order seen in Fig. 405.

We are reminded of the DECUSSATE arrangement of

foliage-leaves by the position of the floral leaves shown in Fig. 406.

The SUPERVOLUTE arrangement is the name given to the folding of the gamosepalous calyx, or the gamopetalous corolla (Fig. 407). Observe whether the overlapping is from right to left, or from left to right, as you stand before the flower. Observe, also, whether the mode of arrangement is the same in the calyx and corolla.

The plaiting of a gamopetalous corolla is shown in Fig. 408.

EXERCISE LXI.

Cymose, or Definite Inflorescence.

It often requires much skill and patience to determine whether a particular panicle, corymb, raceme, or head, is definite or indefinite.

The buttercup, wild columbine, rose, and cinquefoil, are common examples of cymose inflorescence among alternate-leaved plants, while Saint-John's-wort, chickweed, sedum or live-forever, dog-wood, elder, hydrangea, are opposite-leaved examples. Get as many of these as you can, and begin the study with the inflorescence of an alternate-leaved plant. Compare it with Fig. 409. In this plant each shoot terminates in a flower, and the growth is continued by means of branches. Here the main or primary stem (A, A) terminates with a flower which must, of course, be the oldest of the cluster. The branches (B, B, B) continue the growth, blossom, and cease to lengthen. From these branches proceed others (C, C), and so on.

Such a loose, irregular, definite inflorescence is called a *cyme*; but, when the number of branches is greatly increased, and the peduncles acquire such lengths as to give a peculiar outline, the cluster receives a more special name. Fig. 410 represents the cymose inflorescence of an opposite-leaved plant. The main or primary stem terminates in a flower between two branches. These branches, or secondary stems, also terminate in flowers, each one of which is situated between branches of the third order, and so on.

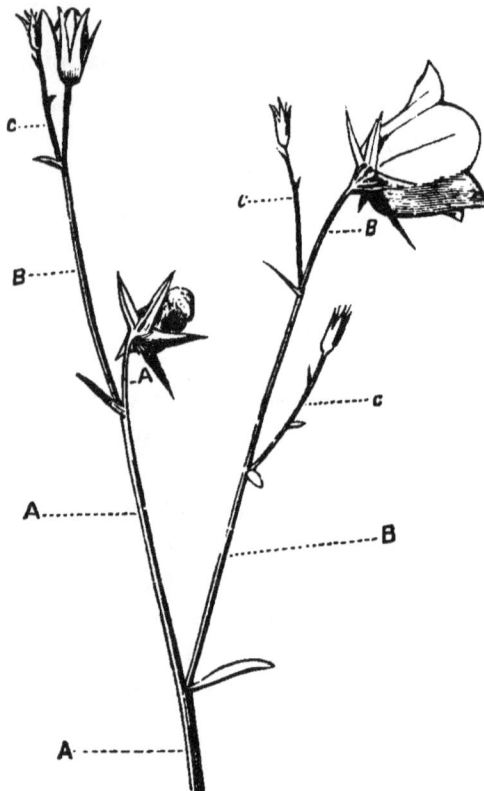

In this way is formed a *forked* or *dichotomous cyme.* If, in place of two, we have three branches, forming a sort of whorl around the primary stem, and each of these branches has another whorl of three tertiary branches, and so on, we get a trichotomous cyme. When the branching is carried forward, as seen in Fig. 411, the cyme becomes *globose.* When the central flower is suppressed, the process of development is not easily traced.

FIG. 409.

Suppose that, at each stage of the branching, one of the divisions is regularly suppressed, as shown in Fig. 412, where the dotted lines take the place of the absent branches, the cyme is apparently changed into a one-sided raceme, and the flowers seem to expand in the same way as in the indefinite raceme. In opposite-leaved plants bearing this kind of inflorescence, the leaf or bract opposite the flower shows that the raceme is definite ; but when, as in Fig. 413, there is no such bract, it is not easy to decide whether the cluster is definite or indefinite. However, the one-sided mode of branching gives the stem a coiled appearance, which is characteristic of the false or cymose raceme, described as *scorpioid*—curved like the tail of a scorpion.

FIG. 410.

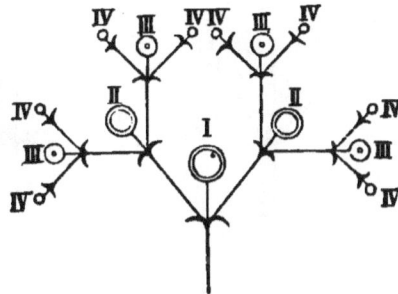

FIG. 411.

You may know a *cymose umbel* by observing that its oldest flowers are in the center of the cluster (Fig. 414), with buds, on short peduncles, surrounding them.

A FASCICLE (Fig. 415) is a cymose cluster of nearly sessile flowers.

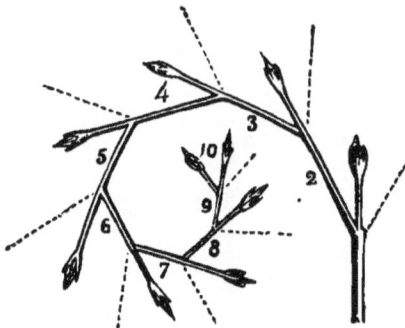

FIG. 412.

FIG. 413.

A GLOMERULE is a cymose cluster of sessile flowers in the axil of a leaf.

What is known as compound inflorescence occurs when the flower-clusters of a plant develop in one way, and the plant itself develops in another way. This state of things is often met with. Compare the development of the sunflower with that of catnip and hoarhound in this respect.

FIG. 414.

FIG. 415.

The indefinite mode of growth is sometimes spoken of as centripetal, because the flowers open first at the circumference; while definite forms are said to be centrifugal, because here the flowers open at the center first.

CHAPTER EIGHTH.

THE COMPOSITÆ.

EXERCISE LXII.

Parts of Flower-Heads.

To illustrate this chapter, gather all the plants you can find that have the inflorescence in a dense head. The dandelion, thistle, aster, marigold, sunflower, daisy, dahlia, burdock, mayweed, bachelor's - button, boneset or thoroughwort, golden - rod, lettuce, saffron, cudweed or everlasting, wormwood, tansy, yarrow, feverfew, camomile, ragweed, tickseed, elecampane, are familiar examples of such plants. For your first observations se-

Florets........

Involucre of Bracts......

FIG. 416.

lect some flower-head in which the parts are well developed, as the marigold, thistle, or dandelion. Fig. 416
shows a thistle-head, with
lines pointing to its principal divisions.

Fig. 417 represents a
marigold, in which the same
parts are shown. In Fig.
418 we look down upon the

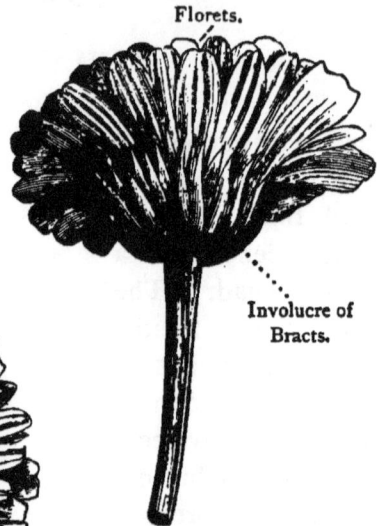

Florets.

Involucre of
Bracts.

FIG. 417.

Disk
Florets.

Ray Florets.

FIG. 418.

Ray Florets...
Disk Florets......

FIG. 419

top of the flower-head,
and observe that it pre-
sents unlikeness of as-
pect, which is still more
plainly shown in section
in Fig. 419.

The parts pointed out in these pictures may be thus
defined :

INVOLUCRE.—The outer green circle of a flower-head,
often mistaken for a calyx.

SCALES.—The bracts forming the involucre of a flower-head.

FLORETS.—The flowers of a flower-head.

RAY FLORETS.—The outer petal-like florets of a flower-head.

DISK FLORETS.—The inner florets of a flower-head.

Observe the bract at the base of the floret in Fig. 421. Observe the chaffy, bract-like bodies growing among the florets in Fig. 420. Examine your specimens, and see if, in any case,

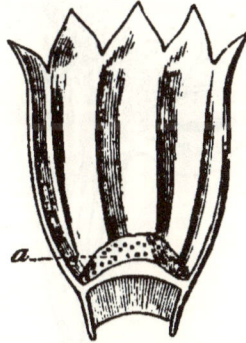

FIG. 420.   FIG. 421.   FIG. 422.

you find such things growing out of the receptacle among the florets. These chaffy bodies are known as *paleæ*. When they are wanting, the receptacle is said to be naked. Separate the naked from the chaffy flower-heads of your collection.

In Fig. 422 you see the convex receptacle at *a*. Observe the different forms presented by the receptacle in the last four figures. Strip away the florets from your flower-heads, and compare them in this respect. Are any conical in shape? Are any columnar? Are any pitted or honey-combed? In Fig. 422 is shown half the involucre of a marigold. Compare the involucres of your collection. They may be hemispherical, conical, inversely conical, squarrose, oblong, cup-shaped, etc. Their scales may be

many or few ; narrow or broad ; in one or several rows ;
loosely or closely imbricate ; chaffy, spinous, or soft ; reflexed, colored, etc.

EXERCISE LXIII.

The Florets.

Let us now examine, with some care, the structure of
florets. The flower-head here dissected is that of the
marigold. If you can not get this plant, take the sunflower, or daisy, or dande-
lion, or thistle, or any other
flower-heads you happen to
have. Of course, it is desirable, at the outset of
study, to get the largest
florets you can find.

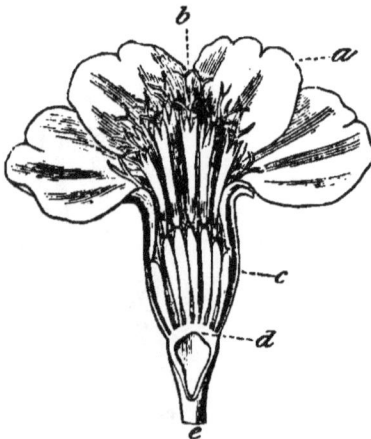

Fig. 423 represents a section of the marigold : *a*, the
ray florets ; *b*, the disk florets ; *c*, the involucre ; *d*,
the receptacle ; and *e*, the
peduncle.

FIG. 423.

Fig. 424 shows one of
the ray florets, with its strap-shaped corolla, *d* the limb,
and *c* the tube. At *e* is seen the forked stigma of the pistil ; *a* is the ovary, and *b* the limb of the calyx. Compare
this picture, or, what is better, a living example, with one
of the florets of a dandelion, and carefully note the differences of structure they present.

Fig. 425 represents a disk floret : *a*, the ovary ; *b*, the
limb of the calyx ; and *c*, the tubular corolla. Compare
this floret with those of the thistle, or any tubular florets
in your collection.

In looking for the limb of the calyx in your specimens,
you have found very various and peculiar appearances.
This part of florets, from its singularity, has received the

special name of *pappus*. In some, you observe, it does not exist at all, the adherent tube of the calyx forming an indistinguishable part of the ovary; in such cases the limb is said to be obsolete. Again, it is a mere rim, or border; sometimes it is cup-shaped, or bristly, or composed of teeth, scales, awns, or beards.

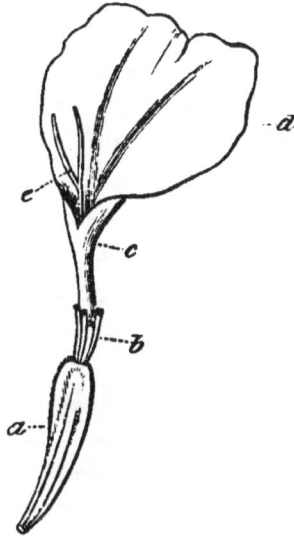

FIG. 424.

FIG. 425.

In the dandelion (Fig. 426) and the thistle it is silky. The cause of this singular condition of the calyx-limb may be that it is starved and stunted while growing, by the constant pressure of the florets against

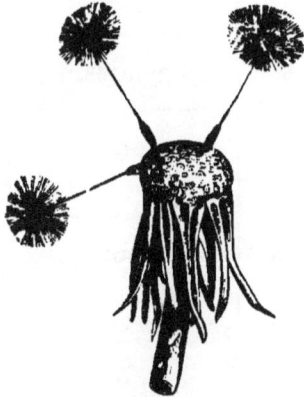

FIG. 426.

FIG. 428.

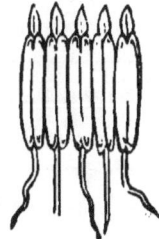

FIG. 427.

each other. In the case of the dandelion, while the seed is maturing, the tube of the calyx is prolonged above the ovary into a kind of stalk, and the pappus is said to be *stipitate* (Fig. 426).

8

But let us return to the florets. We have not yet examined their essential organs. Just below the stigma, in the disk floret (Fig. 425), is a cylindrical body, which, at first, you may not understand. Slit it down, flatten it out, and examine it with your glass. Is not this cylinder composed of slender coherent anthers? Do you not see each anther with its filament, as shown in Fig. 427, and which represents the tube seen in Fig. 428, thus laid open? The stamens of this floret are syngenesious.

The following is a schedule of the ☿ and ♀ florets of the marigold:

SCHEDULE FOURTEENTH.

Organs.	No.	Cohesion.	Adhesion.
Calyx? *Sepals.*	5*	Gamosepalous. Limb of narrow scales.	Superior.
☿ Corolla? *Petals.*	5	Gamopetalous, tubular.	Epigynous.
☿ Stamens?	5	Syngenesious.	Epigynous.
☿ Pistil? *Carpels.*	2	Syncarpous.	Inferior.
♀ Corolla? *Petals.*	5	Gamopetalous, strap-shaped.	Epigynous.
♀ Stamens?	o	o	o
♀ Pistil? *Carpels,*	2	Syncarpous.	Inferior.

* As the corolla is five-lobed, and there are five stamens, the florets seem to be five-merous, and we put the number of sepals as five.

The two carpels are inferred from the two-lobed stigma.

Study the florets of the dandelion. Is there more than one sort in the head? Select a well-developed floret, and describe it. Does your account agree with the following schedule?

SCHEDULE FIFTEENTH.

Organs.	No.	Cohesion.	Adhesion.
Calyx? *Sepals.*	5	Gamosepalous.	Superior.
Corolla? *Petals.*	5	Gamopetalous.	Epigynous.
Stamens?	5	Syngenesious.	Epipetalous.
Pistil? *Carpels.*	2	Syncarpous.	Inferior.
Seeds?		Solitary, erect, exalbuminous.	

In the same way see how many sorts of florets you can find upon the thistle-head, and carefully describe whatever you find. Do the same for all the plants of this family that you have collected. When a flower-head has both disk and ray florets, note whether they are ♂, ♀, ☿, or neutral.

When you have done this, you will be able properly to apply the following terms to inflorescences of this order:

When all the florets of a head are *perfect*, it is said to be *homogamous*.

When part of the florets are *imperfect*, the head is said to be *heterogamous*.

Flower-heads are *discoid* when destitute of ray florets.

EXERCISE LXIV.

Characters of the Compositæ.

Dandelions, daisies, dahlias, thistles, etc., we see, are composed, of many florets, inclosed in a calyx-like involucre. Plants of this kind have, therefore, been named Compositæ, from the compound, or composite, nature of what, to the untaught, seems a single flower. They form one of the most numerous, and, at the same time, one of the most natural and perfect families in the vegetable kingdom. There are about nine thousand different species included in it. They are found in all countries and climates. About $\frac{1}{8}$ of the plants of North America, and $\frac{1}{2}$ of all tropical plants, belong to it; indeed, from $\frac{1}{8}$ to $\frac{1}{10}$ of all the plants in the world are of this order.

Now, why is this order said to be very natural? Why, for instance, is it a more natural group than the rose family? If examples of all these nine thousand species were brought together, they would be seen to have one conspicuous and many important characters in common. In every one of them the inflorescence is a dense head, inclosed in a more or less compact involucre. But, when you have collected all the members of the rose family, you do not see so many features common to all, nor any marked one which stamps them as similar. On the contrary, in all their prominent characters, they are often widely unlike, and only experienced botanists can detect their affinities.

It must not be supposed, however, that all plants with flowers in a head belong to the Compositæ. The case is not quite so simple. Plants are not to be classified by a single character. We must not forget the principle that characters of cohesion and adhesion in the flower are of the first importance in determining affinities.

Now, what are the characters of cohesion and adhesion in which the florets of all the plants named in Ex. LX agree? In the matter of cohesion, you always found the calyx gamosepalous, the corolla gamopetalous, the stamens syngenesious, and the forked style, of which Fig. 429 is a magnified view, seems to imply a syncarpous pistil, although the ovary is one-celled and one-ovuled.

In the matter of adhesion, you always found the calyx-tube adherent to the ovary (Fig. 431), forming the peculiar kind of achenium known as a cypsela, and on further inspection you would find one erect exalbuminous seed (Fig. 430); and, if you were to examine the entire nine thousand species, you would find them all bearing the same characters.

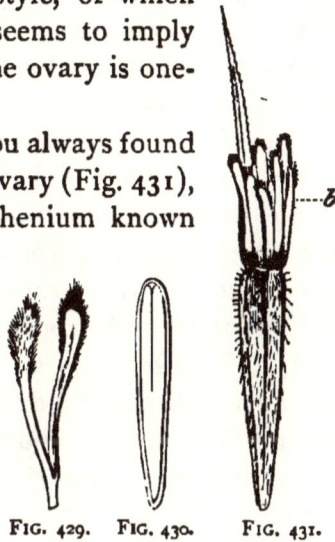

FIG. 429.   FIG. 430.   FIG. 431.

But you need not discover all these characters before you decide that a given plant belongs to the composite order. If you find *syngenesious stamens* in the florets of a dense *flower-head*, it settles the question. The coexistence of the two characters makes sure the inference that the plant has all the above-named characters, and also that it is more or less bitter.

Well, you have now the means of easily recognizing the members of this great family. They differ from all other plants, not in their inflorescence, for many other plants blossom in a head; not in having syngenesious anthers, for in many other plants the anthers are coherent; but they differ from all other plants in possessing both these characters. This circumstance is, therefore, characteristic of the compositæ. It enables you to identify any plant that belongs to the order. (See Flora, pages 185 to 191, where the characteristics of all the principal orders of plants are given.) Observe the distinction between that which characterizes an order and the characters of that

order. The coexistence of the two characters—syngenesious anthers and a flower-head—is sufficient to identify any plant of the order compositæ, or, what is the same
thing, to characterize it; but these, added to the other
characters that invariably accompany them, are the *characters of the order.*

Though all composite plants are alike in certain particulars, called their *ordinal characters*, they differ much
among themselves in other respects. Though they all
have bitter properties, yet some are tonic, some acrid, and
some narcotic. One group will have milky juice, another
will be watery and aromatic, or mucilaginous, or gummy,
or oily. In respect to the structure of flower-heads, you
have already found the dandelion, with all its florets, perfect and ligulate; you found the thistle with perfect tubular florets; you found the marigold with ♀ ligulate disk
florets, and ☿ tubular ray florets; the daisy with ♀ ray
florets, and ♂ disk florets. By referring to this order in
the Flora, you will find that these differences serve in arranging this vast family into sub-families, and these sub-
families are again separated into smaller groups by still
other characters. Differences in the involucre, and in the
conditions of the inferior fruit, serve to separate them into
what are called genera (see Flora, page 242), and then the
species of a genus are found to differ still further in the
characters of leaf and stem, in size, color, etc.

In Order VIII of Chart II, illustrating the Compositæ,
the characters of the dandelion, thistle, marigold, bachelor's-button, and globe amaranth, are given; those of the
dandelion and thistle are presented in full detail, and
much enlarged.

CHAPTER NINTH.

THE CRUCIFERÆ, OR CROSS-BEARERS.

EXERCISE LXV.
Characters of the Cruciferæ.

THE plants of this order bear flowers with a cruciferous corolla. About sixteen hundred species have been discovered, and they are all wholesome. They grow in every zone and country, but chiefly in temperate regions. Both wild and cultivated species are common, and the characters by which they are known are few and obvious, so that you may easily make their acquaintance. Mustard, horse-radish, shepherd's-purse, turnip, cabbage, radish, pepper-grass, cress, and honesty, are familiar examples, which you must often have observed and studied; and I wonder how many of you can recollect certain characters peculiar to these plants. Procure them, and confirm, by direct observation, the following statements :

The flowers of this family of plants have four petals, so placed as to resemble a cross. They have six stamens, four long and two short (Fig. 158)—*tetradynamous stamens*. Their inflorescence is racemose, and *without bracts.* Any plant with these characters is a crucifer. These three characters are alone sufficient to *characterize* a plant as cruciferous ; but they always accompany certain other traits of structure, which you will discover on glancing at the columns of the schedules you have made in describing them. In each case there are four sepals and four petals. There is no cohesion in any of these flowers, unless you except the spuriously syncarpous pistil (Fig. 349). They are also without adhesion. I do not know how successful you may be in observing the embryo, but, with a good magnifying-glass, you should be able to see that the radicle is folded upon the cotyledons, sometimes against their edges, sometimes against the back of one, but always folded. Now, these invariable features are the *ordinal characters* of the Cruciferæ. You may identify any one of the sixteen hundred known species by the three features first named, and, when you have done this, you may safely infer the existence of all the others. You are enabled to do this because botanists have carefully studied and analyzed these plants, and in every case, along with a cruciferous corolla, tetradynamous stamens, and bractless inflorescence, the other features have invariably been found. Compare the *characteristics* of any order of plants, as given at the beginning of the Flora, with the *characters* of that order given in connection with its genera and species in the body of the Flora. In the same way, if any plant have certain ordinal *characteristics*, you may infer the other ordinal characters.

I wish to say a word about the *importance* of the characters by which you determine whether a plant is or is not a crucifer. Some of you may think it strange that such features as the length of stamens and the absence of bracts should be named in describing an order of plants. These points of structure would not be looked upon as ordinal characters but for one circumstance, to be carefully borne in mind. It is their *constancy*, which here gives them value. They take rank from their *permanence*. Permanent or constant characters, no matter how trivial otherwise considered, are of high value in classification.

Order II of Chart I exhibits the characters of the Cruciferæ as here described.

CHAPTER TENTH.

THE UMBELLIFERÆ.

EXERCISE LXVI.

Structure of its Flowers and Fruit.

THE plants of this family blossom in umbels. An umbel, with its pedicels all starting from one point, like the rays of an umbrella, is a feature of plants so striking that it has naturally given its name to the group that bears it. But, as you saw that a plant blossoming in a head did not necessarily belong to the Compositæ, so you are now to find that all umbel-bearing plants are not, therefore, placed among Umbelliferæ. It has been found that certain plants blossoming in umbels are alike in many other respects, and are at the same time unlike all other plants in the structure of their flowers, and particularly of their fruit. These umbelliferous plants constitute the family we are about to examine.

They are "natives chiefly of the northern parts of the northern hemisphere, inhabiting groves, thickets, plains, marshes, and waste places. They appear to be extremely rare in all tropical countries except at considerable elevations, where they gradually increase in number, as the other parts of the vegetation acquire an extra-tropical or mountain character."

At the outset let me warn you that this is an order of plants to be suspected. Though some of its species are excellent food, yet some, when eaten, are deadly poisons, as hemlock, water-parsnip, and fool's-parsley. These poisonous species so strongly resemble esculent ones that only botanists can distinguish them, and many persons have made the fatal mistake of eating their roots. But the carrot, parsnip, parsley, celery, lovage, caraway, coriander, etc., are common cultivated species of this order, and none of the species are poison to the touch.

In your rambles you will be likely to find a large, coarse-looking, hairy or woolly, strong-scented plant, three or four feet high, which grows in moist, cultivated grounds, from Pennsylvania to Labrador, and west to Oregon. It has a thick, furrowed stem, ternate leaves, with large, channeled, clasping petioles, and blossoms in June, bearing huge umbels, often a foot broad. It is a species of cow-parsnip, sometimes called masterwort. Its flowers have white, deeply heart-shaped petals. As its parts are comparatively large, the flower of this plant is here chosen to exhibit the peculiarities of the order. In Fig. 432 it is given in section, and here follows its schedule-description :

SCHEDULE SIXTEENTH.

Organs.	No.	Cohesion.	Adhesion.
Calyx ? *Sepals.*	5	Gamosepalous.	Superior.
Corolla ? *Petals.*	5	Polypetalous.	Epigynous.
Stamens ?	5	Pentandrous.	Epigynous.
Pistil ? *Carpels.*	2	Syncarpous.	Inferior.
Seeds ?		One in each carpel—pendulous, albuminous.	

Now look at an ovary that has attained its full size, and lost its petals and stamens. It has turned brown, the furrows on its sides are deepened, and it separates into two halves, commonly called seeds (caraway-seed, for example). This ovary requires close study. In Fig. 433 you see its two carpels suspended in a peculiar manner. You may see in your specimen this slender, forked *carpophore*.

The fruit of the Umbelliferæ consists of two achenia, called a *cremocarp*, and each achenium, or carpel, is called a *mericarp*. The inner faces of the carpels, which are in contact before ripening, are called the *commissure*.

FIG. 432.

Fig. 434 is a magnified view of the back of a mericarp. Five ridges are seen passing from bottom to top of each mericarp, and often four intermediate or secondary ones, which may be, some, none, or all of them winged. In the

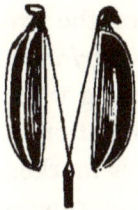

FIG. 433.
Cremocarp of
two Carpels,
each of which
is a Mericarp.

FIG. 434—Mericarp.

FIG. 435.
Cross-section of a Mericarp.

substance of the thin pericarp are little bags of colored oil, called *vittæ*, that give aromatic and stimulating properties to all the plants of this family. Four of these bags are seen in Fig. 434, in the intervals of the ribs. In the cross-section of a mericarp (Fig. 435) the little mouths of the four oil-bags of the back are seen, along with two others in the face of the commissure. If you have difficulty in finding these oil-bags, cut the carpel across, as shown in 435, and look down upon it with your glass, and perhaps their cut ends will be visible to you. A thin section, moistened and seen under a microscope, reveals them very distinctly.

Collect all the plants you can find with this kind of inflorescence and examine their flowers and fruit. In most cases you will need your glass and much patience in doing this; but, if you can not discover all the minute details of structure, you can, at least, tell whether the fruit of the plant is like that of the cow-parsnip or not.

EXERCISE LXVII.

Classification of Umbel-bearing Plants.

The order Umbelliferæ is thus described:

CALYX, superior; LIMB, obsolete, or entire, or a five-toothed border. PETALS, five, mostly with the point inflexed, and along with the five STAMENS, inserted on the outside of a fleshy, epigynous disk at the base of the two styles. FRUIT, consisting of two carpels, called *mericarps*, cohering by their faces, the commissure separating when ripe, and suspended from the summit by a prolongation of the receptacle, called a *carpophore;* each carpel is marked by five primary ribs, and a variable number of intermediate or secondary ones, between which are found oil-tubes, called *vittæ*, filled with aromatic oil. SEEDS, solitary, anatropous, with minute embryo in horny albumen.

HERBACEOUS plants, with hollow, furrowed stems. LEAVES, alternate, mostly compound, usually sheathing at the base (Fig. 436). FLOWERS, in umbels, usually compound, often with involucre and involucels (Fig. 437).

So, you see, we have here a family of fifteen hundred

species, all blossoming in umbels, and named from this circumstance, and yet distinguished from the rest of the vegetable kingdom by quite other characters than the inflorescence. If your notion of the order were founded on its name, or upon the *general aspect* of a few familiar species known to belong to it, you would most likely pronounce an elder-bush an umbelliferous plant. "You would find a large umbel, a small umbel, little, white blossoms,

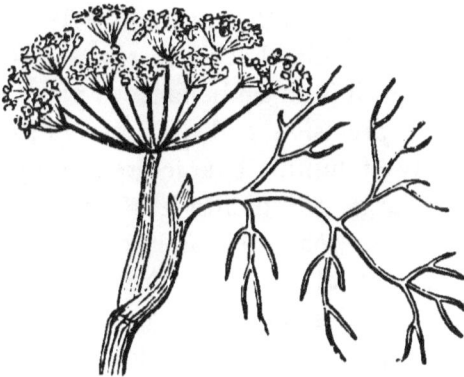 

FIG. 436.   FIG. 437.

an inferior ovary, and five stamens. Yes, it must be an umbelliferous plant. But look again : suppose you study a flower. In the first place, instead of five distinct petals, you find a corolla, with five divisions, it is true, but, nevertheless, with all five joined into one piece ; now, the flowers of umbelliferous plants are not so constructed. Here, indeed, are five stamens, but you see no styles ; you see three stigmas more often than two, and three grains more often than two ; but umbelliferous plants have never either more or less than two stigmas, nor more or less than two grains to each flower. Besides, the fruit of the elder is a juicy berry, while that of umbelliferous plants is dry and hard. The elder, therefore, is not an umbelliferous plant. If you now go back a little, and look more attentively at the way the flowers are disposed, you will also find their arrangement is not like that of umbelliferous plants. The first rays, instead of setting off exactly from the same center, arise, some a little higher and some a little lower ; the little rays originate with still less regularity ; there is nothing like the invariable order you find in umbelliferous plants. In fact, the arrangement of the flowers of the elder is that of a *cyme*, and not of an umbel."

But you need not search for all the characters given in the foregoing description in settling the question whether a plant is or is not umbelliferous. If it bears flowers in umbels, and produces inferior fruit, that when ripe separates into two seed-like bodies, it is an umbelliferous plant. These simple features give precision and distinctness to the order, so that the study of minute characters is only needed in separating this large group into lesser groups with a still greater number of like characters and properties. The number and development of ribs, the presence or absence of vittæ, the form of albumen, etc., are used for this purpose. Hence, although a beginner readily separates the plants of this order from all others, he finds it difficult to tell one genus from another, and, till he acquires skill in observation and has some experience in studying its genera and species with the aid of the Flora, he is quite safe in looking upon all of them with suspicion.

In Order VI, of Chart II, the structure of umbelliferous plants is shown in detail. Enlarged sections of the fruit, with all its peculiarities of structure, are represented in such a way as to reveal the parts with great distinctness.

CHAPTER ELEVENTH.

THE LABIATÆ.

EXERCISE LXVIII.

Characters of the Labiatæ.

CHILDREN who live in or visit the country, and those familiar with market-places, know what mints are, and can easily get peppermint, spearmint, catnip, sage, pennyroyal, thyme, balm, and such like plants, to illustrate this exercise. Compare your specimens with the following description:

Herbs, with square stems and opposite aromatic leaves; flowers, with a more or less two-lipped corolla, didynamous or diandrous stamens, usually with diverging anthers; ovary, deeply four-lobed, on a fleshy disk, four-celled, each cell with one erect ovule forming in fruit four little seed-like nutlets or achenia, around the base of the single style, in the bottom of the persistent calyx. Seeds with little albumen; cotyledons flat. Stamens inserted on the tube of the corolla. Stigma, forked. Flowers, axillary, chiefly in cymose clusters, that are sometimes gathered into spikes or racemes. Leaves, usually dotted with glands, containing a pungent, volatile oil.

Whenever you find a plant that answers to this description, it belongs to the order Labiatæ. The group is named from the two-lipped corolla of its flowers, but you can not know one of these plants by this circumstance alone. There are many plants with labiate flowers that do not belong here. There are many plants with square stems, opposite leaves, and labiate flowers, that still do not belong in this order. Nor do you find in this list of characters any that may not be found elsewhere, as you do in the case of the fruit of Umbelliferæ, for instance. Is it, then, necessary, in every case, to make an extended and minute examination of plants suspected of being in this order before deciding that they really are so? We can best answer this question by carefully observing certain plants. First get a specimen of verbena, a widely-cultivated plant belonging to the family Verbenaceæ, and compare it with any of the labiate plants named in the beginning of this exercise, thus:

The Verbenaceæ are herbs or shrubs with opposite leaves.	The Labiatæ are chiefly herbs, with square stems, opposite, aromatic leaves.
More or less two-lipped or irregular corolla.	More or less two-lipped corolla.
Didynamous stamens.	Didynamous or diandrous stamens.
Two to four celled fruit, dry, or drupaceous, usually splitting, when ripe, into as many one-seeded, indehiscent nutlets.	A deeply four-lobed ovary, which forms in fruit four little seed-like nutlets or achenia surrounding the base of the single style in the bottom of the persistent calyx; each nutlet filled with a single erect seed.
Seeds, with little or no albumen; the radicle of the straight embryo pointing to the base of the fruit.	Albumen, mostly none; embryo, straight; radicle, at the base of the fruit.

The affinities of these orders are so strong that, at first, one almost wonders why botanists regard them as distinct. But we remember that the characters by which they differ, though not conspicuous, are yet very important, being characters of the essential organs and the fruit. The deeply-lobed ovary, with the style growing out from its base, and surrounded in fruit by the four nut-lets, distinctly separates the two groups. But does this structure of the ovary distinguish the Labiatæ from all other plants? Let us see.

There is a family of rough, hairy herbs, known as bo-rages, with flowers in cymose clusters, unrolling as they expand, as described (page 137), which it will be well to study with reference to this point. One of its species, the forget-me-not, is a common, widely-diffused plant of this order, which you may get, and compare with the follow-ing description :

The Boraginaceæ are chiefly rough, hairy herbs, with (not aromatic) alternate, entire leaves.	The Labiatæ are chiefly herbs, with square stems, and op-posite, aromatic leaves.
Symmetrical flowers, with five-parted calyx, and regular five-lobed corolla.	More or less two - lipped co-rolla.
Five stamens inserted on the corolla tube.	Didynamous or diandrous sta-mens.
Ovary, deeply four-lobed, the lobes surrounding the base of the style, and forming in fruit four seed-like nutlets, each with a single seed.	Ovary, deeply four-lobed, form-ing in fruit four seed-like nut-lets around the base of the single style, in the bottom of the persistent calyx, each filled with a single erect seed.
Albumen, none; cotyledons, plano-convex; radicle, point-ing to the apex of the fruit.	Albumen, mostly none; em-bryo, straight ; radicle, at the base of the fruit.

Here, then, is an order of plants, the Boraginaceæ, which is very different from the Labiatæ, except in the characters of the ovary, and in these characters it is almost identical with that order. You have in this instance an ex-ample of the puzzling relationships encountered in classi-fication. The verbenas can not be grouped with the labi-ates, because, though wonderfully like them in many other respects, thcy are so unlike in the characters of the pistil ; the borages, though agreeing essentially with the Labiatæ in the characters of the pistil, can not be classed with them, because of their differences in so many other respects.

At any rate, you now see that the structure of the ovary is not characteristic of the Labiatæ. To identify the members of this group, we have to bear in mind several characters, which you are prepared to do if you have examined and compared the plants named above. When you find a plant with a two-lipped corolla, square stem, and opposite leaves, joined with a deeply-lobed ovary and basic style, you need not hesitate to place it among Labiatæ.

In the same way, at the beginning of the Flora (p. 186) are given the characteristics of all the natural orders of our common flowering plants. Try to refer to its order every flowering plant you meet with, and you will soon find yourself, without effort, referring plants at once to their proper ordinal group.

You have now examined a good many species of plants belonging to four different natural families—the Compositæ, the Cruciferæ, the Umbelliferæ, and the Labiatæ. Can you tell whether their leaves are parallel-veined or net-veined ? Have you ever seen a parallel-veined cruciferous plant ? Have composite plants, as far as you know, parallel-veined or net-veined leaves? Try to find whether the leaves in the plants of these orders are alike in their venation.

Order XII, of Chart III, exhibits the characters of the Labiatæ.

CHAPTER TWELFTH.

THE CONIFERÆ.

Characters of the Coniferæ.

THERE is still another large group of widely-distrib-uted plants that must be specially described. When we speak of evergreens, everybody knows what we mean, and thinks of pines, balsams, hemlocks, spruces, cedars, juni-pers, arbor-vitæs, or whatever species are most familiar. When we speak of cone-bearing trees or shrubs, it is not quite the same group of plants that is thought of, for, although everybody knows what cones are, yet untaught and unobservant people would hardly think of a juniper-berry as in any way allied to a cone. But, although cone-bearing trees are everywhere to be found, and universally known, yet very few people can tell when they flower, what sort of flowers they bear, or what a cone really is ; and yet their structure and habits in respect to flowering and fruiting are even more remarkable than their general appearance. They are monœcious or diœcious, and blos-som in spring. Their flowers are in clusters, usually aments, sometimes in the axils of the leaves, and sometimes at the extremity of the branches. The fruit is two years in ripen-ing, so that the full-grown cones, seen upon them in sum-mer, were blossoms the year before.

To study their flowers, you must begin in the spring, and look carefully for the fertile and sterile aments, which will usually be found on different branches of the same tree. And, while you are searching for their flowers, observe also their remarkable foliage. Fig. 438 shows a fascicle of needle-leaves from the pine. Observe the number of leaves in each fascicle of the specimen you are studying, for the species vary in this respect. Fig. 439 represents the scale-shaped leaves of arbor-vitæ. In evergreens of this sort observe the difference between the foliage on the older and newer parts of the plant. In diœcious species, observe whether the foliage is of the same kind on both ♂ and ♀ plants. When you find awl-shaped leaves upon a young branch, observe them from time to time, and note their gradual passage into scale-shaped, imbricate leaves. Do evergreens shed their foliage? If so, when? and how long does the foliage last?* Can you find young foliage upon old branches?

In the pine the inflorescence of the sterile flowers is a kind of compound spike (Fig. 440). One of the spikelets much magnified is shown in Fig. 441. Each flower of this spikelet consists of a single stamen only, and this stamen has a most peculiar structure. Its filament is so short as to be scarcely discernible. It is really a spikelet of anthers, and their connective. Remove a stamen, and examine its inner face. Compare it with Fig. 442, which is a ♂ flower of the pine. Here you see two anther-cells dehiscing vertically, and Fig. 443 represents a grain of the compound pollen they bear. Seen on the outside, this stamen appears to be all connective. This connective, or scale, as it is usually called, varies in form in different

---

* To find whether evergreens shed their foliage, you have only to watch the ground beneath them for fallen leaves. If you find that their foliage does fall, and wish to learn by observation how long it lasts, notice whether the twigs of the present year keep their foliage all through the coming winter. If they do, observe them again next summer, and if it is still retained, watch them the third season, and so on.

species of evergreens ; but these of the ♂ catkins of the pine are enough like all the others to guide you in search-ing for and studying them. When they have shed their pollen, they wither and dis-appear.

The ♀, or fertile flowers, are also clus-tered, and appear at the same time as the ♂ ones, sometimes on the same, and some-

FIG. 438.

FIG. 440.

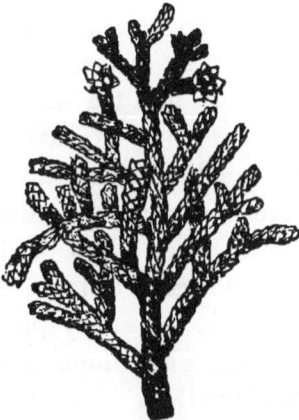

FIG. 439.

FIG. 441.

FIG. 442.

FIG. 443.

times on different branches. It is this ♀ catkin that, in a couple of years, develops into the fruit we call a cone. Fig. 444 represents it when in flower. The fertile flowers are very simple in struct-ure, each one consisting of an open carpellary leaf, or scale. Hitherto, you have always found seeds in seed-vessels, but here you will find them borne upon one side of a scale, and

FIG. 444.

FIG. 445.

hence the Coniferæ are said to be *naked-seeded.* Get one of these ♀ catkins, and detach from it a single flower. Compare it with Fig. 445. Observe the ovules upon its inner surface. These vary in number and position with the species examined. In this specimen of the pine we have two inverted ovules, which, in time, become seeds. Fig. 446 represents a scale from the same kind of catkin after it has become woody, and the seeds have ripened. The left side of this scale shows the cavity from which one winged seed has fallen, while on the other side a seed still remains. You may easily find these seeds in mature cones by breaking them across, or, what is better, by put-ting them in a dry place for a day or two, when the scales will cleave away and so reveal the seeds within.

In some evergreens, as arbor-vitæ and white cedar,

when you examine the small terminal catkins, you will find the ♂ ones composed of several scales or flowers, each scale bearing two to four anther-cells on the lower margin (Fig. 447), while the globular ♀ catkins consist of four rows of scales, each scale or flower bearing one or

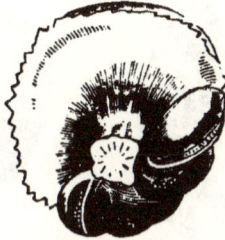

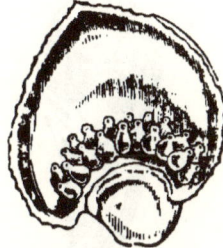

FIG. 446. FIG. 447. FIG. 448.

several erect, bottle-shaped ovules at the base (Fig. 448). The developed cone of the white cedar is scarcely larger than a pea, with scales firmly closed, but opening at maturity.

The juniper or red cedar, common on dry, sterile, rocky hills, both northward and southward, blossoms in April. The various species are mostly diœcious, and the catkins are very small. Observed only when in fruit, you would scarcely regard the juniper as a coniferous plant, but the ♀ catkin, when in flower, is seen to consist of from three to six scales, bearing a variable number of ovules precisely in the same manner as the pine. But, in ripening, these scales grow together, turn purple, and form a berry-like fruit as large as a pea. Fig. 449 represents one of these berries with its scaly bracts underneath, while Fig. 450 shows one of its enlarged bony seeds. The berries ripen the second year from the flower.

The ground-hemlock is another coniferous plant with a berry-like fruit. Its ♀ flower is more simple than those we have been examining, for it consists of a single ovule, without even an accompanying scale. This straggling bush, two or three feet high, is found in shady places, along streams, on thin, rocky soils, from Canada to Pennsylvania and Kentucky, and south along the Alleghanies. Its linear leaves are nearly an inch in length, in two opposite rows, along the branches. It blossoms in April. Fig. 451 represents its axillary ♂ inflorescence, consisting of six scale-like connectives, bearing the anther-cells on their

FIG. 449.   FIG. 450.   FIG. 451.

inner faces. Fig. 452 represents its solitary fertile flower. You see it is a single, erect, sessile ovule, surrounded by scaly bracts. At its base is a cup-shaped disk, that becomes pulpy, red, and berry-like, as the ovule ripens and turns black. Fig. 453 represents a vertical section of this fruit. The embryo of a coniferous seed is shown in Fig. 454. It is said to be *polycotyledonous*.

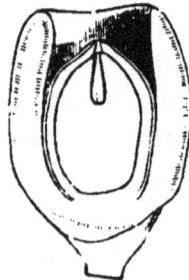

FIG. 452.   FIG. 453.   FIG. 454.

The lower half of Chart IV is devoted to the Coniferæ. Examples of the leading genera of this order are given, showing the foliage, fruit, and seed, the latter much magnified, and all colored from Nature.

CHAPTER THIRTEENTH.

THE ORCHIDACEÆ.

EXERCISE LXX.

Characters of the Orchidaceæ.

THERE is a widely-distributed and well-known plant, with showy flowers, blossoming in early summer, and called the lady's-slipper, or sometimes the moccasin-flower (Fig. 455). It is an orchid ; and, though unlike other orchids in some respects, it has the chief traits of the order to which it belongs. ·

Provide yourself with some of these plants, and compare them with the following description : Herbs with parallel-veined leaves and irregular flowers. Perianth of six parts in two sets ; the three outer ones nearly alike, and petaloid in structure and appearance ; the three inner ones unlike. One of these, differing much in shape and direction from the others, is called the *lip.* In Fig. 455 the lip is the sac or slipper, which gives the plant its common name. The lip varies much in different orchids, but in all its appearance is singular and striking. It is seen spurred and lobed, and assumes many fantastic forms.

Examine, now, the stamens and pistil of your flower. Lift up the little, drooping organ opposite the lip, and compare the structure beneath with Fig. 456. You have here the stamens and pistil consolidated into one organ, and known as the *column.* The fertile anthers are shown at *a, a,* while a sterile stamen back of the stigma is marked

9

st. The stigma is marked *stig.* The fertile anthers are sessile upon the style. In most orchids there is but one an-

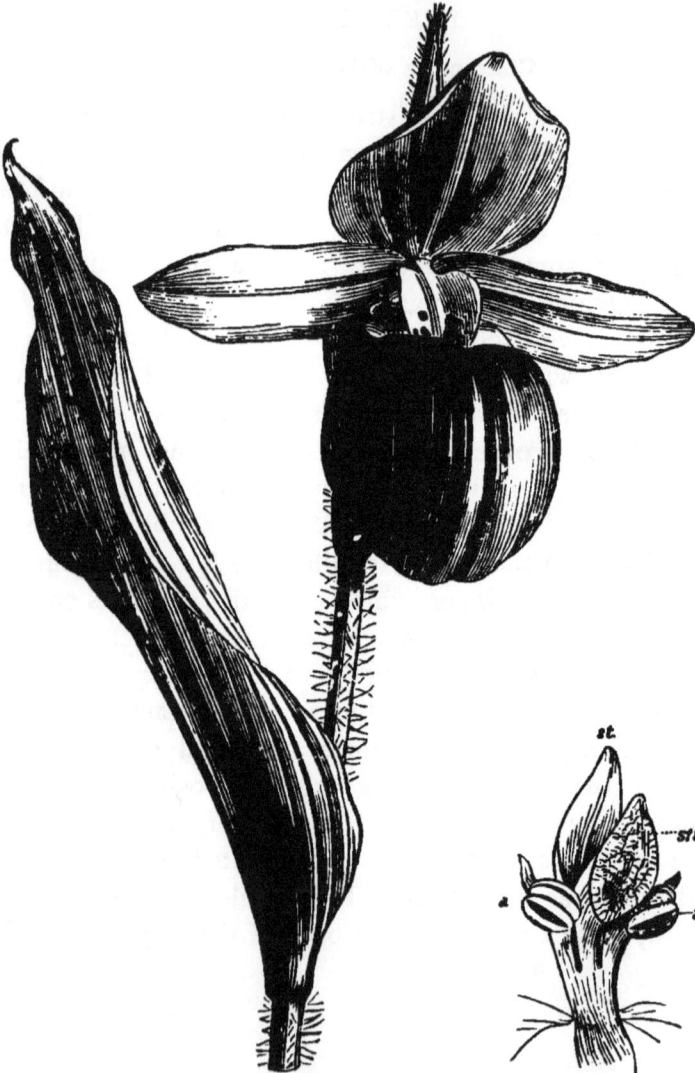

FIG. 455.

FIG. 456.

ther, which is fertile, and placed behind the stigma, in the position of the sterile stamen of the lady's-slipper. Examine the pollen. Instead of being dry and powdery, you find it pulpy-granular. In many orchids it coheres into coarse grains, held together in one mass by cobwebby tissue, and known as *pollinia* (Fig. 457). You find just such pollen masses, or pollinia, in the gynandrous stamens of the milk-weed (Fig. 458). The ovary of the lady's-

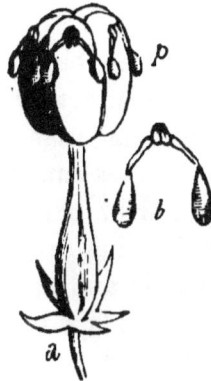

FIG. 457.   FIG. 458.

slipper is inferior, forming in fruit a one-celled pod, with innumerable minute seeds borne on parietal placentæ. In some orchids you find it so twisted as to alter the position of the petals.

The characters of the Orchidaceæ will be better understood by comparing them with other groups of parallel-leaved plants. Provide yourself with lilies of any sort, and specimens of blue flag, or flower-de-luce. Compare your lilies with the following description :

Herbs with simple, sheathing or clasping, parallel-veined leaves. *Flowers* regular, perfect. Perianth of six parts in two circles of similar color and form. *Stamens* six, inserted on the leaves of the perianth ; anthers introrse. Ovary free, three-celled, with numerous ovules on axile placentas; the styles united into one.

What number have you found prevailing in the lilies you have examined ? What number occurred oftenest in describing the Compositæ ? The Labiatæ ? The Umbelliferæ ? The Cruciferæ ? Point out the affinities of the lady's-slipper and the lily.

Compare flower-de-luce, or blue flag, with the following description :

Herbs with parallel-veined, equitant, two-ranked leaves and perfect flowers. Tube of the perianth coherent with the three-celled ovary ; limb petal-like and six-parted ; convolute in the bud in two sets. *Stamens* three, monadelphous or distinct, with extrorse anthers. Pod three-celled, loculicidal, many-seeded.

What affinities can you point out between the flower-de-luce and lily ? between the lady's-slipper and flower-de-luce ? In what respect are these three plants alike ?

The genera and species of these orders are described in their proper place in the Flora.

We think of a SPECIES as made up of individuals that have descended from a common ancestor, or that are so nearly alike that they might have done so, like the individual plants in a bed of pansies or a field of wheat. All the dandelions scattered everywhere constitute such a species. They are of common descent, and they produce plants like themselves from their seed.

A GENUS (the singular of genera) is an assemblage of species that resemble each other much more nearly than they resemble any other plants. All the species of oak form an oak genus ; the species of clover, white, red, etc., a clover genus ; the roses, a rose genus, and so on.

On Chart V several orders of parallel-leaved plants are given, and their characters are so magnified that they may be easily seen and compared.

CHAPTER FOURTEENTH.

THE GRAMINEÆ.

EXERCISE LXXI.

Characters of the Gramineæ.

THERE is a large group of plants blossoming in peculiar-looking spikes, heads, and panicles, the flowers of
which are furnished with green or brown scales, called
glumes, whence the entire group is known as the Glumaceæ. They constitute a twelfth part of the described species of flowering plants, and at least nine tenths of the individuals composing the vegetation of the world. They grow
everywhere. All grasses and all the cultivated crops of
grain belong among them, besides many other plants not so
important to man. They have true flowers, but no calyx
or corolla. The Glumaceæ are divided into two groups;
one group—the sedges—having solid stems, while the
other—the grasses—has hollow stems. The flowers of
both these groups have a special structure, which your
previous study will not enable you to understand.

From this large class we will select examples that belong to the family of grasses or Gramineæ, the members
of which have hollow stems, and the sheaths of their ligulate leaves are split in front.

Gather specimens of wheat, if possible, in blossoming-
time, when the stamens are to be seen (Fig. 459). Along
the rachis are rows of peculiar-looking bundles. The
number of these rows varies in different kinds of wheat.

Break the spike at about the middle, and take off a bundle from the top of the lower half.   Observe whether it is
attached by its side or its end, and whether any of its
scales adhere to the rachis either wholly or in part.

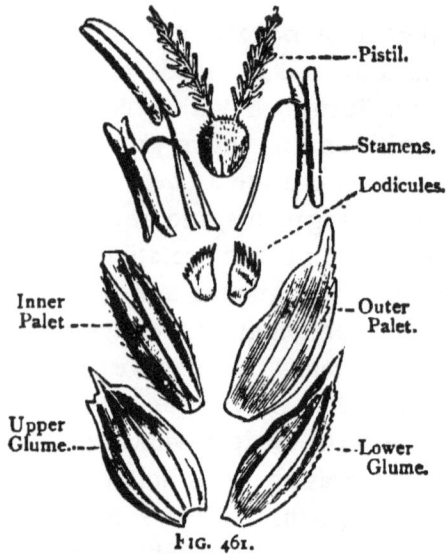

Pistil.

Stamens.

Lodicules.

Inner
Palet

Outer
Palet.

Upper
Glume.

Lower
Glume.

FIG. 461.

FIG. 459.   FIG. 460.   FIG. 462.

Remove the first two of these scales : there is no trace
of either pistil or stamens within them.   They are quite
empty.   What do you find next?   Are there not two or

three separate flowers forming a sort of spikelet within these two outer scales (Fig. 460). Examine one of them.

In Fig. 461 a single flower is shown, with the two glumes found at the base of the spikelet, and called the *lower* and *upper glumes*. What remain are the parts of a single flower. Beginning with the outermost of these at the right, you see a scale called the *outer palet*. Does the outer palet, in the specimen you are studying, terminate in a bristle?

At the left you see a peculiar scale, folded at the sides, and called the inner *palet*. Then come the scales. Look carefully at your flower for these minute bodies, which are thought to be a sort of perianth, the outer and inner scales being of the nature of bracts. We next come upon the stamens, with their versatile anthers, and the pistil, with its plumose stigmas—the unmistakable flower. The peculiar features of this inflorescence, then, are—

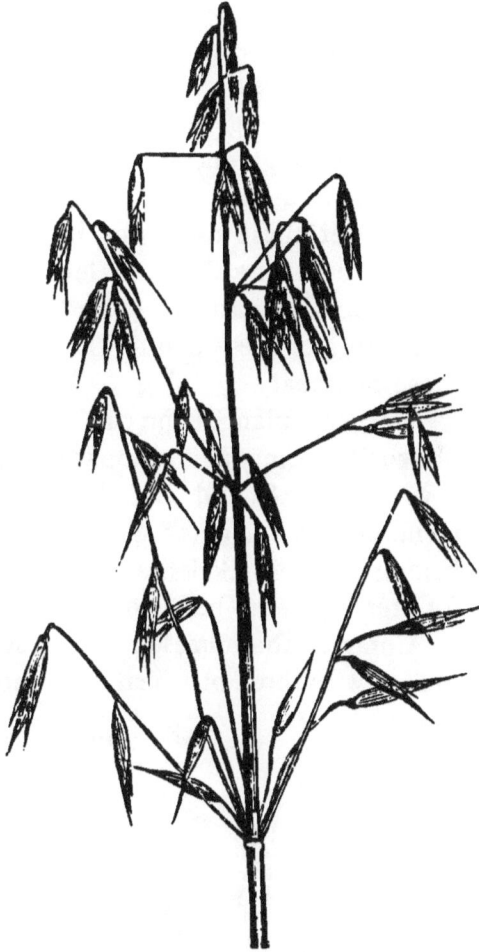

FIG. 463.

GLUMES.—Scales of the spikelet, and exterior scales of the flower.

PALETS.—Chaffy, inner scales of the flower.

AWN.—The beard or bristle of a scale.

SQUAMULA.—One of the minute scales at the base of the ovary of grasses.

The following questions, which form a schedule for this group of plants, are answered as if asked concerning Figs. 459 and 461 :

Answer these questions in regard to the heads of barley and rye. Compare the culm* and leaves of these plants with those of wheat.

Gather a plant of the oat in blossoming-time, and compare it with Figs. 463 and 464. Remember that the outer glumes

Inflorescence ?	Spike.
Glumes ?	2.
Outer palet ?	1.
Inner palet ?	1.
Lodicules ?	2.
Stamens ?	3.
Styles ?	2.

belong to the spikelet, and not to the flower. Look out for sterile flowers below or above the perfect ones.

Compare the culm, leaves, and stipules of the oat with those of wheat, rye, and barley.

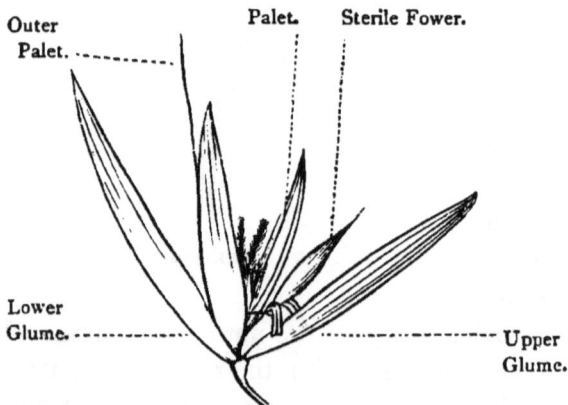
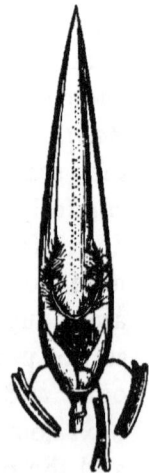

Outer Palet.

Palet.

Sterile Fower.

Lower Glume.

Upper Glume.

FIG. 464.

FIG. 465.

* *Culm :* a straw ; the stem of grasses and sedges.

In Fig. 465 are seen the palet, squamulæ, stamens, and pistil. The oat may be thus described :

Compare a plant of Indian corn, when in blossom, with the following description : ♂ flowers in a terminal panicle of racemes known as the *tassel ;* spikelets two-flowered; glumes herbaceous, palets membranous ; anthers three, linear. ♀ flowers in

Inflorescence ?	Panicle.
Glumes ?	2.
Outer palet ?	1.
Palet ?	1.
Lodicules ?	2.
Stamens ?	3.
Styles ?	2.

an axillary spike, partially imbedded in the rachis, known as the *cob*, the bracts forming its spathe being the *husks ;* lower flower of each spikelet consisting of two palets, abortive ; glume broad, thick, membranous, obtuse ; styles, very long, filiform, exserted and pendulous, forming the *silk ;* kernels in eight, ten, twelve, or some even number of rows.

The further study of orders, genera, and species may be pursued with the aid of the Flora.

The *botanical name* of a plant is the name of its genus followed by that of the species. The generic name may be compared to the surname or family name of a person, as Jones or Smith, and the specific name to the given or baptismal name, as James or William. The botanical names are in Latin, that the botanists of all countries may have a common language. The name of the genus comes before that of the species. Thus, in naming the members of the oak genus, the scientific name of the white oak is *Quercus alba*, *Quercus* being the generic name, and *alba* that of the species. The red oak has the name *Quercus rubra ;* the black oak, *Quercus nigra*. The name of the genus is a substantive, and that of the species an adjective. ORDERS are commonly named from their most representative genus, thus : *Ranunculaceæ*, from the genus *Ranunculus ; Rosaceæ*, from the genus *Rosa*.

CHAPTER FIFTEENTH.

FLOWERLESS PLANTS.

EXERCISE LXXII.

Ferns.

You have often seen dense, green patches of plants, more or less resembling Fig. 466, and called *brakes*, or *ferns*. They seem, when growing, to be all leaf and no stem; but you see in the figure that the stem is a short, underground rhizoma. In some ferns the rhizoma takes a vertical direction, and bears a whorl or tuft of foliage at the top. Here it gives off single leaves as it advances. Although, in our climate, the stems of ferns are found creeping underground, yet in the warm climates of the tropics they rise in the air, sometimes forming trees, forty or fifty feet in height.

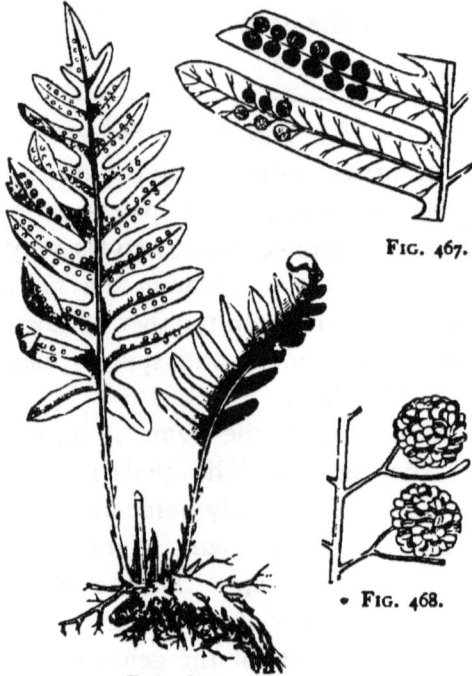

FIG. 467.

FIG. 468.

FIG. 466.

Did you ever see any flowers upon this sort of plant? anything that looked like fruit? Since studying the Coniféræ, you are aware how very simple and obscure flowers may become, and you will, of course, look very carefully at a plant before deciding that it has none. Gather as many kinds of ferns as you can find, and search for the seed-bearing portions. Meantime you can learn the terms

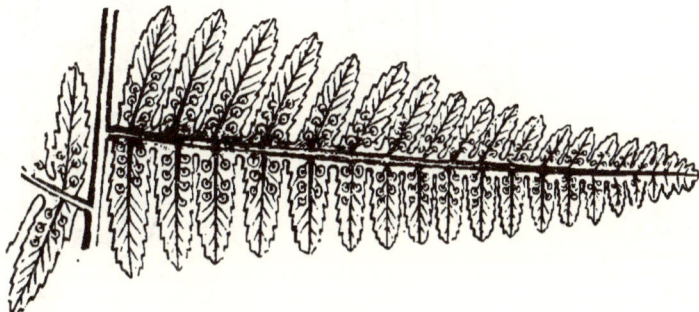

FIG. 469.

by which their parts are distinguished. They are the following:

The leaf of a fern is called a *frond*. The stalk or petiole of a frond is called a *stipe*. Point out the frond and stipe in the specimens you have gathered. The lobes of a frond are called *pinnæ* (Fig. 467). Subdivisions of pinnæ are called *pinnules* (Fig. 469). Point out the pinnæ in your specimens. Have you· found any in which the pinnæ are divided or lobed by pinnules? Observe the differences of stipe in your specimens. What kind of soil did you find them in? Were they growing in shady or sunny places? Did you observe the way the young fronds were folded in the bud?

EXERCISE LXXIII.

Reproduction of Ferns.

Did you find anything that you could fancy to be a flower, in your examination of ferns? Look them over

once more on all sides, and note all appearances that are
repeated on different specimens. Observe carefully the
under side of the frond, along the veins and the margin.

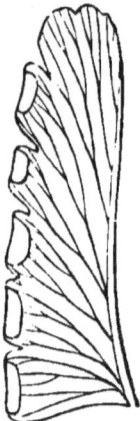

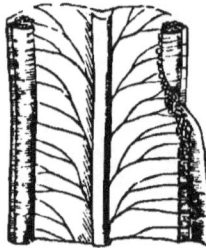

FIG. 471.

FIG. 470.   FIG. 472.

Do you not anywhere find little brown patches resembling
the spots seen in Fig. 467, representing magnified pinnæ,
or the pinnules of Fig. 469? In Figs. 470 and 471 you see
how these spots may be concealed under folds of the margin of fronds.

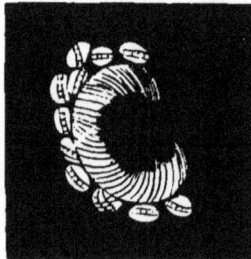

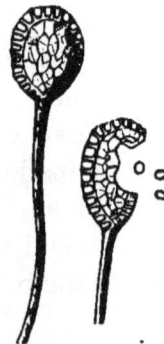

FIG. 473.   FIG. 474.   FIG. 474 A.

These brown patches certainly look very little like
flowers. Examine them never so carefully with your microscope, you will not find stamens or pistils. And yet

these little brown patches answer, in a certain way, to
seeds. It is from them that new ferns arise. They are
the reproductive parts of this class of plants, and the
fronds that bear them are said to be *fertile.* Examine
these spots carefully with your magnifying-glass, and compare them with Fig. 468 or Fig. 472. The small, brownish clusters of fruit-dots seen on the under surface of
fronds, in rows along the veins, or on the margin of the
pinnæ, are called *sori*, and a single cluster a *sorus.* The
scale or protective covering of a sorus, seen in Fig. 472,
but absent in Fig. 468, is called an *indusium.* This organ
is still more plainly seen in Fig. 473.

In the sorus (Fig. 473) you see little, peculiar-looking
bodies escaping from beneath the indusium. Each of

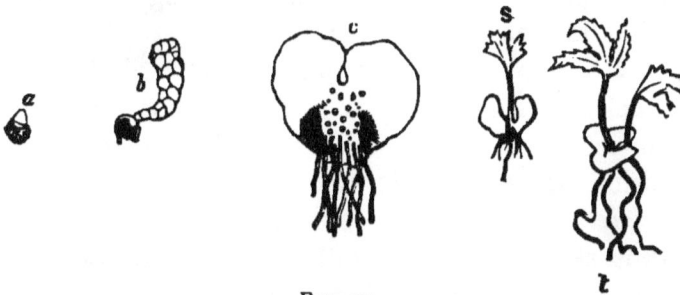

FIG. 475.

these cell-like bodies, of which the sorus is composed, is
known as a *spore-case, sporange,* or *theca.* They are sometimes stalked, as seen in Fig. 474. The singular-looking
band around them is an elastic membrane, which bursts
when they are mature, and thus the spores contained in
the spore-case escape (Fig. 474, A). It is from spores that
ferns arise, but by a process more like budding than like
the sprouting of a seed. When a spore commences to
grow, appearances like those represented in Fig. 475 may
be observed. The growth begun by a spore, as at *a*, and
seen more advanced at *b*, is shown at *c*, expanded into a
leaf-like body, called a *prothallus*, which gives off roots at

the under surface. Among these roots may be found certain bodies, analogous to the stamens and pistils of flowers,
and called the *antheridia* and *pistillidia.* It is not until
these bodies have matured and done their work that the
young fern appears. If there is anything like flowering
in the history of ferns, it is the prothallus produced from
the spore that bears the flowers, and from these produces
the young fern as seen at *s,* and the same, still more developed, at *t.*

EXERCISE LXXIV.

Mosses.

In place of flowers, mosses have *antheridia* and *pistillidia.* These plants may be either monœcious or diœcious. Fig. 476 represents a moss having its antheridia
and pistillidia on different plants.

At *a* you notice a moss-plant with sessile leaves and
unbranched stem, ending in a sort of rosette, which is
seen in section at *b,* where you may observe the peculiar
cylindrical bodies growing among the leaves. These are
antheridia. One of these bodies, detached and much magnified, is seen at *c.* The stalk-like bodies accompanying
the antheridia (*h*) are called *paraphyses.* They are not
well understood, but are thought to be abortive states
of the *antheridia.* At first these little organs contain mu-
cilage, but, when mature, their contents, seen escaping at
c, are granular, and each of the little ejected cellules sets
free an active *antherozoid.* Sometimes the leaves that sur-
round the antheridia grow together into a kind of cap
called a *perigone,* and in monœcious mosses, the antheridia
and pistillidia are often found within the same perigone.

The *archegone* or *pistillidia* of mosses also arise in clusters of leaves, and are cell-like bodies, having a cap or
epigone of the same nature as the perigone of antheridia.
But the pistillidia bursts its cap, leaving part of it as a
sheath below, and is carried up on a stalk (*d*), at the top

of which is seen an urn-shaped body of curious structure, called a *sporange* (*c*).

SETA.—The stalk of a sporange (*d*).

VAGINULE.—The collar or sheath at the base of the seta, resulting from the bursting of the epigone.

CALYPTRA. — The cap or hood of a sporange, shown at *f*, and seen in place at *e*.

OPERCULUM.—The lid of the sporangē (*g*), seen when the calyptra is removed.

PERISTOME. — A single or double fringe of teeth around the mouth

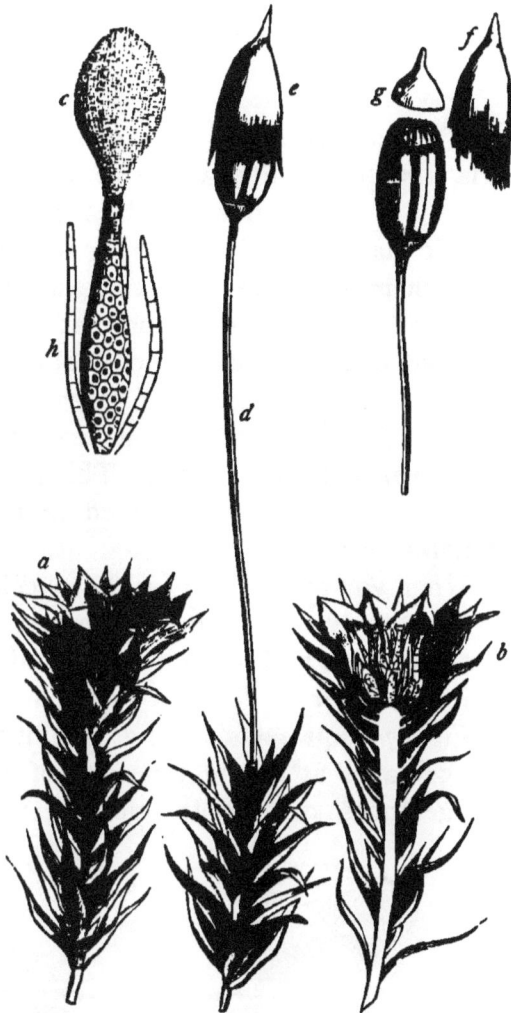

FIG. 476.

of a sporange. It is sometimes altogether absent. These teeth vary very much in number, but are always either four or some multiple of four.

ANNULUS.—An elastic ring sometimes found in the mouth of a sporange.

SPORES.—The ripened contents of the sporange.

EXERCISE LXXV.

Fungi.

The common mushroom, or toadstool, as children call it, is a well-known example of this group of flowerless plants. It is found everywhere growing upon decaying organic matter. If, in gathering specimens for study, you break them off above the surface of the ground, you will leave the plant itself behind, and bring only the fruit. The part concealed in the rich mold, or spread on its surface, is a tangled mass of filaments that you might mistake for fibrous roots; but it answers to the root, stem, and leaves of higher plants. This portion of the plant is called the *mycelium*, represented by the root-like fibrous portion of Fig. 477.

When you are looking for the mycelium of mushrooms, observe the *young* fruit just appearing above the surface. You may often find it in clusters, in all stages of growth, in rich mold, or on decayed logs or stumps.

Fig. 477 represents a full-grown mushroom and several younger ones at different periods of development. The younger ones are smooth, globular masses, but, as they

Fig. 477.

Fig. 478.

get larger, the outer wrappage breaks, as you see at the right in the figure, and reveals a stem with an umbrella-like cap. The ring around the stalk, seen in the full-grown specimen, shows where this covering, called the *volva*, was attached. The stout stem is called a *stipe*, and its cap the *pileus*. Along the under surface of the pileus you see numerous thin plates, called *gills*, and it is within these plates that the spores are found, many thousands occurring on the gills of a single mushroom.

Puff-balls are mushrooms without the stem and pileus. The " smoke " which escapes when they are broken consists of spores, which are so exceedingly small that they may penetrate everywhere. A few species of fungi are good to eat, but many are poisonous, and to be avoided. Yeast, mildew, smut, mold, and dry rot, all belong to this group of plants.

The gray, yellow, or greenish, crust-like layers that are seen on stones and the bark of trees, on old walls, and in rocky places, are a low form of vegetation, called *lichens* (Fig. 478). They have little distinction of parts, except that of upper and under surface, and certain specialized places in which spores are formed. *Algæ*, or the sea-weed family, is another order of flowerless plants, which contains many fresh-water species. The green scum seen on the surface of stagnant water is one of the lowest forms of fresh-water algæ, called *conferva*.

EXPLANATION OF TABLE.

IN the analysis of flowering plants, on the opposite page, the first division, it will be seen, is into Angiosperms (plants with seeds formed in an ovary) and Gymnosperms (naked-seeded plants, see page 164).

The Angiosperms are again separated into Dicotyledons and Monocotyledons, and these groups are divided and redivided by characters with which the use of the Flower Schedule has made you familiar.

The last groups reached in this analysis are used to divide the orders into sections; so that, by finding the section to which a plant belongs, you at once limit the number of groups with which it must be compared in determining the order to which it belongs.

SYSTEMATIC BOTANY.

PRIMARY ANALYSIS OF FLOWERING PLANTS.

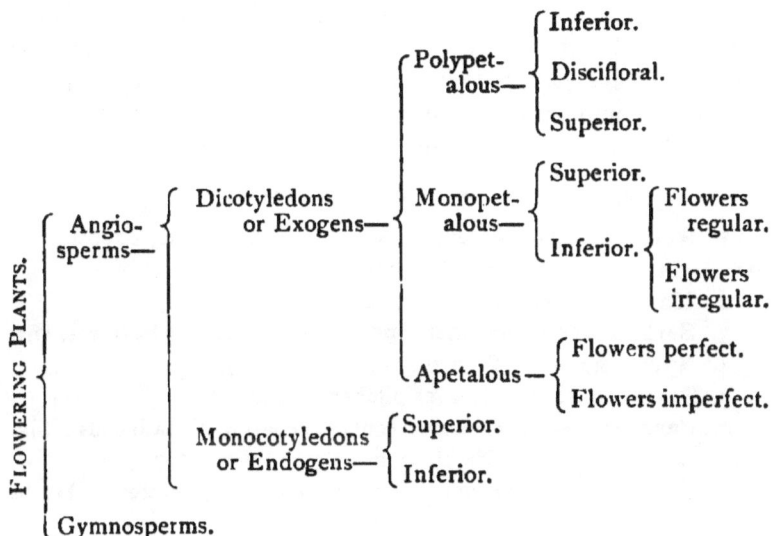

FLOWERING PLANTS.
- Angiosperms—
  - Dicotyledons or Exogens—
    - Polypetalous—
      - Inferior.
      - Discifloral.
      - Superior.
    - Monopetalous—
      - Superior.
      - Inferior.
        - Flowers regular.
        - Flowers irregular.
    - Apetalous—
      - Flowers perfect.
      - Flowers imperfect.
  - Monocotyledons or Endogens—
    - Superior.
    - Inferior.
- Gymnosperms.

CHARACTERISTICS OF THE NATURAL ORDERS OF FLOWERING PLANTS.

In Chapters Eighth to Twelfth it was shown that a natural order may be identified by a few characteristic features, without reference to its entire group of ordinal characters.

The orders in the following list are first separated into eleven groups, on the basis of characters indicated in the above table. The characteristics by which any order may be known are clearly and concisely given, and they follow each other in a natural sequence, according to their increasing divergence from the ordinal characters of the Ranunculaceæ. The same thing is seen in passing from first to last through the orders shown on the Botanical Charts.

To find the order of a plant, first determine the group to which it belongs ; then compare it with the characteristics of the orders of that group, one after another, until the one is reached to which the plant belongs. Its page in the body of the Flora is indicated by the figures. Now refer to this order, and look through the analysis of its genera to find the genus of the plant. If it is a common one, you will find it described and named among the species of this genus.

Hypogynous or Inferior Polypetalous Exogens.

1. Ranunculaceæ.—Stamens ∞; often apetalous, sepals colored ; carpels distinct. Herbs, rarely shrubs. 191.

2. Magnoliaceæ.—Like 1, but trees. 194.

3. Berberidaceæ.—Stamens opposite petals, opening by valves. 195.

4. Nymphæaceæ.—Stamens ∞; aquatic. 196.

5. Sarraceniaceæ.—Leaves pitcher-shaped. 197.

6. Papaveraceæ.—Flowers regular ; sepals 2-3, caducous ; placentæ parietal. Herbs with milky or colored juice. 197.

7. Fumariaceæ.—Flowers irregular ; sepals 2, persistent. Herbs with watery juice. 198.

8. Cruciferæ.—Flowers cruciate ; stamens tetradynamous ; ovary syncarpous, 2-celled. Herbs. 199.

9. Capparidaceæ.—Like 8, but stamens not tetradynamous ; ovary 1-celled. 202.

10. Resedaceæ.—Flowers irregular ; stamens ∞, on a lateral disk ; placentæ parietal, pod opening before maturity. Herbs. 202.

11. Violaceæ.—Flowers irregular ; stamens 5, on a disk ; pod 3-valved, with 3 parietal placentæ. Herbs. 203.

12. Cistaceæ.—Stamens ∞; pods 3-5-valved ; placentæ parietal. Shrubs or herbs. 204.

13. Polygalaceæ.—Flowers irregular ; stamens 6-8, with apical pores ; pod 2-seeded. 204.

14. Caryophyllaceæ.—Flowers regular ; ovary usually 1-celled, with free central placentæ. Herbs with opposite leaves. 205.

15. Portulacaceæ.—Like 14, but with 2 sepals. Leaves fleshy. Low herbs. 207.

16. Hypericaceæ.—Flowers regular ; stamens ∞, polyadelphous ; leaves entire, opposite, with pellucid dots. Herbs or shrubs. 208.

17. Malvaceæ.—Flowers regular ; stamens ∞, monadelphous ; leaves alternate, palmate. Herbs or shrubs. 209.

18. Tiliaceæ.—Flowers regular ; stamens ∞, in 5 clusters ; fruit a 1-2-seeded nut. Trees. 210.

Discifloral Polypetalous Exogens.

19. Linaceæ.—Flowers regular; stamens 5; petals convolute; capsule 8–10-seeded. Herbs with entire leaves. 210.

20. Geraniaceæ.—Flowers regular; ovary 3–5-lobed; receptacle elongated; glands outside the stamens; leaves variously cut. Herbs, or shrubby. 211.

21. Rutaceæ.—Flowers regular; ovary on a disk; leaves punctate, compound, or divided. Herbs, shrubs, or trees. 212.

22. Aquifoliaceæ.—Flowers 4–8-merous; stamens on base of petals; drupe berry-like. Trees or shrubs. 213.

23. Rhamnaceæ.—Stamens opposite the petals, and on a disk; a single erect ovule in each of the 2–5 cells; leaves simple, alternate. Shrubs or trees, with bitter principle. 213.

24. Vitaceæ.—Stamens opposite the caducous petals. Climbing shrubs. 214.

25. Sapindaceæ.—Flowers unsymmetrical; stamens 5–10, on a disk; ovary 2–3-celled, with 1–2 seeds in each cell. Trees, shrubs, or herbaceous climbers. 214.

26. Anacardiaceæ.—Flowers small, symmetrical; ovary 1-seeded, with 3 styles. Trees or shrubs, with acrid juice. 215.

Perigynous and Epigynous or Superior Polypetalous Exogens.

27. Leguminosæ.—Flowers papilionaceous or regular; odd sepal anterior; stamens usually 10. Fruit a legume. 216.

28. Rosaceæ.—Flowers regular, odd sepal posterior; stamens ∞, perigynous; usually stipules. Endosperm. 217.

29. Saxifragaceæ.—Flowers regular, much like 28, but having albumen in the seeds, and no stipules. 225.

30. Crassulaceæ.—Flowers symmetrical; pistils usually separate, ovules ∞. Leaves fleshy. 226.

31. Droseraceæ.—Flowers regular. Herbs with gland-bearing leaves. 227.

32. Hamamelaceæ.—Flowers in close clusters; carpels 2, united below. Shrubs or trees. 228.

33. Halorageæ.—Flowers small; ovaries 1–4-celled; styles distinct. Aquatic plants. 228.

34. Melastomaceæ.—Anthers opening by pores; leaves opposite, 3–7-ribbed, without stipules. 228.

35. Lythraceæ.—Calyx inclosing the 1–∞-celled ovary; anthers open lengthwise; style 1, seeds many. Leaves opposite, without stipules. 229.

36. **Onagraceæ.**—Flowers in fours, epigynous.; ovary ∞-seeded ; style 1.   Herbs or shrubs, without stipules.   229.

37. **Cucurbitaceæ.**—Flowers imperfect. epigynous ; anthers united, tortuous.   Fruit a pepo.   Climbing or prostrate herbs.   231.

38. **Cactaceæ.**—Floral parts numerous ; ovary 1 - celled, with parietal placentæ.   Fleshy and usually leafless plants.   232.

39. **Umbelliferæ.**—Inflorescence umbellate ; ovary decarpellary, splitting in two when ripe.   Endosperm.   232.

40. **Araliaceæ.**—Like 39, but with more than two styles, and the fruit a drupe.   236.

41. **Cornaceæ.**—Style 1 ; ovule 1 ; fruit a 1–2-seeded drupe. Woody plants, with opposite leaves.   237.

Superior Gamopetalous Exogens.

42. **Caprifoliaceæ.**—Ovary 2–5-celled ; leaves opposite, without stipules.   Shrubs, rarely herbs.   238.

43. **Rubiaceæ.**—Like 42, but with stipules, or the leaves in whorls.   240.

44. **Dipsaceæ.**—Flowers in heads, with involucres ; stamens 4 ; ovary 1-celled.   Endosperm.   Herbs.   241.

45. **Compositæ.**—Flowers in heads, surrounded with a common involucre ; anthers syngenesious ; ovary 1-celled, 1-seeded ; style bifid. No endosperm.   Herbs.   241.

46. **Campanulaceæ.**—Corolla 5-lobed, and stamens alternate ; ovary 2–5-celled ; style 1 ; ovules ∞.   Herbs, rarely shrubs, with alternate leaves and milky juice.   251.

Inferior Gamopetalous Exogens with Regular Flowers.

47. **Ericaceæ.**—Stamens hypogynous, anthers opening by terminal pores ; ovary 2–12-celled ; style simple.   252.

48. **Plumbaginaceæ.**—Flowers 5-merous ; stamens opposite petals ; ovary 1-celled ; styles 3–5.   The single ovule pendulous.   Herbs. 255.

49. **Primulaceæ.**—Flowers like 48, with a single style and ovary, free central placenta ; seeds ∞.   256.

50. **Oleaceæ.**—Stamens 2, epipetalous. ovary 2-celled.   Trees or shrubs.   257.

51. **Apocynaceæ.**—Corolla convolute ; ovaries 2, separate ; styles and stigmas united ; leaves opposite.   Plants with milky juice.   258.

52. **Asclepiadaceæ.**—Like 51, with anthers united to stigma, and pollen in masses.   259.

53. **Gentianaceæ.**—Cymose 4–5-merous flowers ; ovary 1-celled.

2 placentæ, seeds ∞. Herbs with opposite, entire leaves, without stipules. 260.

54. Polemoniaceæ.—Flowers 5-merous, ovary 3-celled, and style 3-cleft. Herbs with alternate leaves, without stipules. 261.

55. Boraginaceæ.—Inflorescence scorpioid; flowers 5-merous, with 4-lobed ovary becoming 4 achenes; style 1. 262.

56. Convolvulaceæ.—Flowers 5-merous, corolla convolute, ovary 2-celled, 1-styled; embryo large, crumpled. Twining or trailing herbs. 264.

57. Solanaceæ.—Flowers 5-merous, on bractless pedicels; ovary 2-celled, ∞-seeded; pod or berry. Mostly narcotic herbs. 265.

Inferior Gamopetalous Exogens with Irregular Flowers.

58. Scrophulariaceæ.—Flowers 2-lipped, stamens 2 or 4, didynamous; ovary 2-celled, single style, placentæ axile, seeds ∞. Chiefly herbs. 267.

59. Lentibulaceæ.—Stamens 2, ovary 1-celled. Floating herbs. 272.

60. Bignoniaceæ.—Stamens 2 or 4; seeds ∞, winged. Trees or shrubs with opposite leaves. 272.

61. Verbenaceæ.—Stamens 2-4, ovary 4-celled. Herbs or shrubs with opposite leaves. 273.

62. Labiatæ.—Stamens didynamous, ovary 4-lobed, becoming 4 nutlets. Herbs with square stems and opposite leaves. 274.

63. Plantaginaceæ.—Flowers spiked, 4-merous; stamens epipetalous, anthers versatile. 280.

Apetalous Exogens with Perfect Flowers.

64. Nyctaginaceæ.—Perianth petaloid, united at base, inclosing the fruit; ovary 1-celled, 1-seeded utricle. 281.

65. Amarantaceæ.—Flowers in dense clusters; stamens united at base; ovary 1-celled; fruit a utricle. 281.

66. Chenopodiaceæ.—Like 65, perianth green; filaments not united. Leaves often mealy. 282.

67. Phytolaccaceæ.—Inflorescence racemose; stamens hypogynous. Leaves alternate, glabrous. 283.

68. Polygonaceæ.—Ovary 1-celled, styles 2-3, and 1 orthotropous seed. Leaves alternate, with membranous sheathing stipules. 283.

69. Aristolochiaceæ.—Ovary 4-6-celled, seeds ∞. Herbs or shrubs, often twining. 285.

70. Lauraceæ.—Sepals 4-6 in two rows, anthers opening by valves. Aromatic trees or shrubs with punctate leaves. 285.

71. Thymeleaceæ.—Flowers in close clusters; ovary with 1 pendulous ovule in each cell. Shrubs or trees with tough bark. 286.

72. Santalaceæ.—Ovary inferior, 1-celled, 1-3 ovules on a free central placenta. Usually parasitic. 286.

Apetalous Exogens with Imperfect Flowers.

73. Euphorbiaceæ.—Ovary usually 3-celled, with 1-2 seeds hanging in each cell. Plants with milky juice. 287.

74. Urticaceæ.—Flowers monœcious or diœcious ; calyx free from 1-seeded fruit. Leaves with stipules. 288.

75. Juglandaceæ.—Male flowers in catkins, stamens ∞ on a bract ; ovary 1-celled, 1 ovule. Trees with pinnate leaves. 290.

76. Myricaceæ.—Like 75, but stamens few ; no calyx. Leaves simple. 291.

77. Cupuliferæ.—Ovary 2-3-celled ; fruit 1-seeded nut. Trees with simple leaves. 292.

78. Salicaceæ.—Perianth of scales ; ovary 1-celled ; capsule 2-4-valved, seeds plumed. Shrubs or trees with alternate leaves. 294.

Superior Endogens.

79. Orchidaceæ.—Flowers very irregular ; stamens 1-2, confluent with style, pollen in masses. 296.

80. Iridaceæ.—Flowers regular, 3-merous ; stamens 3, anthers extrorse ; leaves 2-ranked, equitant. 299.

81. Amaryllidaceæ. — Flowers regular ; stamens 6. Bulbous herbs. 300.

Inferior Endogens.

82. Alismaceæ.—Flowers regular, 3-merous ; anthers extrorse ; ovaries 3, distinct. Marsh herbs. 301.

83. Naiadaceæ.—Perianth scale-like ; aquatic plants ; stems jointed, with sheathing stipules. 302.

84. Typhaceæ.—Flowers imperfect, in close clusters ; ovules 1-2, pendulous. Aquatic plants with linear leaves. 303.

85. Araceæ.—Flowers on a spadix ; berry with few seeds ; leaves mostly reticulate. Plants with acrid juice. 303.

86. Liliaceæ.—Perianth regular ; ovary 3-celled ; seeds ∞. 305.

87. Smilaceæ.—Flowers regular, imperfect. Climbing plants, with tendril-stipules. 309.

88. Juncaceæ.—Flower structure like 87, perianth glumaceous. Herbage grass-like. 309.

89. Pontederiaceæ.—Flowers perfect, petaloid from a spathe. Aquatic herbs. 310.

90. Cyperaceæ.—Flowers spiked with perianth of bristles; fruit an achene. Grass-like plants, solid stems, and tristichous leaves. 310.

91. Gramineæ.—Flower spikelets in chaffy glumes; perianth rudimentary, stamens 3; ovary 1; style 2; fruit a grain. Stems solid at joints, distichous leaves. 315.

Gymnosperms.

92. Coniferæ.—Male flowers in catkins, female in catkins or cones; leaves acicular. Trees or shrubs. 323.

HYPOGYNOUS OR INFERIOR POLYPETALOUS EXOGENS.

Order I.—RANUNCULACEÆ (*Crowfoot Family*).

A large family of herbs, or climbers, with a colorless, acrid juice. Leaves usually divided, with stalk dilated at the base; stipules none. The sepals, 3–15, distinct and hypogynous; petals, 3–15, or wanting, rarely united; stamens, usually many; embryo, minute; fruit, dry pods, achenia, or berries.

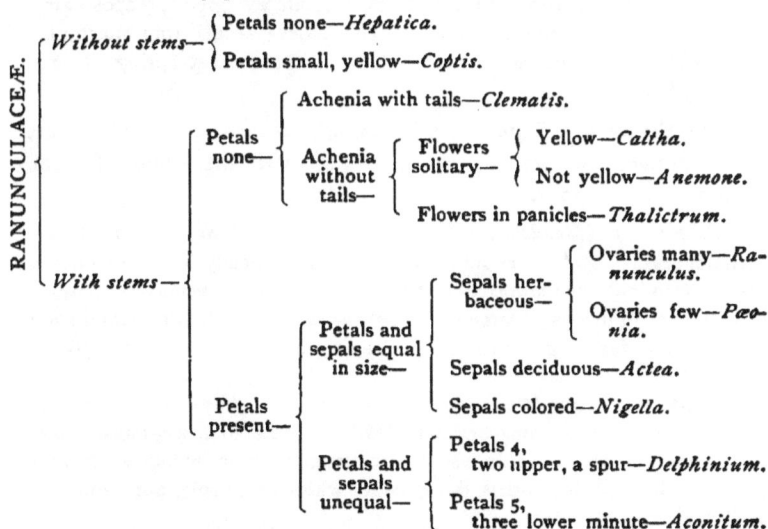

RANUNCULACEÆ.

Without stems—
 Petals none—*Hepatica.*
 Petals small, yellow—*Coptis.*

With stems—
 Petals none—
  Achenia with tails—*Clematis.*
  Achenia without tails—
   Flowers solitary—
    Yellow—*Caltha.*
    Not yellow—*Anemone.*
   Flowers in panicles—*Thalictrum.*
 Petals present—
  Petals and sepals equal in size—
   Sepals herbaceous—
    Ovaries many—*Ranunculus.*
    Ovaries few—*Pæonia.*
   Sepals deciduous—*Actea.*
   Sepals colored—*Nigella.*
  Petals and sepals unequal—
   Petals 4, two upper, a spur—*Delphinium.*
   Petals 5, three lower minute—*Aconitum.*

1. Clematis.—Sepals 4–8, colored. Petals small or none. Anthers linear, extrorse. Achenia many in a head, terminating in long, plumose tails. Herbs or vines climbing by clasping leaf-stalks. ♃

1. C. VIRGINIANA (*Virgin's Bower*).—Leaves ternate, smooth; leaflets ovate, somewhat cordate at base; flowers in panicles, often diœcious, white; fruit covered with long, feathery tails, which in autumn cause the plant to appear at a distance as if in bloom. A vine, in thickets and low grounds, 10–15 feet long. *July–August.*

2. **Anemone.**—Involucre distant from the flower, composed of 3 incised leaves. Sepals 5–15, petaloid. Ovaries collected into roundish or oval heads. Achenia mucronate. ♃

1. A. NEMOROSA (*Wood Anemone*).—Stem simple, smooth; leaves ternate; leaflets 3–5-lobed; involucre of 3 leaves, on short petioles near the top of the stem; flowers solitary, nodding, white, purplish outside; heads of achenia small, globose. A pretty little plant, growing from 3′–6′ high, in old woods and thickets. *April–May.*

2. A. VIRGINIANA (*Wind-flower*).—Stem pubescent; leaves ternate; leaflets petiolulate, 2–3-cleft, ovate, dentate; flowers greenish white, with the sepals pubescent beneath, on long peduncles; stem 2–3 feet high, with a 3-leaved involucre. A coarse-looking plant. Pastures and fields. *June–July.*

3. **Hepatica.**—Involucre, very near the flower, of 3 ovate, obtuse bracts. Sepals petaloid, 6–9 in number, in 2–3 rows. Petals none. Achenia awnless. ♃

1. H. TRILOBA (*Heart Liverwort*).—Acaulescent; leaves broadly cordate, mostly 3-lobed, with the lobes entire, smooth, evergreen; scapes hairy, 1-flowered, flower nodding; involucre green, hairy; sepals from bluish purple to nearly white, oblong, obtuse. One of the earliest spring flowers. *March–April.*

4. **Thalictrum.**—Sepals 4–5, petaloid, caducous. Peta's none. Stamens numerous. Achenia 4–15, tipped with the stigma, falcate. Flowers often diœcious. ♃

1. T. CORNUTI (*Meadow Rue*).—Stem erect, branching, smooth; leaves triternate, sessile; leaflets rhomboidal, 2–3-lobed, mostly smooth; flowers in large, terminal, compound panicles, white, small, diœcious or polygamous; sepals caducous; filaments somewhat clavate. A tall plant in wet meadows, 3–8 feet high, with large leaves and a hollow stem. *June–August.*

2. T. ANEMONOIDES (*Rue Anemone*).—Stem low, simple, smooth; radical leaves biternate, on long petioles; leaflets rhomboidal, 2–3-lobed; stem leaves 3, ternate, nearly sessile, verticillate, resembling an involucre; flowers large, 3–6, in umbels; sepals 6–10, oval, white or purple, not caducous. *April–May.*

5. **Ranunculus.**—Sepals 5. Petals 5, with a scale or gland at the base of each petal inside. Flowers solitary; stamens indefinite. Achenia ovate, pointed, crowded into heads. ♃

1. R. ABORTIVUS (*Small-flowered Crowfoot*).—Stem erect, branching, very smooth; radical leaves reniform, upper ones 3-5-lobed; flowers small, yellow; fruit in heads; style very short, straight. This species grows 10′–18′ high. Damp woods. *May–June.*

2. R. RECURVATUS (*Wood Crowfoot*).—Stem erect, hairy; leaves deeply 3-parted, segments ovate, dentate, pubescent; flowers small, greenish yellow; petals narrow, scarcely equal to the reflexed sepals; achenia in globose heads, tipped with the minute, hooked styles. *June.*

3. R. ACRIS (*Buttercups*).—Stem erect, branching, often hollow hairy; leaves generally pubescent, deeply 3-parted; segments deeply incised, divisions of the upper ones linear; flowers large, of a burnished golden yellow; calyx spreading; heads globose; achenia, beak short, recurved. Common, 1-2 feet high. *June–September.*

4. R. REPENS (*Creeping Crowfoot*).—Stem stoloniferous, mostly smooth; leaves on long petioles, trifoliate, segments deeply 3-lobed; peduncles furrowed; flowers yellow; sepals spreading; stems 1–3 feet long, with dark leaves. Wet grounds. Common. *May–July.*

5. R. AQUATILIS (*White Water Crowfoot*).—Stem floating; submerged leaves filiformly dissected, the emersed, when present, 3-parted; flowers rather small, dull white, about 1′ out of water; petals dull white, with yellow claws. *May–August.*

6. Caltha.—Sepals 5–9, resembling petals. Petals none. Ovaries 5–10. Follicles 5–10, compressed, erect, many-seeded. ♃

1. C. PALUSTRIS (*Marsh Marigold*).—Plant very smooth; stem erect, hollow, leaves reniform, crenate; flowers large, bright yellow, pedunculate, in umbellate clusters of 3-5. *April–May.*

7. Coptis.—Sepals 5–7, petaloid, deciduous. Petals 5–7. Stamens 15–25. Follicles 5–10, stipitate, somewhat stellately diverging, 4-8-seeded. ♃

1. C. TRIFOLIA (*Goldthread*).—Leaves radical, ternate; leaflets sessile, wedge-obovate, on petioles 1′–2′ long; peduncles twice as long, slender, 1-flowered, with a single, minute bract above the middle; petals minute, yellow; stamens white; root slender, creeping, of a golden yellow. A delicate little plant, growing in bogs. *May.*

8. Aquilegia.—Sepals 5, petaloid. Petals 5, tubular, extending below into long, spurred nectaries. Follicles 5, erect, many-seeded, tipped with the style. ♃

1. A. CANADENSIS (*Columbine*).—Stem erect, smooth, branching; lower leaves biternate; leaflets on long petioles, 3-lobed; flowers large, nodding, scarlet; sepals ovate-oblong; petals scarlet below, each with a straight spur, with honey. A beautiful plant, growing 1-2 feet high, in the clefts of rocks. *May.*

9. Delphinium.—Sepals 5, irregular, the upper with a spur. Petals 4, irregular, the upper two forming spurs. Pistils 1–5; many-seeded pods.

1. D. Consolida (*Larkspur*).—Stem erect, smooth ; leaves palmately divided into fine segments ; flowers few, in loose racemes, but from blue to white, often double ; pedicels longer than the bracts ; carpels smooth. Cultivated in gardens. *July–August.*

2. D. exaltatum (*Tall Larkspur*).—Stem erect, glabrous below, pubescent near the summit ; leaves deeply 3–5 parted, lobes cuneiform, 3-cleft ; petioles not dilated at base ; racemes straight ; flowers large, of a rich purplish blue. Native, and common in cultivation. *June–August.*

10. Aconitum.—Sepals 5, petaloid, upper one large, vaulted. Petals 2—the 3 lower ones minute, expanded into a sac, or short spur at the summit. ♃

1. A. napellus (*Monk's-Hood*).—Stem straight, erect ; leaves deeply 5-cleft ; leaflets pinnatifid ; upper sepal arched at the back, somewhat resembling a monk's cowl ; ovaries smooth ; flowers large, blue, in long racemes. Plant 4 feet high in gardens. *August.*

11. Actæa.—Sepals 4–5, roundish, deciduous. Petals 4–8, spatulate. Stamens indefinite. Anthers 2-lobed. Stigma capitate, sessile. Berry globose, 1-celled. Seeds many, compressed. ♃

1. A. spicata, var. rubra (*Red Baneberry*).—Stem erect, smooth ; leaves ternately decompound ; leaflets ovate, serrate, smooth ; flowers small, white, in a dense hemispherical raceme, pedicellate, followed by red berries. Quite common in rocky woods, 1–2 feet high. *May.*

12. Pæonia.—Sepals 5, unequal, leafy, persistent. Petals 5. Stamens numerous. Ovaries 2–5. Stigmas sessile, double, persistent. Follicles many-seeded, opening above. ♃

1. P. officinalis (*Peony*).—Stem erect, herbaceous, smooth ; lower leaves bipinnately divided, coriaceous ; leaflets ovate-lanceolate, incised ; fruit downy, nearly straight. Flowers 2′–3′ in diameter, generally double, and varying from red to rose-color and white. *May–June.*

Order II.—Magnoliaceæ (*Magnolia Family*).

Trees or shrubs with the leaf-buds covered with stipules. Flowers large, solitary, polypetalous, hypogenous. Pistils many, closely covering the long receptacle. Seeds 1-2 in each carpel ; embryo minute.

Sepals 5—*Magnolia.*
Sepals 3—*Liriodendron.*

1. Magnolia.—Sepals 3. Petals 6–9. Carpels 1–2-seeded, forming a strobile-like fruit. Seeds baccate, suspended by a long funiculus. Trees.

1. M. glauca (*Small Magnolia*).—Leaves oval, entire, obtuse, glaucous beneath ; flowers solitary, 2′ in diameter, fragrant ; sepals 3 ; petals obovate, concave, narrowed at base, erect. *July.*

2. Liriodendron.—Sepals 3, caducous. Petals 6. Carpels 1–2-seeded, indehiscent, imbricated in a cone. Trees.

1. L. TULIPIFERA (*Tulip-tree*).—Leaves dark green, very glabrous, truncate; flowers large, solitary, fragrant, orange within. A noble tree. *May–June.*

Order III.—BERBERIDACEÆ.

Herbs or shrubs. Petals opposite the sepals. Stamens definite, as many as the petals, and opposite them; or else twice as many. Anthers extrorse, usually opening by recurved valves. Ovary of a single carpel, forming in fruit a one-celled berry or capsule. Seeds few.

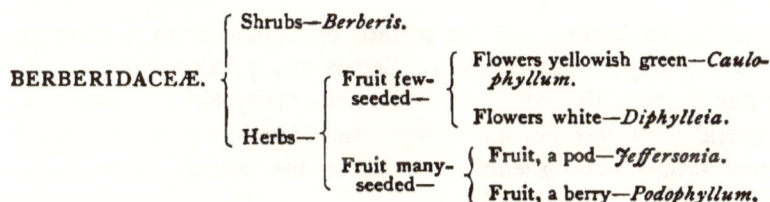

BERBERIDACEÆ.

Shrubs—*Berberis.*

Herbs—

Fruit few-seeded—
- Flowers yellowish green—*Caulophyllum.*
- Flowers white—*Diphylleia.*

Fruit many-seeded—
- Fruit, a pod—*Jeffersonia.*
- Fruit, a berry—*Podophyllum.*

1. Berberis.—Petals 6, with 2 glands at the base of each. Stamens 6; filaments flattened. Stigma sessile. Fruit a 2–3-seeded, oblong berry. Shrubs.

1. B. VULGARIS (*Barberry*).—Leaves oval or obovate, sharply serrate, in clusters, with 3 spines at base; flowers yellow, in nodding racemes; berries oblong, red, very acid. The stamens are irritable, springing against the pistil when touched at the base. *June.*

2. Caulophyllum.—Petals 6, with nectariferous scales at the base within. Stamens 6. Pericarp membranaceous, 2–4-seeded. Seeds erect, globose.

1. C. THALICTROIDES (*Blue Cohosh*).—Very glabrous; leaves biternate and triternate; leaflets ovate, irregularly lobed, terminal one broadest; stem simple, smooth, bearing 2 leaves; flowers greenish yellow; seeds deep blue, soon bursting the ripe fruit, resembling berries. Young plant is purple, and somewhat resembles a fern. *April–May.*

3. Diphylleia.—Sepals 6, fugacious. Petals 6, larger than sepals. Ovules 5–6, attached to side of ovary. Seeds without aril. Root-stock perennial; leaves umbrella-like.

1. C. CYMOSA (*Umbrella-leaf*).—Root-leaves 1–2 feet in diameter, 2-cleft, and the divisions much lobed. Berries blue. Wet places. *May.*

4. Jeffersonia.—Sepals 4, fugacious. Petals 8. Stamens 8. Stigma 2-lobed. Pod pear-shaped, opening with a lid. Seeds with aril on the sides.

1. J. DIPHYLLA (*Twin-leaf*).—A glabrous herb with fibrous roots. Leaves long, petioled, divided into two leaflets. Flowers white. *April-May.*

5. **Podophyllum.**—Sepals 3, caducous. Petals 6-9, obovate. Stamens 9-18. Anthers linear. Ovary ovate, thick, sessile, peltate. Fruit a large, ovoid berry, 1-celled, and tipped with the stigma. *Per.*

1. P. PELTATUM (*May-apple*).—Stem smooth, 1 foot high, with 2 leaves, and a nodding flower between them; leaves from 6'-10' in diameter, peltate, often cordate at base, palmately 5-7-lobed; flower white, 1'-2' in diameter; petals curiously veined; fruit ovoid, yellowish, subacid and eatable. *May.*

Order IV.—NYMPHÆACEÆ (*Water-lily Family*).

Aquatic herbs. Leaves peltate, or cordate, from a rhizoma. Flowers large, often fragrant. Sepals and petals several, imbricated in rows, the latter inserted in the fleshy disk. Stamens in several rows, with introrse anthers, and filaments petaloid. Ovary many-celled, many-seeded, crowned by the radiate stigma, indehiscent.

NYMPHÆACEÆ.
{
  Flowers small.
     Petals 3-4 — *Brasenia.*

  Flowers large. { Stamens on ovary—*Nymphæa.*
     Petals many— { Stamens under ovary—*Nuphar.*
}

1. **Nymphæa.**—Sepals 4-5. Petals and stamens indefinite, passing into each other. Stigma surrounded by rays. Pericarp many-celled, many-seeded.

1. N. ODORATA (*White Pond-lily*).—Leaves orbicular, cordate at base, entire, floating, with prominent veins beneath; sepals 4, equaling the petals, which are lanceolate, white, often tinged with purple; filaments yellow. *July.*

2. **Nuphar.**—Sepals 5-6, somewhat petaloid. Petals numerous, nectariferous on the back, stamens linear, stigma surrounded with rays.

1. N. ADVENA (*Yellow Pond-lily*).—Leaves oval, smooth, entire, cordate at base, with an open sinus, on long petioles; flowers large, dull yellow; sepals 6; petals many; stigma 12-15-rayed, with a crenate margin. *June-July.*

3. **Brasenia.**—Sepals 3-6, colored within, persistent. Petals 3-4. Stamens 18-36. Ovaries 6-18. Carpels oblong, 1-2-seeded.

1. B. PELTATA (*Water-shield*).—Stem floating, branched; leaves elliptical and peltate, entire, smooth, floating; flowers just rising above the water; dull purple, ½' in diameter. *July.*

Order V.—Sarraceniaceæ (*Pitcher-plant Family*).

Perennial, acaulescent bog-herbs. Leaves pitcher-shaped. Flowers large, solitary, nodding. Sepals 5, persistent, with 3 bracts at base. Petals 5. Stamens indefinite, hypogynous, style single, stigma large, petaloid. Seeds numerous.

1. **Sarracenia.**—Petals, deciduous. Stigma peltate, 5-angled, persistent. Capsule 5-celled, 5-valved, many-seeded.

1. S. PURPUREA (*Pitcher-plant*).—Leaves with tubular, inflated, gibbous petioles, which are winged on the inside, ending in a broadly cordate, erect lamina, hairy on the inside. Scapes 12'–30' high, with a single, large, nodding, dark-purple flower; petals inflected over the stigma. *June.*

Order VI.—Papaveraceæ.

Herbs, usually with a milky or colored juice. Leaves alternate. Stipules none. Flowers solitary, on long peduncles, never blue. Sepals 2, rarely 3, caducous. Petals 4, rarely 6, regular, hypogynous. Stamens usually some multiple of 4. Ovary 1-celled, forming a pod with 2 or 3 parietal placentæ, or a capsule with several.

PAPAVERACEÆ.
- Without stem—*Sanguinaria.*
- With stem—
  - Juice yellow—
    - Flower-buds droop—*Chelidonium.*
    - Flower-buds erect—*Argemonia.*
  - Juice white—*Papaver.*

1. **Sanguinaria.**—Sepals 2. Petals 8–12. Stamens numerous. Stigma 1–2-lobed, sessile. Capsule oblong, 1-celled, 2-valved, many-seeded.

1. S. CANADENSIS (*Bloodroot*).—Acaulescent; rhizomas creeping, fleshy, of a red color. Leaves reniform, 3–7-lobed, smooth. Flower white, inodorous, and of brief duration. *April–May.*

2. **Chelidonium.**—Perennial herb. Sepals 2. Petals 4. Stamens numerous. Stigma sessile, bifid. Capsule pod-like, linear, 2-valved.

1. C. MAJUS (*Celandine*).—Stem erect, branching, very smooth; leaves pinnate, 5–7-foliate; leaflets ovate; flowers yellow, very fugacious; sepals orbicular; petals elliptical. A branching, pale-green weed. *May–August.*

3. **Agremone.**—Petals 4–6. Style none; pod prickly, valved. Seeds crested, juice yellow, leaves sessile.

1. A. MEXICANA (*Prickly Poppy*).—Leaves with prickly teeth; flowers yellow. Waste places. *July–September.*

4. Papaver.—Sepals 2. Petals 4. Style none. Capsule obovate, opening by pores under the broad, persistent stigma.

1. P. SOMNIFERUM (*Opium Poppy*).—Caulescent, plant very smooth and glaucous; leaves clasping; sepals smooth; capsule globose; 1-2 feet high; flowers about 3′ in diameter. *June-July.*

Order VII.—FUMARIACEÆ.

Smooth herbs, with a watery juice. Stems brittle. Leaves alternate, much divided. Flowers irregular. Sepals 2. Petals 4, in pairs, the outer ones spurred or saccate at base; the inner cohering at apex. Stamens 6, diadelphous. Ovary simple, becoming a 1-celled, 2-valved pod, or an indehiscent, persistent, globular capsule. Seeds with an aril. Embryo minute. Albumen fleshy.

FUMARIACEÆ.
- Corolla 2-spurred—
  - Petals slightly united, seed crested—*Dicentra.*
  - Petals permanently united, seed crestless—*Adlumia.*
- Corolla 1-spurred—
  - Ovary several-seeded—*Corydalis.*
  - Ovary 1-seeded—*Fumaria.*

1. Dicentra.—Sepals 2, minute. Petals 4, 2 outer saccate. Stamens in 2 sets of 3 each. Capsule pod-shaped, 2-valved, many-seeded.

1. D. CUCULLARIS (*Dutchman's Breeches*).—Scape from a bulb. Raceme simple; corolla with two long spurs.
2. D. CANADENSIS (*Squirrel-Corn*).—Acaulescent; leaves very finely dissected, glaucous beneath; flowers pale purple, with short, rounded spurs. *May-June.*

2. Adlumia.—Sepals minute. Petals united into a persistent, monopetalous corolla, 4-toothed at apex. Capsule pod-shaped, linear-oblong, many-seeded.

1. A. CIRRHOSA (*Mountain Fringe*).—Stem herbaceous; leaves biternately decompound; leaflets rhomboidal; flowers numerous, in axillary, nodding, racemose clusters, pale purple. *June-August.*

3. Corydalis.—Sepals minute. Petals only 1-spurred at base. Stamens 6, in 2 sets. Capsule pod-shaped, 2-valved, many-seeded.

1. C. GLAUCA (*Pale Corydalis*).—Stem erect, branching; leaves bipinnately decompound, glaucous; flowers in erect racemes, large, rose-colored and yellow. *May.*

4. Fumaria.—Sepals caducous. Petals unequal, 1 only spurred at base. Fruit a 1-seeded, ovoid, or globose valveless nut.

1. F. OFFICINALIS (*Common Fumitory*).—Stem erect or decumbent, branching; leaves biternately dissected, segments linear; flowers small, rose-colored, in loose racemes. Sepals ovate-lanceolate, acute; nut globose, retuse. *June—September*.

Order VIII.—CRUCIFERÆ.

Herbs. Leaves alternate. Stipules none. Flowers usually yellow or white, in racemes or corymbs, destitute of bracts. Sepals 4, deciduous. Corolla of 4 usually regular, unguiculate petals, in the form of a cross. Stamens 6, 2 shorter than the other 4. Ovary consisting of 2 carpels united by a membranous partition, usually a 2-celled pod, called a silique, or a silicle. Seeds destitute of albumen. Embryo variously folded, with the cotyledons on the radicle.

A large, important, and very natural order, containing some very beautiful and fragrant flowers.

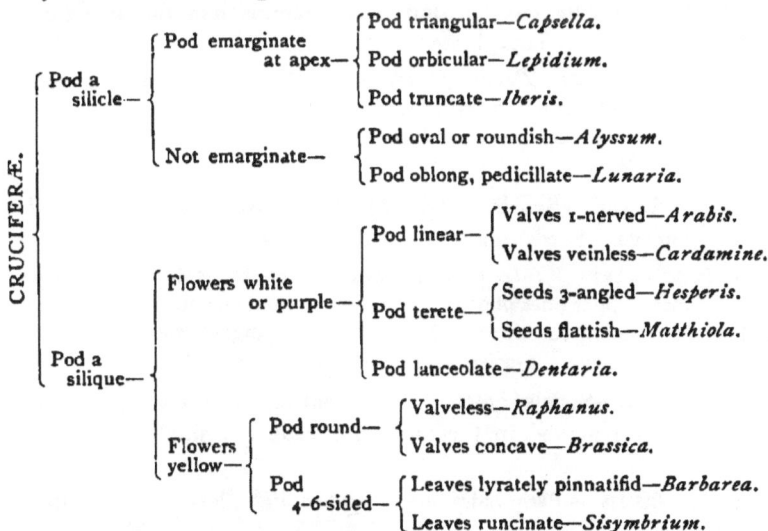

CRUCIFERÆ.
- **Pod a silicle—**
  - **Pod emarginate at apex—**
    - Pod triangular—*Capsella.*
    - Pod orbicular—*Lepidium.*
    - Pod truncate—*Iberis.*
  - **Not emarginate—**
    - Pod oval or roundish—*Alyssum.*
    - Pod oblong, pedicillate—*Lunaria.*
- **Pod a silique—**
  - **Flowers white or purple—**
    - **Pod linear—**
      - Valves 1-nerved—*Arabis.*
      - Valves veinless—*Cardamine.*
    - **Pod terete—**
      - Seeds 3-angled—*Hesperis.*
      - Seeds flattish—*Matthiola.*
    - Pod lanceolate—*Dentaria.*
  - **Flowers yellow—**
    - **Pod round—**
      - Valveless—*Raphanus.*
      - Valves concave—*Brassica.*
    - **Pod 4-6-sided—**
      - Leaves lyrately pinnatifid—*Barbarea.*
      - Leaves runcinate—*Sisymbrium.*

1. Capsella.—Silicles triangular-cuneiform, obcordate. Valves wingless. Cells small, many-seeded. Style short. ①

1. C. BURSA-PASTORIS (*Shepherd's-purse*).—Stem erect, branching; radical leaves pinnatifid, on short petioles, growing in a flat tuft, upper ones linear-lanceolate, auriculate at base; flowers very small, white, in long racemes; capsules obcordate, tipped with the short style. A common weed, 1–2 feet high. *June–October*.

2. Lepidium.—Sepals and petals ovate. Silicles nearly orbicular, emarginate. Valves carinate, dehiscent. Cells 1-seeded. ①

1. L. Virginicum (*Wild Peppergrass*).—Stem erect, branching, smooth; leaves linear-lanceolate, dentate, acute, smooth; flowers minute, white, in terminal racemes; silicle orbicular, not winged. In dry soil, about 1 foot high. Leaves with peppery taste, like the garden peppergrass. *June–October*.

3. **Alyssum.**—Calyx equal at base. Petals entire. Stamens toothed. Silicle orbicular, or oval, with valves flat, or convex in the center. 2͵

1. A. maritimum (*Sweet Alyssum*).—Stem suffruticose and procumbent; leaves linear-lanceolate, hoary; flowers small, white, fragrant; pods oval, smooth. Gardens. *June–October*.

4. **Lunaria.**—Sepals somewhat bi-saccate at base. Petals nearly entire. Silicles pedicellate, elliptical, or lanceolate. Valves flat. ②—2͵

1. L. biennis (*Honesty*).—Pubescent; stem erect; leaves cordate, with obtuse teeth; flowers lilac; silicles oval, obtuse at both ends. A garden plant, 3–4 feet high. *May–June.* ②

5. **Iberis.**—The two outside petals larger than the two inner. Silicles compressed, emarginate; cells 1-seeded. ①

1. I. umbellata (*Purple Candy-tuft*).—Stem herbaceous, smooth; leaves linear-lanceolate; lower ones serrate; flowers purple, terminal, in simple umbels; silicles acutely 2-lobed. A common garden plant, 1 foot high. *June–July.*

6. **Barbarea.**—Sepals nearly equal at base. Silique 4-sided; valves concave. Seeds in a single series. ②—2͵

1. B. vulgaris (*Winter Cress*).—Stem smooth, branching above; lower leaves lyrate; upper ones obovate, pinnatifid at base, crenate; flowers small, yellow, in terminal racemes; siliques obtusely 4-angled, slender. Brooksides and damp fields; 1–2 feet high, dark green. *May–June.*

7. **Arabis.**—Sepals erect. Petals entire, with claws. Silique linear, valves 1-nerved in the middle. Seeds in a single row in each cell. ②

1. A. hirsuta.—Plant hairy, erect, 1–2 feet high; leaves arrow-shaped; flowers small, greenish white; pods upright. Rocky places. *May.*

8. **Cardamine.**—Calyx spreading. Silique linear. Valves flat, veinless, opening elastically. Seeds ovate, with slender stalks. 2͵

1. C. hirsuta (*Bitter Cress*).—Stem erect, branching; leaves pinnate pinnatifid; flowers small, white, racemose; silique long, slender, erect, tipped with a short style. A variable, dark-green plant, common in wet grounds. *June.*

2. C. rhomboida.—Stem upright, from a tuber; flowers white or purple; leaves toothed, and lower ones heart-shaped. Wet places. *April–May.*

9. Dentaria.—Sepals converging. Silique lanceolate ; valves flat, nerveless, opening elastically. Seeds in a single row, without margins. ♃

2. D. DIPHYLLA (*Pepper-root*).—Rhizoma elongated, with a pungent flavor ; stem erect, with 2 opposite ternate leaves half-way upon the stem ; leaflets ovate, serrate ; flowers large, white, or very pale purple, in a terminal raceme. Woods and meadows. *May.*

10. Hesperis.—Calyx closed, furrowed at base. Petals linear or obovate, bent obliquely. Silique nearly round, or 4-sided. Stigmas 2, erect. Seeds 3-sided, without margins. ♃

1. H. MATRONALIS (*Rocket*).—Stem simple, erect ; leaves ovate-lanceolate ; stem with scattered, bristly hairs ; flowers large, purple, racemed ; siliques erect, 2'-4' long, smooth. A garden perennial, 3-4 feet high. *May-July.*

11. Sisymbrium.—Sepals equal at base. Petals with claws, entire. Silique terete ; valves concave. Style short. Seeds ovate. ①

1. S. OFFICINALE (*Hedge Mustard*).—Stem erect, very branching ; leaves hairy ; flowers in slender racemes, small, yellow ; siliques sessile, erect. A common weed, 1-3 feet high. *June-September.*

12. Matthiola.—Calyx closed. Sepals bi-saccate at base. Silique terete. Stigmas connivent, thickened, or carinate on the back. ①—♃

1. M. ANNUA (*Ten-weeks Stock*).—Stem herbaceous, branched ; leaves hoary, lanceolate, obtuse, toothed ; flowers large, variegated ; silique sub-cylindrical. A garden plant, 2 feet high, with soft stellate pubescence. *June-August.*

13. Brassica.—Sepals equal at base. Petals obovate. Silique compressed, with concave valves. Style short, obtuse. Seeds globose. ②

1. B. NIGRA (*Mustard*).—Stem erect, smooth, branching ; lower leaves lyrate ; upper, linear-lanceolate, entire ; flowers yellow, ¼' in diameter, racemose ; siliques smooth, 4-sided, nearly 1' long ; seeds small, nearly black. Cultivated grounds, 3-6 feet high. *June-July.*

2. B. RAPA (*Turnip*).—Stem and leaves deep green ; radical leaves lyrate, rough ; lower stem-leaves incised ; upper, entire, clasping ; flowers yellow ; seeds reddish brown. Common in cultivation as an esculent vegetable, and for feeding stock. *June.*

3. B. OLERACEA (*Cabbage*).—Leaves smooth and glaucous, fleshy, toothed, or lobed, sub-orbicular ; flowers yellow, in paniculate racemes. Native of sea-shores and cliffs in Europe, where it shows no appearance of a head like that of the esculent varieties, thus indicating the great power of cultivation. The cauliflower, broccoli, as well as the cabbage, are varieties of this species. *June.*

14. Raphanus.—Calyx erect. Petals obovate. Silique terete, valveless, many-celled. Seeds sub-globose, in a single series. ① ②

1. R. RAPHANISTRUM (*Wild Radish*).—Stem erect, branching, terete ; leaves lyrate ; flowers bright yellow, large, racemose ; petals turning white, purple, or blue ; silique round, jointed, 1-celled when mature. Road-sides and fields. *July.*

Order IX.—Capparidaceæ (*Caper Family*).

Herbs ; or, in the Tropics, shrubs or trees. Leaves alternate, without true stipules. Flowers solitary, or in racemes. Sepals 4. Petals 4, cruciform, unguiculate, unequal. Stamens usually some multiple of 4. Ovary of 2 united carpels. Styles united into one. Fruit a 1-celled pod, without any partition ; or baccate. Seeds usually many, reniform. Embryo curved, with leafy cotyledons.

An order much resembling Cruciferæ, but usually distinguished by the number of its stamens and the structure of its fruit. The plants are often more acrid than the Cruciferæ ; and their roots, bark, and herbage are nauseously bitter, and sometimes poisonous.

CAPPARIDACEÆ. { Ovary and pod stalked—*Cleome.*
{ Ovary and pod sessile—*Polanisia.*

1. Cleome.—Calyx 4-cleft. Petals 4. Torus minute or nearly round. Stamens 6, rarely 4. Pod subsessile or stipitate. ①

1. C. PUNGENS (*Spiderwort*).—Glandular-pubescent ; stem smooth ; peti-oles prickly ; leaves digitate, 5–9, long-petiolate ; leaflets lanceolate, slightly toothed ; flowers purple, racemed ; petals on filiform claws ; stamens 6, twice as long as the petals. A common garden flower, 3 to 4 feet high. *July–August.*

2. Polanisia.—Petals 4, with claws ; stamens many, unequal. Receptacle short ; pod linear. Fetid annuals.

1. P. GRAVEOLENS.—Leaves of 3 leaflets. Flowers small, styles short. Gravelly shores. *June–August.*

Order X.—Resedaceæ (*Mignonette Family*).

Herbs, with a watery juice. Leaves alternate, without stipules, but often with 2 glands at base. Flowers small, fragrant, in terminal racemes. Sepals 4–7, united at base. Petals 2–7, usually unequal and lacerated, with nectariferous claws. Stamens 3–40, on the fleshy glandular disk. Ovary 1-celled, 3–6-lobed at summit, usually many-seeded, with 3–6 parietal placentæ, opening before maturity. Seeds reniform, with no albumen.

1. **Reseda.**—Sepals and petals many. Torus large, fleshy, bearing the ovary with several stamens and styles. ①

1. R. ODORATA (*Mignonette*).—Leaves 3-lobed, or entire ; sepals shorter than the petals. A plant widely known and admired for its unsurpassed fragrance.

Order XI.—VIOLACEÆ (*Violet Family*).

Herbs ; sometimes shrubby. Leaves alternate, or radical, petiolate, with stipules. Flowers irregular, solitary. Sepals 5, persistent, often auricled at base. Petals 5, unequal ; the lower one large, with a spur at base. Stamens 5, inserted on the disk. Filaments broad ; 2 of them usually furnished with a spur-like process. Anthers united in a ring. Ovary 1-celled, composed of 3 united carpels, with 3 parietal placentæ. Fruit a 3-valved, many-seeded capsule.

1. **Viola.**—Perennials, with 1-flowered peduncles. The spurred petal, which is really the upper one, appears to be the lower, on account of the reversed position of the flower. ♃

1. V. CUCULLATA (*Hooded Violet*).—Leaves cordate, crenate, often cucullate at base ; flowers blue, marked with dark lines ; stipules linear ; lower petals bearded. Everywhere in meadows and wet ground, and sometimes even in dry soils. *May.*

2. V. PEDATA (*Bird-foot Violet*).—Nearly smooth ; leaves pedate, 5–7 parted ; segments lanceolate, entire or 3-parted ; flowers pale, brilliant blue, rarely white ; petals beardless ; stigma large and thick, margined, obliquely truncate. Borders of dry woods, and sandy fields. *April–May.*

3. V. BLANDA (*White Violet*).—Leaves cordate, slightly pubescent, on short, pubescent petioles ; flowers small, white, on scapes longer than the leaves ; petals marked with blue lines, greenish at base ; rhizoma creeping. Meadows ; common. *May.*

4. V. CANADENSIS (*Canadian Violet*).—Stem nearly purple ;. radical leaves reniform ; cauline ones cordate ; serrate, with pubescent veins ; flowers pale blue or white ; stipules ovate-lanceolate, entire ; peduncles shorter than the leaves ; petals yellowish at base ; upper ones purple outside ; lateral ones bearded. In woods, 6'–12' high. *May–June.*

5. V. PUBESCENS (*Large Yellow Violet*).—Stem erect, pubescent ; leaves broad, cordate, obtuse, toothed, pubescent ; stipules ovate ; flowers large, yellow ; lateral petals bearded ; upper ones with brown lines ; peduncles pubescent, shorter than the leaves ; sepals oblong-lanceolate ; spur very short. In rich woods, 3'–6'. Very variable. *May–June.*

6. V. TRICOLOR (*Pansy*).—Stem angular ; leaves ovate, obtuse ; stipules lyrate, very large ; flowers large, on long axillary peduncles ; spur thick, obtuse, very short. A beautiful and very variable species, cultivated in gardens.

Order XII.—CISTACEÆ (*Rock Rose Family*).

Herbs, or low shrubs. Leaves simple, the lowest commonly opposite. Sepals mostly 5, the 3 inner convolute in prefloration, the 2 outer minute, or wanting. Petals 5, rarely 3, regular, in prefloration usually convolute in a direction contrary to that of the sepals. .Stamens indefinite, hypogynous, distinct. Ovary of 3–5 united carpels. Style 1. Fruit a 1-celled, or imperfectly 3–5-celled capsule. Seeds few, or numerous.

CISTACEÆ. { Petals 3, persistent—*Lechea.*
{ Petals 3–5, fugacious—*Helianthemum.*

1. **Lechea.**—Sepals 5 ; 2 outer much smaller. Petals 3, lanceolate, small. Stamens 3–12. Stigmas 3, nearly sessile. Capsule 3-valved, 3-celled. Placentæ each 1–2-seeded. ♃

1. L. MINOR (*Small Pinweed*).—Stem nearly smooth, very branching ; leaves linear ; often nearly oblong, scattered, sometimes verticillate ; flowers dull purple, in nearly simple racemes, separate, pedicellate ; capsule globose. In dry grounds, 3'–5' high. *June–September.*

2. **Helianthemum.**—Petals 5, or rarely 3, sometimes wanting, fugacious. Stamens numerous. Stigmas 3, more or less united. Capsule triangular, 3-valved. ♃

1. H. CANADENSE (*Rock Rose*).—Stem simple, pubescent ; leaves oblong, entire, usually alternate, acute, paler beneath ; flowers of two kinds, the earlier ones large, few, bright yellow, fugacious, terminal ; petals large, thin, nearly orbicular, emarginate, twice as long as the calyx ; later ones apetalous, or with very small petals, axillary, sessile, nearly solitary, very small ; capsules smooth, shining ; those of the apetalous flowers very small. *June–September.*

Order XIII.—POLYGALACEÆ (*Milk-wort Family*).

Herbs, or somewhat shrubby plants. Leaves simple. Flowers perfect, irregular. Sepals 5, distinct, very irregular ; 3 exterior and smaller ; the 2 lateral, interior ones larger, and petaloid. Petals irregular, the anterior one (*the keel*) larger than the others. Stamens 6–8, hypogynous. Filaments united into a tube, which is split on the upper side. Ovary compound, free from the calyx, consisting of 2 united carpels.

1. **Polygala.**—Embryo large, with broad cotyledons. Some species bear concealed flowers, near the ground. ①—♃

1. P. VERTICILLATA (*Green-flowered Polygala*).—Stem erect, branched ; leaves linear, in whorls ; spikes linear, slender ; flowers very small, crested,

greenish white; bracts deciduous. A slender plant, with inconspicuous flowers, 3'-4' high. On dry hills. *July-October*.

2. P. PAUCIFOLIA (*Fringed Polygala*).—Stem erect, simple, leafy at summit; leaves ovate, entire, petiolate; flowers 2-3, large and handsome, ½' long, deep rose-color, on pedicels nearly ½' long, crested; radical flowers wingless; crest inconspicuous, purple; rhizoma creeping and branching. A handsome plant, in low woods and swamps. *May*.

Order XIV.—CARYOPHYLLACEÆ (*Pink Family*).

Herbs. Leaves opposite, entire, sometimes verticillate. Flowers regular. Sepals 4, or 5, sometimes coherent in a tube. Petals 4, or 5, or none. Stamens as many, or twice as many, as the petals, rarely only 2 or 3. Ovary mostly 1-celled. Styles 2-5. Fruit a 1-celled utricle, or a capsule, 2-5-valved, or opening at top by twice as many valves, or teeth, as there are stigmas.

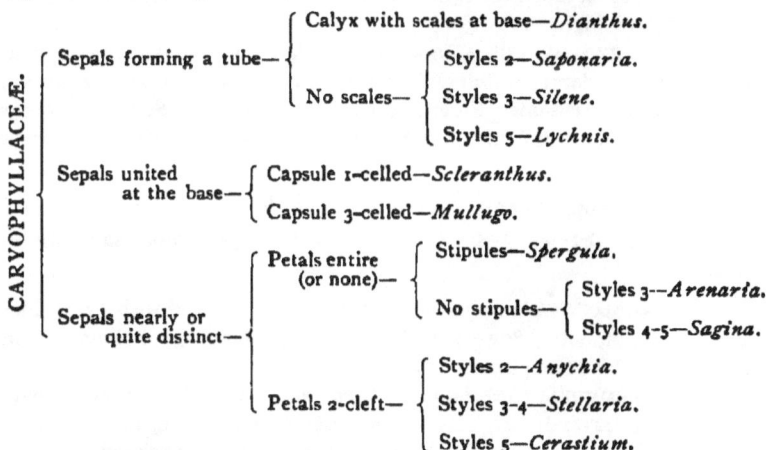

CARYOPHYLLACEÆ.

Sepals forming a tube—
  Calyx with scales at base—*Dianthus*.
  No scales—
    Styles 2—*Saponaria*.
    Styles 3—*Silene*.
    Styles 5—*Lychnis*.

Sepals united at the base—
  Capsule 1-celled—*Scleranthus*.
  Capsule 3-celled—*Mullugo*.

Sepals nearly or quite distinct—
  Petals entire (or none)—
    Stipules—*Spergula*.
    No stipules—
      Styles 3—*Arenaria*.
      Styles 4-5—*Sagina*.
  Petals 2-cleft—
    Styles 2—*Anychia*.
    Styles 3-4—*Stellaria*.
    Styles 5—*Cerastium*.

1. **Silene.**—Calyx tubular, swelling, 5-toothed. Petals 5, unguiculate. Stamens 10. Styles 3. Capsule 3-celled, many-seeded, opening at the top by 6 teeth. ①—♃

1. S. ARMERIA (*Garden Catchfly*).—Very glabrous; stem erect, branching, glutinous; leaves ovate-lanceolate; flowers purple, in cymes, numerous; petals obcordate, crowned; calyx clavate, 10-striate. A common garden flower, 12'-18' high. *July-September*.

2. **Lychnis.**—Calyx tubular, 5-toothed. Petals 5, unguiculate; claws slender. Stamens 10. Styles 5. Capsule 1-celled or 5-celled at base. ①—♃

1. L. GITHAGO (*Corn-cockle*).—Stem dichotomous, hirsute; leaves pale green, sessile, soft-hairy; flowers few, light purple, on long peduncles. A handsome weed, 1-3 feet high, in cultivated grounds. *July*.

2. L. CHALCEDONICA (*Scarlet Lychnis*).—Nearly smooth; leaves ovate-lanceolate, dark green; flowers in terminal, dense fascicles, scarlet; petals 2-lobed. Flowers varying to white, sometimes double. Gardens, growing 6' high. *June-July.*

3. **Saponaria.**—Calyx tubular, 5-toothed. Petals 5, with claws as long as the calyx. Stamens 10. Styles 2. Capsule 1-celled. ♃

1. S. OFFICINALIS (*Bouncing Bet*).—Plant smooth, somewhat fleshy; leaves ovate-lanceolate, smooth, sessile; flowers large, pale rose-color, in paniculate fascicles; crown of the petals linear; flowers often double. Common in waste places, 1-2 feet high. Introduced. *July-August.*

4. **Dianthus.**—Calyx tubular, 5-toothed, with one or more pairs of opposite, imbricated scales at base. Petals 5, with long claws. Stamens 10. Styles 2. Capsule 1-celled. ♃

1. D. BARBATUS (*Sweet-William*).—Stem branching; leaves lanceolate; flowers red, often much variegated, in dense fascicles. A well-known garden flower, 10'-18' high. *May-July.*

2. D. CARYOPHYLLUS (*Carnation*).—Glaucous; leaves linear, channeled; flowers large, solitary, fragrant; scales short, ovate; petals very broad, beardless, crenate. Stem 1-3 feet high, branched. From this species, under the influence of cultivation, have been derived all the splendid varieties of carnations.

5. **Stellaria.**—Sepals 5, connected at the base. Petals 5, 2-cleft. Stamens 10. Styles 3-4. Capsule 1-celled, 3-valved, many-seeded. ①—♃

1. S. MEDIA (*Chickweed*).—Stem procumbent, marked with hairy lines; leaves ovate, smooth; flowers small, white; petals oblong; stamens varying from 3-10. In waste places. *March-November.*

2. S. LONGIFOLIA (*Stitchwort*).—Stem weak, with rough angles, slender and brittle; leaves linear, sessile, 1-nerved; flowers white, in long cymes, with lanceolate, scarious bracts; petals deeply cleft. In meadows, 8'-10' high. *June-July.*

6. **Cerastium.**—Sepals 5, somewhat united at base. Petals 5, bifid. Stamens 10, rarely fewer. Styles 5. Capsule roundish, 1-celled, 10-toothed. Seeds numerous. ①—♃

C. NUTANS (*Mouse-ear Chickweed*).—Clammy pubescent; leaves oblong; loosely flowered; pods long, nodding. In moist ground.

7. **Arenaria.**—Sepals 5. Petals 5, entire. Stamens 10, rarely fewer. Styles 3. Capsule 3-valved; valves usually 2-parted.

A. SERPYLLIFOLIA (*Sandwort*).—Stem diffuse, dichotomous, pubescent; leaves small, ovate, sessile, acute; flowers small, white, numerous; sepals lanceolate, hairy, striate; capsule ovate, 6-toothed. In cultivated grounds 3'-4' high. Introduced. *May-June.*

8. Sagina.—Sepals 4–5, united at base. Petals 4–5, entire, or none. Stamens 4–5. Capsules 4–5-valved, many-seeded. ①

S. PROCUMBENS (*Pearlwort*).—Glabrous; stem slender, procumbent; leaves linear; flowers small, white or green, axillary; petals sometimes wanting; stamens, sepals, and petals 4–5. In wet, springy grounds. *May–July.*

9. Anychia.—Sepals 5, ovate-oblong, connivent, subsaccate at the apex. Petals none. Stamens 2–5, inserted on the base of the sepals. ①

A. DICHOTOMA (*Forked Chickweed*).—Stem smooth, dichotomously branched, slender; leaves oval, sessile; flowers minute, white, axillary, solitary, or in terminal clusters of 3. A delicate, very branching plant, on hill-sides, 4'–8' high. *June–August.*

10. Spergula.—Sepals 5, nearly distinct. Petals 5, entire. Stamens 5–10. Styles 3–5. Capsules ovate, 3–5-valved, many-seeded. ①

1. S. ARVENSIS (*Corn Spurrey*).—Stem branching, somewhat viscid; leaves linear, verticillate, 10–20 in a whorl, dark green; stipules minute; flowers in terminal cymes; petals white; stamens 10; styles 5; seeds uniform. In cultivated grounds, 1 foot high. *May–August.*

2. S. RUBRA (*Red Corn Spurrey*).—Stem decumbent, much branched, smooth; leaves narrow-linear, somewhat fleshy. Stipules ovate, cleft; flowers small, solitary, axillary, red, or rose-color, on hairy peduncles. A variable little plant, in dry soils.

11. Scleranthus.—Sepals 5, united at base. Petals none. Stamens 10 or 5, inserted at the throat of the calyx. Styles 2. ①

S. ANNUUS (*Knawel*).—Stem procumbent, branching, tufted; leaves numerous, narrow-linear, acute, opposite; flowers small, greenish, nearly sessile, in leafy clusters. In dry soils, 2'–4' long. *June–July.*

12. Mollugo.—Sepals 5, united at base. Petals none. Stamens 5, sometimes 3–10. Styles 3. Capsule 3-valved, 3-celled, many-seeded. ①

M. VERTICILLATA (*Carpet-weed*).—Stem branched; leaves spatulate, entire, in verticils of 5; flowers greenish white, axillary; stamens mostly 3. A weed in cultivated grounds, spreading flat on the earth. *July–September.*

Order XV.—PORTULACACEÆ (*Purslane Family*).

Succulent or fleshy-herbs. Leaves entire. Flowers showy, opening in the sunshine. Sepals 2, often cohering to the ovary. Petals 5, rarely more, ephemeral. Stamens sometimes as many as the petals, and opposite them. Ovary 1-celled. Styles 2–8. Fruit a pyxis, or a loculicidal' capsule, with as many valves as there are stigmas.

PORTULACACEÆ. { Stamens 5—*Claytonia.*
{ Stamens more than 5—*Portulaca.*

1. **Portulaca.**—Sepals 2, united ; upper portion deciduous. Petals 4–6, equal. Stamens 8–20. Styles 3–6, cleft at apex. Capsule a pyxis, many-seeded. ①

P. OLERACEA (*Purslane*).—Stem thick, very branching, prostrate, spreading ; leaves sessile ; flowers pale yellow, sessile ; petals 5, cohering at base ; foliage reddish green. A common and troublesome weed. Introduced. *June–July.*

2. **Claytonia.**—Sepals 2, persistent. Petals 5, hypogynous. Stamens 5, on the claws of the petals. Stigma 3-cleft. Capsule 3-valved, 2–5-seeded. ♃

C. VIRGINICA (*Spring Beauty*).—Stem simple, glabrous ; leaves opposite, linear ; flowers white, veined with purple, in a raceme ; sepals rather acute. A handsome little plant 4′–8′ high, arising from a root. Low grounds. *April–May.*

Order XVI.—HYPERICACEÆ (*St.-John's-wort Family*).

Herbs or shrubs. Leaves opposite, entire, without stipules, punctate with black glands and transparent dots. Flowers mostly regular. Sepals 4–5, persistent. Petals as many as the sepals, and alternate with them, twisted in prefloration. Stamens hypogynous, usually numerous, and cohering by their filaments in 3 or more sets ; sometimes definite, and monadelphous, or distinct. Anthers versatile. Ovary composed of 2–5 united carpels. Fruit a many-seeded capsule, with septicidal dehiscence, either 1-celled. or more or less completely 2–5-celled.

HYPERICACEÆ. { Petals convolute—*Hypericum.*
{ Petals imbricated—*Elodes.*

1. **Hypericum.**—Sepals 5, connected at base. Petals 5, oblique. Stamens numerous, sometimes few, united at base into 3–5 parcels, occasionally distinct. Styles 3–5, separate, or united, persistent. ①—♃

1. H. PERFORATUM (*St.-John's-wort*).—Stem erect, 2-edged, smooth, branching ; leaves elliptical, obtuse, sessile, punctate with pellucid dots ; flowers bright yellow, in panicles ; sepals lanceolate, shorter than the obovate petals. A troublesome plant, in pastures and dry grounds. Introduced. *June–July.*

2. H. MUTILUM (*Small St.-John's-wort*).—Stem erect, smooth, 4-angled ; leaves oval, obtuse, entire, sessile, 5-veined ; flowers very small, greenish yellow, in leafy cymes ; sepals lanceolate, a little longer than the petals ; stamens 6–12, distinct ; capsule ovate, conical. A small species in wet grounds. *July–August.*

2. Elodea.—Sepals 5, equal, somewhat united. Petals 5, deciduous. Stamens in 3 parcels, which alternate with 3 hypogynous glands. Styles 3, distinct. Capsule 3-celled. ♃

E. VIRGINICA (*Marsh St.-John's-wort*).—Stem smooth, branching; leaves sessile, clasping, oblong, obtuse, glaucous beneath; flowers large, dull orange purple, in racemes; petals obovate, marked with reddish veins; stamens united below the middle, 3 in a set. In swamps and ditches, 8'–16' high. *July–August.*

Order XVII.—MALVACEÆ (*Mallow Tribe*).

Herbs, shrubs, or trees. Leaves alternate, stipulate. Flowers regular, axillary. Sepals 5, somewhat united. Petals alternate with sepals, hypogynous. Stamens numerous, monadelphous, hypogynous, united to the petals at base. Anthers uniform, 1-celled, bursting transversely. Pollen hispid. Ovary 1, several-celled; or ovaries several, arranged circularly round a common axis. Fruit a several-celled capsule, or consisting of several separate or separable 1–2-seeded carpels.

MALVACEÆ.
{ Calyx without involucel—*Abutilon.*
{ Calyx with involucel—{ Involucel 3-leaved—*Malva.*
{ Involucel 6–9 cleft—*Althæa.*
{ Involucel many-cleft—*Hibiscus.*

1. **Malva.**—Calyx 5-cleft, with an involucel of 3 leaves. Carpels several, 1-celled, 1-seeded, dry, indehiscent, circularly arranged. ♃

M. ROTUNDIFOLIA (*Low Mallow*).—Stems prostrate, branching; leaves on long petioles, obtusely 5-lobed, crenate; flowers axillary, pale pink, or whitish; petals deeply notched; involucre 3-leaved; fruit spherical, depressed in the center, mucilaginous. *May–September.*

2. **Althæa.**—Calyx with a 6–9-cleft involucel. Carpels numerous, indehiscent, arranged around the axis, separating when ripe. ♃

A. ROSEA (*Hollyhock*).—Stem erect, hairy; leaves rough, cordate, 5–7-angled; flowers large, axillary, sessile. Flowers red, purple, white, or yellow, often double. Stem 6–8 feet high.

3. **Hibiscus.**—Calyx 5-cleft, surrounded by a many-leaved involucel. Stigmas 5. Carpels 5, united into a 5-celled capsule. ①—♃

H. SYRIACUS (*Althæa*).—Leaves 3-lobed, toothed; flowers delicate, large, purple, axillary, solitary; white, red, and variegated flowers. Shrub, 5–10 feet high. *July–September.*

4. **Abutilon.**—Calyx 5-cleft, without an involucel. Ovary 5-celled, several-seeded. Capsule of 5 or more carpels, 2-valved, 1–3-seeded. ①

A. AVICENNÆ (*Indian Mallow*).—Stem erect, with spreading branches ; leaves orbicular, cordate, acuminate, velvety ; flowers large, orange-yellow, on peduncles, solitary ; carpels about 15, 3-seeded, inflated, 2-beaked, hairy. In waste places, 2–5 feet high. *July–September*.

Order XVIII.—TILIACEÆ (*Linden Family*).

Trees, or shrubby plants. Leaves alternate, stipulate, deciduous. Sepals 4–5, deciduous. Petals 4–5, with glands at base. Stamens indefinite, distinct, hypogynous. Ovary with 2–10 united carpels. Styles united. Stigmas as many as the carpels. Fruit a 2–5-celled capsule.

Tilia.—Sepals 5, united, colored. Petals 5. Stamens numerous, in several parcels, mostly 5 in each set, with a petaloid scale. Ovary globose, 5-celled. Cells 1–2-seeded.

T. AMERICANA (*Bass-wood*).—Leaves alternate, obliquely cordate, sharply serrate, abruptly acuminate, glabrous ; flowers dull white, with a heavy odor, in dense, pendent cymes. Peduncle united to the mid-vein of an oblong bract ; petals truncate, or obtuse ; fruit greenish, as large as peas. The inner bark is very mucilaginous, and its fiber is extremely strong. A tall, elegant tree of regular growth. *June.*

DISCIFLORAL POLYPETALOUS EXOGENS.

Order XIX.—LINACEÆ (*Linen Family*).

Herbs, sometimes suffruticose. Leaves sessile, entire, alternate. Flowers regular, symmetrical. Sepals 3–5. Petals as many as the sepals, and alternate with them. Stamens 3–5, with 5 processes resembling teeth. Styles as many as the stamens. Ovaries of 3–5 united carpels. Stigmas capitate. Capsule globose, 3–5-celled. Carpels 2-valved at apex, 2-seeded. Seeds without albumen.

Linum.—Herbs with tough bark. Leaves simple, sessile, and exstipulate. All flower-circles regularly 5-merous. Carpel 5-celled ; seeds flat, mucilaginous.

L. USITATISSIMUM (*Flax*).—Glabrous ; leaves linear-lanceolate, very acute ; flowers large, blue, in a corymbose panicle ; sepals ovate, 3-nerved at base. Cultivated for the seed and fiber, the basis of the linen fabric.

Order XX.—GERANIACEÆ.

Herbs, sometimes somewhat suffruticose. Leaves usually palmately veined and lobed, the lower ones opposite. Sepals 5, persistent. Petals 5, unguiculate. Stamens 10, hypogynous, united by their broad filaments. Ovary with 5 2-ovuled carpels. Styles attached to the base of a prolonged axis. Fruit consisting of 5 1-seeded carpels, which separate from the axis by curving back from their base.

GERANIACEÆ.
- Flowers regular, herbs—
  - Styles 5—*Oxalis.*
  - Style 5-cleft—*Geranium.*
- Flowers somewhat irregular, shrubby—*Pelargonium.*
- Flowers very irregular—
  - Stamens 8, climbing herbs—*Tropæolum.*
  - Stamens 5, erect herbs—*Impatiens.*

1. Geranium.—Sepals 5, equal. Petals 5, equal. Stamens 10; alternate ones larger, with a nectariferous gland at base. Styles persistent. Fruit beaked. ♃

G. MACULATUM (*Cranesbill*).—Stem erect, dichotomous, angular; leaves palmately 5-7-parted; peduncles dichotomous, 1-3-flowered; flowers large, light purple; sepals awned; petals entire. Woods, fields, and thickets, 1-2 feet high. *May–June.*

2. Pelargonium.—Sepals 5; upper one terminating in a nectariferous tube, extending down the peduncle. Petals 5, irregular, larger than the sepals. Filaments 10; 3 of them sterile. ♃

1. P. ZONALE (*Horseshoe Geranium*). — Stem thick, shrubby; leaves orbicular, with shallow lobes, dentate, marked with a colored zone near the margin; flowers bright scarlet, in umbels with long peduncles.

2. P. INQUINANS (*Scarlet Geranium*).—Stem erect, with downy branches; leaves round-reniform, scarcely lobed, crenate, viscid; flowers bright scarlet, in many-flowered umbels.

3. P. PELTATUM (*Ivy-leaved Geranium*).—Stem long, climbing; leaves 5-lobed, with the lobes entire, fleshy, smooth, peltate; flowers purplish, in few-flowered umbels.

3. Oxalis.—Sepals 5, distinct or united at base. Petals 5, much longer than the calyx. Capsule oblong, or subglobose. Carpels 5. ♃

O. STRICTA (*Wood-sorrel*).—Stem simple, smooth, leafy; leaves trifoliate, on long petioles; leaflets obcordate; flowers yellow, in umbels; capsules hirsute, leaves acid to the taste. Fields, from 3'-6' high. *April–September.*

4. Impatiens.—Sepals 5, colored, apparently 4, from the union of the 2 upper ones; lowest spurred. Petals 4, apparently 2. Anthers cohering at apex. Capsule often 1-celled. ①

1. I. FULVA (*Jewel-weed*).—Stem succulent; leaves rhombic-ovate, obtuse, coarsely serrate, with mucronate teeth; flowers deep orange, spotted with brown dots, very irregular in form; spur longer than the petals, recurved. In wet grounds, 1-3 feet high. *June–September.*

2. I. BALSAMINA (*Balsamine. Touch-me-not*).—Stem succulent; leaves lanceolate, serrate, lower ones opposite; flowers large, in axillary clusters; spur shorter than the flower.

5. Tropœolum.—Herbs climbing by leaf-stalks, exstipulate. Sepals 5, united at base in a long spur. Petals 5, with claws. Stamens 8, unequal.

T. MAJUS (*Nasturtium*).—Leaves peltate, orbicular; petioles long; flowers large, orange-colored, with darker spots; petals obtuse; the 2 upper distant from the 3 lower, which are fimbriate at base.   *June–November.*

Order XXI.—RUTACEÆ.

Herbs, shrubs, or trees. Leaves punctate, without stipules. Flowers perfect. Sepals 4–5. Petals 4–5. Stamens as many, or twice as many as the petals, inserted on a hypogynous disk. Ovary 3–5-lobed, 3–5-celled. Fruit usually separating into 3–5 few-seeded carpels.

RUTACEÆ. { Herbs—*Ruta*. — Shrubs or trees— { Flowers perfect—*Citrus.* Flowers not all perfect—*Zanthoxylum.* }

1. Citrus.—Sepals 5, united. Petals 5. Stamens arranged in clusters of 5 each. Filaments dilated at base. Fruit a berry, 9–18-celled. ♃

1. C. LIMONUM (*Lemon*).—Leaves, or rather leaflets, oval, acute, toothed; petioles somewhat winged; flowers white, fragrant; stamens 35; fruit pale yellow, oblong-spheroidal, rind thin, pulp very acid. A low tree.

2. C. AURANTIUM (*Orange*).—Leaflet oval, acute; petioles winged; stamens 20; berry globose, with a thin rind, and sweet pulp; flowers white, very fragrant. A middle-sized tree.

2. Ruta.—Sepals 4–5, united at base. Petals 4–5, concave, obovate, distinct. Stamens 10. Capsule lobed. ♃

R. GRAVEOLENS (*Rue*).—Nearly smooth, suffruticose; leaves bi- and tripinnately divided; segments all entire, or incised, punctate with conspicuous dots; flowers yellow, terminal, corymbose; petals entire. Plant 3–4 feet high.

3. Zanthoxylum.—Polygamous. Perfect flowers: Sepals 5. Petals none. Stamens 3–6. Pistils 3–5. Carpels 3–5, 1-seeded. ♃

Z. AMERICANUM (*Prickly Ash*).—Branches armed with stout, hooked prickles; leaves pinnate; leaflets 5–7, ovate, mostly entire, sessile; flowers small, greenish, in umbels. Bark bitter, aromatic.

Order XXII.—AQUIFOLIACEÆ (*Holly Family*).

Shrubs, or trees, Leaves simple, alternate, or opposite, often evergreen, exstipulate. Flowers small, white or greenish, axillary, clustered or solitary, often diœcious or polygamous. Calyx-tube free from the ovary; limb 4–6 cleft. Corolla regular, 4–6-parted. Stamens as many as the segments of the corolla, alternate with them, inserted on its base. Anthers opening longitudinally. Ovary 2-celled, with 1 ovule in each cell. Stigmas 2–6. Fruit drupaceous, with 2–6 stones.

Ilex.—Calyx 4–5-toothed. Corolla 4–5 parted, somewhat rotate. Stamens 4–5. Stigmas 4–5, united or distinct. Berry 4–5-seeded.

1. I. OPACA (*American Holly*).—Leaves evergreen, oval, flat, tapering at both ends, coriaceous, smooth and shining, armed with strong, spiny teeth on the margins; flowers small, greenish white, in loose clusters, which are axillary, or situated at the base of the young branches; calyx-teeth acute; berry ovate, red when ripe. A tree. *June.*

2. I. VERTICILLATUS (*Black Alder*).—Leaves deciduous, oval, or wedge-lanceolate, serrate; flowers white, axillary, on very short peduncles; fertile ones closely aggregated; sterile ones somewhat umbelled; pedicels 1-flowered; berries roundish, scarlet, persistent through the winter. In swamps, 4–8 feet high. *June.*

Order XXIII.—RHAMNACEÆ.

Shrubs, or trees, often with spinose branches. Leaves simple. Flowers small. Sepals 4–5, united at base. Petals 4–5, inserted in the throat of the calyx. Stamens 4–5, opposite the petals. Ovary of 2–4 united carpels, 2–4-celled, usually more or less free from the calyx, sometimes immersed in the fleshy disk surrounding it. Fruit, a berry, or a capsule with dry carpels.

RHAMNACEÆ. { Fruit berry-like—*Rhamnus.*
{ Fruit a hard pod—*Ceanothus.*

1. Rhamnus.—Calyx urceolate, 4–5-cleft. Petals 4–5, perigynous. Ovary free from the calyx, 2–4-celled. Styles, 2–4. Fruit drupaceous. ♃

R. CATHARTICUS (*Buckthorn*).—Leaves ovate, doubly serrate, acute, strongly veined, nearly smooth, alternate, in fascicles at the ends of the branches; flowers polygamous, in fascicles, mostly tetrandrous; petals entire; fruit black, globose, nauseous, and cathartic. *June.*

2. Ceanothus.—Shrubs. Flowers in umbels. Petals 5, hood-shaped. Ovary 3-celled, becoming a hard, 3-seeded pod.

C. AMERICANUS (*Jersey Tea*).—Young branches pubescent; leaves ovate, serrate, white-downy beneath, flowers small, white, numerous, in dense, axillary, thyrsoid panicles. A small shrub, with a profusion of white flowers, growing in woods, preferring a rather dry soil. Stem 2–3 feet high. *June.*

Order XXIV.—Vitaceæ.

Woody plants, climbing by tendrils. Flowers small, often polygamous, or diœcious. Calyx small, entire, or with 4 or 5 teeth, lined
by a disk. Petals 4 or 5, inserted on the margin of the disk, often
cohering by their tips, and caducous. Stamens 4 or 5, opposite
the petals, and inserted with them. Ovary 2-celled. Style short,
or none. Fruit a globose, usually pulpy berry, often 1-celled, and
1 or few seeded.

VITACEÆ. $\begin{cases} \text{Calyx on a fleshy disk—}Vitis. \\ \text{No fleshy disk—}Ampelopsis. \end{cases}$

1. Vitis.—Calyx nearly entire. Petals 4–5. Ovary surrounded
and partly inclosed in the elevated torus, 2-celled; cells 2-ovuled.
Berry 1-celled, 1–4 seeded.

1. V. LABRUSCA (*Wild Grape*).—Leaves broadly cordate, 3-lobed,
toothed, tomentose beneath; flowers diœcious, small, green, in compact, oblong panicles; fruit large, globose, black or reddish-purple, pleasant and
eatable. In low grounds, with very long stems, which often reach the tops
of the highest trees, climbing by means of its tendrils. *June.*

2. V. CORDIFOLIA (*Frost Grape*).—Young branches mostly smooth;
leaves cordate, often 3-lobed; racemes loose, many-flowered; berries small,
black, late, very acid. In low grounds and woods.

2. Ampelopsis.—Calyx 5-toothed. Petals thick. Disk none.
Leaves digitate, with 5 leaflets. Tendrils with disks at tips.

A. QUINQUEFOLIA (*Woodbine*).—Stem climbing, smooth; leaflets oblong, serrate, acuminate, petiolate, smooth; flowers greenish, in dichotomous, many-flowered panicles; berries dark blue, as large as a small pea,
with crimson peduncles and pedicels. A rapidly growing and spreading vine.
Along fences and borders of woods. *July.*

Order XXV.—Sapindaceæ (*Soapberry Family*).

Trees, shrubs, or herbaceous climbers, with compound leaves,
and irregular and often polygamous or diœcious flowers. Petals
imbricated in the bud on a disk. Ovary 2–3-celled, with 1–3
ovules in each cell. Fruit a pod or samara.

SAPINDACEÆ. $\begin{cases} \text{Fruit bladdery pod—}Staphylea. \\ \text{Fruit winged, leaves simple—}Acer. \\ \text{Fruit leathery pod—}Æsculus. \end{cases}$

1. Acer.—Flowers polygamous. Calyx 5-cleft. Petals 5, or none.
Stamens 7–10. Styles 2. Samaras 2-winged, united at base.

1. A. DASYCARPUM (*Silver or White Maple*).—Leaves deeply 5-lobed,
white and smooth beneath; sinuses obtuse; lobes acute, entire toward the

ba..e ; flowers small, yellowish green, in crowded, simple umbels ; pedicels short and thick ; petals none ; fruit tomentose when young, nearly smooth when old, with very large upwardly dilated diverging wings. Wood, white, soft. *April.*

2. A. SACCHARINUM (*Sugar-Maple*).—Leaves palmately lobed, cordate at base ; sinuses obtuse and shallow ; lobes acuminate, with a few coarse, repand teeth ; flowers pale yellow, on long, pendulous, filiform, villous pedicels ; sepals bearded inside ; petals none ; fruit yellowish, with wings 1' long. A noble tree of elegant foliage and growth, often cultivated. Its sap yields maple-sugar. *May.*

2. Æsculus.—Calyx campanulate, tubular, 5-toothed. Petals 4–5, more or less unequal. Stamens 6–8, on a disk ; ovary of 3 united carpels.

Æ. HIPPOCASTANUM (*Horse-Chestnut*).—Leaves digitate ; leaflets 7, obovate, abruptly acute, serrate ; flowers large, in pyramidal thyrses or racemes, pink and white ; fruit large, dark chestnut-colored, not eatable. Tree, 4c-:o feet high, of elegant growth. *June.*

3. Staphylea.—Flowers perfect. Sepals 5, colored, persistent, erect. Petals 5. Stamens 5. Styles 3. Capsules membranaceous, 3-celled, 3-lobed. ⚥

S. TRIFOLIA (*Bladdernut*).—Leaves ternate, opposite ; leaflets ovate, finely serrate ; stipules caducous ; flowers white, in nodding, axillary racemes ; petals narrow-obovate ; fruit composed of 3 inflated, united, 1–several-seeded carpels. *May.*

Order XXVI.—ANACARDIACEÆ.

Trees, or shrubs, with a resinous, gummy, milky, or acrid juice. Leaves alternate, not dotted, destitute of stipules. Flowers small, often polygamous, or diœcious. Sepals 5, persistent. Petals as many as the sepals, sometimes none. Stamens as many as the petals, alternate with them, inserted on the base of the calyx. Ovary 1-celled, nearly or quite free from the calyx. Ovule solitary. Styles 3–5, distinct or united. Fruit a drupe, or a bony, 1-seeded nut.

Rhus.—Sepals 5, united. Petals 5. Stamens 5. Styles 3. Stigmas capitate. Fruit a dry drupe, with a bony, 1-celled nut. Flowers often diœcious.

1. R. GLABRA (*Sumach*).—Leaves pinnate, 6–15-foliate ; leaflets lance-oblong, acuminate, smooth ; flowers small, greenish, in dense, terminal, thyrsoid panicles, followed by small drupes covered with crimson hairs, of a sour taste. A shrub 6–10 feet high, in pastures and thickets. *June–July.*

2. R. VENENATA (*Dogwood, Poison Sumach*).—Very glabrous ; leaflets 7–13, oval, entire ; flowers very small, green, mostly diœcious, in loose panicles ; drupes smooth, greenish, as large as peas. A shrub, in swamps, 10–15 feet high, and exceedingly poisonous to most persons. *June.*

3. R. TOXICODENDRON (*Poison Oak, Poison Ivy*).—Stem erect, or decumbent; leaves pubescent, ternate; leaflets broad oval; flowers green, in racemose panicles; drupes sub-globose, smooth, pale brown. A low shrub 1–3 feet high, poisonous, but less so than the last.

SUPERIOR POLYPETALOUS EXOGENS.

Order XXVII.—LEGUMINOSÆ (*Pea Family*).

Herbs, shrubs, or trees. Leaves alternate, often compound. Stipules present. Calyx consisting usually of 5, more or less united sepals. Petals 5, either papilionaceous or regular. Stamens perigynous, sometimes hypogynous, diadelphous, monadelphous, or distinct. Ovary single, and simple. Fruit a legume, assuming various forms, sometimes divided into several 1-seeded joints, when it is called a loment. Seeds solitary, or several, destitute of albumen. A very large and important order, distributed throughout the world, except in Arctic countries.

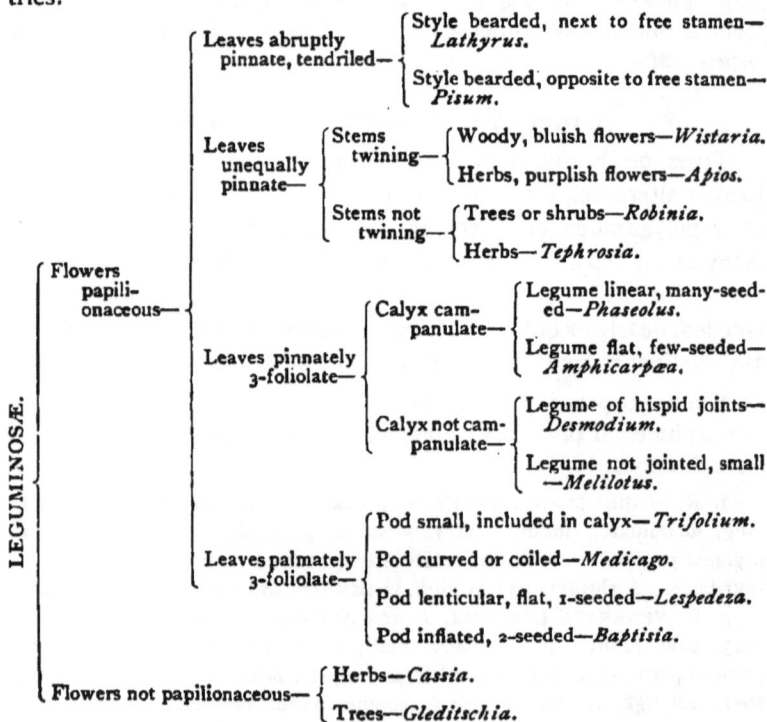

LEGUMINOSÆ

- **Flowers papilionaceous—**
  - Leaves abruptly pinnate, tendriled—
    - Style bearded, next to free stamen—*Lathyrus.*
    - Style bearded, opposite to free stamen—*Pisum.*
  - Leaves unequally pinnate—
    - Stems twining—
      - Woody, bluish flowers—*Wistaria.*
      - Herbs, purplish flowers—*Apios.*
    - Stems not twining—
      - Trees or shrubs—*Robinia.*
      - Herbs—*Tephrosia.*
  - Leaves pinnately 3-foliolate—
    - Calyx campanulate—
      - Legume linear, many-seeded—*Phaseolus.*
      - Legume flat, few-seeded—*Amphicarpæa.*
    - Calyx not campanulate—
      - Legume of hispid joints—*Desmodium.*
      - Legume not jointed, small—*Melilotus.*
  - Leaves palmately 3-foliolate—
    - Pod small, included in calyx—*Trifolium.*
    - Pod curved or coiled—*Medicago.*
    - Pod lenticular, flat, 1-seeded—*Lespedeza.*
    - Pod inflated, 2-seeded—*Baptisia.*
- **Flowers not papilionaceous—**
  - Herbs—*Cassia.*
  - Trees—*Gleditschia.*

1. **Lathyrus.**—Calyx campanulate, 5-cleft. Style flat, dilated above, ascending, pubescent, or villous on the inside next the stamen. ①—♃

L. PALUSTRIS (*Marsh Vetch*).—Glabrous; stem ascending, winged; leaves 4–6-foliate; leaflets narrow-oblong, sessile; stipules minute; peduncles axillary, 3–5 flowered; flowers light purple; legumes broad-linear, compressed. In meadows, 1–2 feet high. *June.*

2. **Pisum.**—Calyx-segments leafy; 2 upper shortest. Banner reflexed. Stamens 10, in 2 sets, 9 and 1. Style compressed. Legume oblong, many-seeded. ①

P. SATIVUM (*Pea*).—Glaucous, smooth; stem nearly simple, climbing; leaves 4–6-foliate; leaflets ovate, entire; stipules ovate, semi-cordate at base; flowers large, white, 2–5 on axillary peduncles. A cultivated plant, 2–5 feet high, climbing by its tendrils. *May–June.*

3. **Phaseolus.**—Calyx campanulate, 5-toothed, 2 upper teeth more or less united. Keel, together with the stamens and style, spirally twisted. Legume linear, or falcate, more or less compressed.

1. P. VULGARIS (*Bean*).—Stem twining; leaflets ovate, acuminate; racemes solitary, shorter than the leaves; pedicels in pairs; legume pendulous. Cultivated in gardens.

2. P. MULTIFLORUS (*Scarlet Pole-Bean*).—Stem twining; leaflets ovate, acute; flowers large, scarlet, very ornamental, in solitary racemes; pedicels opposite; legumes pendulous; seeds reniform. Cultivated.

4. **Apios.**—Calyx 2-lipped; standard broad, reflexed; stamens diadelphous. Pod many-seeded. ♃

A. TUBEROSA (*Ground-Nut*).—Stem twining; leaves 5–7-foliate; leaflets ovate-lanceolate, entire; flowers dark purple, of a peculiar leathery appearance. The root bears numerous nutritious tubers. Low grounds and thickets. *July–August.*

5. **Wistaria.**—Calyx campanulate; upper lip with 2 short teeth. Banner with 2 callosities. Wings and keel falcate. Legume many-seeded. ♃

W. FRUTESCENS (*Common Wistaria*).—Stem long, climbing; leaves 9–13-foliate; leaflets ovate-lanceolate, acute, slightly pubescent; racemes long, pendulous, with large, colored bracts; flowers lilac-colored. Common in cultivation. *May.*

6. **Amphicarpæa.**—Calyx tubular, campanulate, 4, sometimes 5-toothed; segments nearly equal. Petals oblong. Banner with appressed sides. Stigma capitate. Ovary stipitate. Legume flat, 2–4 seeded. Flowers of two kinds. ①

A. MONOICA (*Wild Pea-vine*).—Stem slender, hairy, twining; leaves pinnately trifoliate; leaflets ovate, acute, smooth; upper and perfect flowers nodding, in axillary racemes ; lower imperfect flowers on radical peduncles ; cauline legumes 3-7-seeded ; radical ones often beneath the surface, 1-seeded. A very delicate vine in low woods and thickets. *July–September.*

7. **Robinia.**—Calyx short, campanulate, 5-toothed. Banner large. Keel obtuse. Stamens diadelphous. Style bearded on the inside. Legume compressed, many-seeded. Trees and shrubs. ♃

R. PSEUDACACIA (*Locust-Tree*). — Branches with stipular prickles ; leaves 9-19-foliate ; leaflets ovate, or oblong-ovate, thin, smooth ; flowers large, white ; legumes smooth. A beautiful tree with elegant foliage. *May–June.*

8. **Tephrosia.**—Calyx equally 5-cleft ; standard roundish. Pod linear. Hoary herbs. ♃

T. VIRGINIANA (*Goat's Rue*).—Villous ; stem simple, erect ; leaves 17-29-foliate ; leaflets oblong, softly villous beneath ; racemes terminal, oblong ; flowers large, pale yellow and purple ; legumes falcate, villous. In sandy soils, 1-2 feet high. *July.*

9. **Trifolium.**—Calyx tubular, campanulate, 5-cleft, persistent. Petals withering. Banner larger than the wings, reflexed. Wings oblong. Legume small, membranaceous, indehiscent. Seeds sub-globose. Leaves trifoliate. Flowers in dense heads.

1. T. REPENS (*White Clover*).—Smooth ; stem creeping, spreading ; leaflets obcordate, denticulate ; petioles long ; stipules narrow-lanceolate ; heads globose ; corollas white, becoming pale brown, very fragrant, reflexed when past flowering ; legume 4-seeded. In damp soils.

2. T. ARVENSE (*Rabbit's-foot Clover*).—Silky-pubescent ; stem erect, branching ; leaflets oblong-obovate, minutely 3-toothed at apex ; petioles very short ; flowers pale red or whitish, in cylindrical, very hairy heads. A hairy plant, 3'-8' high, in pastures and dry soils. *July–August.*

3. T. PRATENSE (*Red Clover*).—Stems hairy, slightly pubescent ; leaflets ovate, with a large, lighter-colored spot in the center ; stipules ovate-lanceolate, membranaceous, strongly nerved ; flowers red, in dense, short, fragrant heads. Cultivated for hay.

10. **Melilotus.**—Calyx tubular, persistent, 5-toothed. Corolla deciduous. Keel-petals completely united, cohering with the wings. Stamens in 2 sets (9 and 1). Legume coriaceous. 1-few seeded. ♃

M. OFFICINALIS (*Yellow Mellilot Clover*).— Stem erect, branching, smooth ; leaves pinnately 3-foliate ; leaflets obovate-oblong, obtuse ; flowers yellow, in loose, axillary racemes ; legume ovate, 2-seeded. In alluvial soils, 2-3 feet high. *June–August.*

11. **Medicago.**—Calyx 5-cleft, somewhat cylindric. Keel of the corolla remote from the standard. Legume falcate, or spirally coiled, usually many-seeded. ②

M. LUPULINA (*None-such*).—Stem procumbent, angular; leaves trifoliate; leaflets obovate; stipules lanceolate, acute; flowers small, yellow, in small ovate heads, on slender, pubescent peduncles, longer than the petioles; legumes reniform, 1-seeded. In fields and roadsides. *May–October.*

12. Desmodium.—Calyx with 2 bracteoles at base, bilabiate, 5-cleft. Corolla inserted on the calyx at the base. Banner roundish. Keel obtuse. Style filiform. Stigma capitate. Legume compressed, of several, 1-seeded, separate joints. Leaves pinnately trifoliate. ♃

1. D. NUDIFLORUM (*Scape Trefoil*).—Stem erect, leafy at summit; leaflets ovate; flowers small, in racemes, purple; stamens monadelphous; legume with obtusely triangular joints. *July–August.*

2. D. ROTUNDIFOLIUM (*Creeping Trefoil*).—Stem prostrate, hairy; petioles hairy; leaflets orbicular, hairy; terminal one largest; stipules large, reflexed; racemes with long peduncles; flowers light purple; legumes with 3–5 rhomboidal hispid joints. *August.*

13. Lespedeza.—Calyx 5-cleft, with 2 bracteoles at base. Keel-petals very obtuse, on slender claws. Legume lenticular, small, reticulated, unarmed, indehiscent, 1-seeded. ♃

1. L. CAPITATA (*Headed Bush Clover*).—Stem erect, villous; leaves crowded; leaflets elliptical, silky beneath; flowers in axillary racemes; corolla white. Dry fields and hills, 2–3 feet high. *August–September.*

2. L. VIOLACEA (*Bush Clover*).—Stem erect or diffuse, branching, pubescent, leaves on short petioles; leaflets hairy beneath; flowers in axillary, few-flowered racemes, slender ones violet-purple; apetalous flowers glomerate and subsessile in the axils of the leaves; legumes much longer than the calyx. *August–September.*

14. Baptisia.—Calyx 4–5 toothed; standard reflexed. Stamens 10, distinct. Leaves 3-foliolate. ♃

B. TINCTORIA (*Indigo Weed*).—Stem erect, smooth, branching; leaves palmately trifoliate, on short petioles; leaflets roundish; flowers rather small, yellow, in few-flowered racemes, terminating the branches. In dry fields and woods, 2–4 feet high. *July–September.*

15. Cassia.—Sepals 5, slightly united at base, nearly equal. Petals 5, unequal. Stamens 10, distinct. Legume many-seeded. ♃

C. MARILANDICA (*Wild Senna*).—Glabrous; stem erect; leaves 12–18-foliate; leaflets oblong-lanceolate; flowers bright yellow, numerous, in axillary racemes and terminal panicles; legumes linear; anthers blackish. In alluvial soils, 4–6 feet high. *July–August.*

16. Gleditschia.—Polygamous. Sepals 3–5, equal, united at base. Petals 3–5, distinct, opposite the sepals. Stamens 3–5. Styles short. Stigma pubescent. Legume compressed, often interrupted between the seeds by sweet pulp.

G. TRIACANTHOS (*Honey Locust*).—Branches armed with stout, most-ly branching, triple thorns ; leaves abruptly pinnate, or bipinnate ; leaflets alternate, oblong-lanceolate, obtuse ; flowers small, white, spicate ; pods long, flat, curved, pendulous. A handsome, thorny tree. *June.*

Order XXVIII.—ROSACEÆ (*Rose Family*).

Trees, shrubs, or herbs. Leaves alternate. Stipules usually present. Flowers regular, usually perfect. Sepals 5, rarely less, more or less united, commonly persistent. Petals 5, perigynous, sometimes wanting. Stamens numerous, rarely few, inserted on the calyx, distinct. Ovaries several, or 1, often adherent to the calyx-tube, and to each other. Styles distinct, or united. Fruit a drupe, pome, achenium, or follicle.

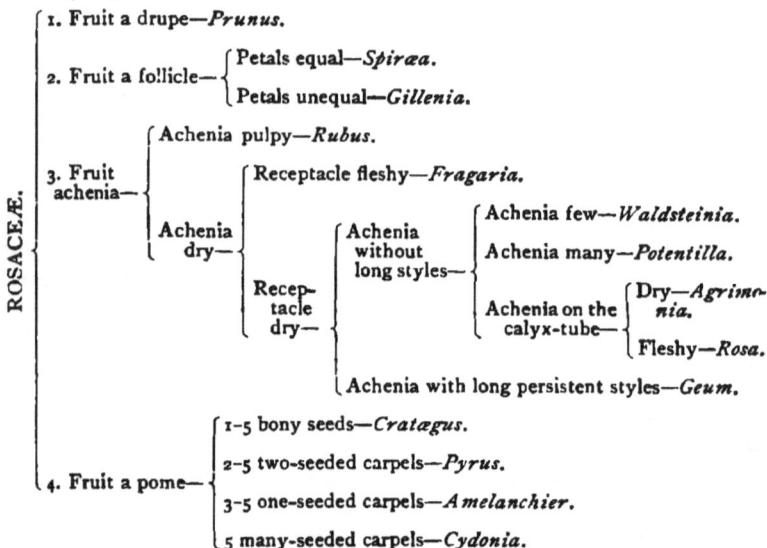

ROSACEÆ.

1. Fruit a drupe—*Prunus.*

2. Fruit a follicle—
- Petals equal—*Spiræa.*
- Petals unequal—*Gillenia.*

3. Fruit achenia—
- Achenia pulpy—*Rubus.*
- Achenia dry—
  - Receptacle fleshy—*Fragaria.*
  - Receptacle dry—
    - Achenia without long styles—
      - Achenia few—*Waldsteinia.*
      - Achenia many—*Potentilla.*
      - Achenia on the calyx-tube—
        - Dry—*Agrimonia.*
        - Fleshy—*Rosa.*
    - Achenia with long persistent styles—*Geum.*

4. Fruit a pome—
- 1-5 bony seeds—*Cratægus.*
- 2-5 two-seeded carpels—*Pyrus.*
- 3-5 one-seeded carpels—*Amelanchier.*
- 5 many-seeded carpels—*Cydonia.*

1. Prunus.—Calyx 5-cleft, regular, deciduous. Petals spreading. Stamens 25–30. Drupe globose, fleshy, destitute of bloom. ♃

1. P. CERASUS (*Cherry*).—Leaves oblong-ovate, hairy beneath ; flowers large, white, in umbels ; drupe ovoid. Cultivated, and esteemed for its delicious fruit. *May.*

2. P. AMERICANA (*Wild Plum*).—Leaves oval, varying to ovate, doubly serrate ; flowers white, in sessile umbels ; drupe roundish-oval, red and orange, very smooth, sweet and pleasant, with yellow pulp, and a thick, tough skin. A straggling shrub, 8–10 feet high, with somewhat thorny branches. *May.*

3. P. PERSICA (*Peach*).—Flowers rose-colored ; calyx bell-shaped. Fruit globular, with thick pulp. A small tree, largely grown for its excellent fruit.

4. P. DOMESTICA (*Plum*).—Leaves oval or ovate-lanceolate; flowers white; drupe round, to ovoid. A small tree, 10–15 feet high, with unarmed branches, long cultivated for its delicious fruit, sometimes black, sometimes white. *May.*

5. P. SEROTINA (*Wild Black Cherry*).—Leaves oval or oblong, smooth, finely serrate, shining above; petioles with 2 or more glands; flowers white, in spreading racemes; drupe black, somewhat bitter; bark bitter, tonic. A tall, elegant tree. *May.*

6. P. VIRGINIANA (*Choke-Cherry*).—Leaves obovate, to oval, sharply serrate; petiole with 2 glands; flowers sessile, in short, erect, spreading racemes; drupes sub-globose, dark red; extremely bitter and astringent. Small tree, 5–20 feet high. *May.*

2. **Spiræa.**—Calyx 5-cleft, persistent. Petals 5, roundish, equal. Stamens 10–50. Carpels 3–12, distinct, follicular, 1-celled, 1–10-seeded.

1. S. TOMENTOSA (*Hardhack*).—Stem shrubby, brittle, woolly-tomentose, and rust-colored; leaves ovate, or oblong, numerous, ferruginous-tomentose beneath; racemes short, dense, numerous, in a dense panicle; flowers pale purple, very small, numerous. A shrub, 2–3 feet high, in pastures and low grounds. *July.*

2. S. SALICIFOLIA (*Meadow-sweet*).—Nearly smooth; leaves oblong-obovate, sharply and doubly serrate; flowers white, often tinged with red, in dense, terminal panicles; carpels 5, smooth. A common shrub, in meadows, 2–4 feet high. *July-August.*

3. **Gillenia.**—Calyx tubular-campanulate, 5-cleft. Petals 5, linear-lanceolate. Stamens 10–15. Carpels 5. Styles filiform, terminal. Follicles 3, 2-valved, 2–4-seeded. ♃

G. TRIFOLIATA (*Indian Physic*).—Stem shrubby at base, slender, and nearly smooth, branching; leaves 3-foliate; leaflets ovate-oblong; stipules linear; flowers rose-color, or nearly white, on long pedicels, in corymbose panicles; root emetic and cathartic. In woods, 2–3 feet high. *June-July.*

4. **Agrimonia.**—Calyx-tube armed with hooked bristles, contracted at the throat, with a 5-cleft limb. Petals 5. Stamens 12–15. Ovaries 2. Styles terminal. Achenia included in rim of the calyx. ♃

A. EUPATORIA (*Agrimony*).—Stem erect, hirsute, branching; leaves 5–7-foliate, upper ones 3-foliate; leaflets ovate, oval, coarsely toothed; stipules large; flowers yellow, in spikes, on very short pedicels. A hairy plant, 2–4 feet high. *July.*

5. **Geum.**—Calyx deeply 5-cleft, with 5 alternate segments. Petals 5. Stamens numerous. Achenia numerous on the conical, dry receptacle. ♃

G. RIVALE (*Water Avens*).—Stem erect, pubescent; radical leaves interrupted and lyrately pinnate; cauline ones 3-foliate, or 3-lobed; flowers few, purple, nodding; petals purplish-yellow, broad obovate. In bogs and wet meadows. *June.*

6. Potentilla.—Calyx 4–5-cleft, with 4–5 alternate, exterior segments. Petals 4–5, dentate, deciduous. Stamens numerous, with slender filaments. Ovaries numerous, in a head, on a dry receptacle. Styles deciduous. Achenia numerous. ♃

1. P. CANADENSIS (*Five-finger*).—Hirsute-pubescent; procumbent; leaves palmately 3–5-foliate; leaflets obovate, silky beneath; stipules 2–3-cleft, or entire; flowers yellow, on axillary, solitary pedicels. A very common and variable species, sporting into apparently distinct varieties under the influence of different soils. *April–August.*

Var. PUMILA.—Very small and delicate. In dry soils. *April–May.*

2. P. ARGENTEA (*Silvery Cinquefoil*).—Stems ascending, hoary-tomentose; leaves palmately 3–5-foliate; leaflets oblong, incised, entire at base, with a revolute margin, silvery beneath; flowers small, yellow, in crowded corymbs; petals longer than the canescent sepals. *June–September.*

7. Fragaria.—Calyx concave, deeply 5-cleft. Petals 5, obcordate. Stamens numerous. Achenia dry, smooth, scattered on the enlarged, pulpy, deciduous receptacle. Leaves trifoliate. Stems stoloniferous. ♃

1. F. VIRGINIANA (*Field Strawberry*).—Plants pubescent; leaflets oval, coarsely serrate; flowers white, on few-flowered scapes; fruit roundish-ovoid; achenia imbedded in pits on the receptacle; calyx spreading in fruit. A well-known plant, in fields and meadows, universally a favorite for its delicious fruit. *April–May.*

2. F. VESCA (*English Strawberry*).—Plant pubescent, leaflets oval, coarsely serrate, or dentate; flowers white, on scapes longer than the leaves; achenia scattered in the surface of the conical or semi-spherical fruit, which is not pitted. Common in cultivation. *April–May.*

8. Waldsteinia. — Calyx 5-cleft, with 5 alternate bracteoles. Petals sessile, deciduous. Stamens numerous, inserted into the calyx. Achenia few, dry, on a short receptacle. ♃

W. FRAGARIOIDES (*Dry Strawberry*).—Leaves trifoliate, with pubescent petioles; leaflets crenately toothed, and incised; scapes many-flowered; flowers yellow; petals obovate. In shady and hilly woods. *June.*

9. Rubus.—Calyx 5-parted, spreading. Petals 5, deciduous. Stamens numerous. Ovaries numerous, with 2 ovules. Achenia pulpy, aggregated on a conical, juicy receptacle. Receptacle deciduous. ♃

1. R. VILLOSUS (*High Blackberry*).—Stem mostly erect, angular, armed with stout, curved prickles; young branches and peduncles villous and glandular; leaves palmately 3–5-foliate; leaflets ovate; flowers white, in leafless racemes of 20–25. Fruit oblong, large, black, sweet and delicious.

2. R. HISPIDUS (*Bristly Blackberry*).—Stem slender, prostrate, clothed

with retrorse bristles ; leaves 3-foliate ; leaflets obovate, mostly obtuse, smooth, entire toward the base, coarsely serrate ; peduncles corymbose, with several flowers, often bristly ; flowers small, white ; petals obovate. Fruit small, dark purple, or blackish, sour. *May-June.*

3. R. IDÆUS (*Garden Raspberry*).—Stem hispid ; leaves pinnately 3-5-foliate ; leaflets broad-ovate, unequally serrate, hoary-tomentose beneath ; flowers white, corymbosely paniculate ; petals entire, shorter than the tomentose, acuminate calyx. Fruit red. *May.*

4. R. STRIGOSUS (*Raspberry*).—Stem unarmed, shrubby, strongly hispid ; leaves pinnately 3-5-foliate ; leaflets oblong-ovate, acuminate, serrate, hoary-tomentose beneath ; peduncles 3-6-flowered, hispid ; petals white. Fruit light red, juicy, of a peculiar and very pleasant flavor. *May.*

5. R. OCCIDENTALIS (*Black Raspberry*).—Stem shrubby, glaucous, armed with hooked prickles ; leaves pinnately 3-foliate ; leaflets ovate, coarsely and doubly serrate, hoary-tomentose beneath ; peduncles 1-3-flowered, with short pedicels ; petals white. Fruit dark purple, covered with a glaucous bloom, pleasant flavor. *May.*

10. **Rosa.**—Calyx-tube contracted at the mouth, at length fleshy, with 5 segments. Petals 5. Achenia numerous, bony, hairy, attached to the inside of the fleshy calyx-tube. Leaves pinnate. Shrubs.

1. R. CAROLINA (*Swamp Rose*).—Stem smooth, armed with stout-hooked prickles ; leaflets 5-9, oblong, serrate, pale beneath ; petioles somewhat prickly ; flowers 3-5, in leafy clusters at the ends of the branches, light red ; calyx and peduncles glandular-hispid. Common in swamps, 4-8 feet high. *June-July.*

2. R. LUCIDA (*Wild Rose*).—Stems armed with scattered prickles ; leaflets 5-9, elliptical, sharply serrate, smooth and shining above ; flowers 1-3, pale red ; peduncles and appendaged calyx-segments glandular-hispid ; fruit depressed, globose, small, red, hispid. A shrub, in dry fields, 1-4 feet high. *June-July.*

3. R. SETIGERA (*Prairie Rose*).—Branches elongated, glabrous, with a few, stout, somewhat hooked prickles ; leaflets 3-5, large, ovate, smooth and shining above, sharply serrate ; flowers in very large, corymbose clusters, nearly scentless, of a changeable reddish color ; styles united ; fruit globose. A climbing species, 10-20 feet high. *June-July.*

4. R. RUBIGINOSA (*Sweet Brier*).—Stem smooth, armed with stout, recurved prickles ; leaflets 5-7, roundish-oval, sharply serrate, and with the petioles and stipules clothed with ferruginous glands beneath ; flowers light red, or white, fragant, mostly solitary ; fruit ovate, or obovate, reddish-orange when full grown. Common in fields ; often cultivated. *June.*

11. **Cratægus.**—Calyx-tube urceolate, with a 5-cleft limb. Petals 5, spreading. Stamens numerous. Styles 1-5. Pome fleshy, containing 1-5 long, 1-seeded carpels.

1. C. COCCINEA (*White Thorn*).—Leaves roundish-ovate, 5-9-lobed, sharply incised, thin, acutely serrate, on slender petioles ; flowers white,

corymbed ; calyx and pedicels usually smooth ; styles 3–5 ; fruit large, globose, bright red. A small tree, 10–20 feet high. *May.*

2. C. TOMENTOSA (*Black Thorn*).—Leaves oval, or elliptic-ovate, doubly serrate, dentate toward the apex, nearly smooth above, tomentose beneath ; flowers large, fragrant, white, in large, leafy corymbs, with calyx and pedicels villous-tomentose ; styles 3–5 ; fruit pyriform, ovary red, eatable, but rather insipid. In wet thickets, 10–15 feet high. *May.*

12. Pyrus.—Calyx urceolate, with a 5-cleft limb. Petals 5, roundish. Styles 2–5, often united at base. Pome closed, fleshy, 2–5 carpeled. Carpels cartilaginous, 2-seeded.

1. P. MALUS (*Apple*).—Leaves ovate, or oblong-ovate, serrate, acute, or briefly acuminate, tomentose beneath, petiolate ; flowers large, light rose-color, fragrant, in corymbs ; petals short ; styles 5, united and villous at base ; pome globose. A fruit-tree, cultivated, and almost naturalized, 20–40 feet high. *May.*

2. P. COMMUNIS (*Pear*).—Leaves ovate-lanceolate, smooth above, pubescent beneath ; flowers smaller than in the apple, white, in racemose corymbs, styles 5, distinct and villous at base ; pome usually pyriform. A tree, 20–50 feet high. *May.*

3. P. ARBUTIFOLIA (*Choke-berry*).—Leaves oblong-ovate, obtuse, or acute, smooth and shining above ; flowers in compound, terminal corymbs ; calyx and pedicels tomentose when young ; fruit pyriform, dark red, or purple when ripe, astringent. A shrub, 2–5 feet high, in low grounds. *May.*

4. P. AMERICANA (*Mountain Ash*). — Leaves pinnate, 13–15-foliate, smooth ; leaflets oblong-lanceolate, sharply serrate ; flowers white, in compound cymes ; fruit bright red, or scarlet, globose, sour. A small tree, 15–25 feet high, common in damp woods in mountainous districts. *May–June.*

13. Amelanchier.—Calyx 5-cleft. Petals oblong-ovate. Stamens short. Styles 5, more or less connected. Pome 3–5-celled, cells partly divided by a false dissepiment.

A. CANADENSIS (*Shad-flower*).—Leaves ovate, or oval, softly tomentose when very young, smooth when fully grown, sharply serrate ; flowers white, racemose ; berries purple, eatable. A common and variable shrub in damp, rocky woods, and low grounds, 6–15 feet high. *May.*

Var. OBLONGIFOLIA. — Leaves oval-oblong, serrate, with short acute teeth, tomentose on the lower surface during flowering ; flowers smaller than in normal form ; petals obovate-oblong. Apparently well marked while in flower, but difficult to distinguish afterward.

14. Cydonia.—Calyx urceolate, with a 5-cleft limb. Petals 5. Styles 5. Pome with 5 cartilaginous, many-seeded carpels. Seeds covered with a mucilaginous pulp.

C. VULGARIS (*Quince*).—Leaves oblong-ovate, entire, smooth above, woolly beneath ; flowers large, solitary, on woolly peduncles ; calyx woolly ; petals white ; pome soft, downy, obovoid, yellow. A large shrub of straggling growth. *May.*

Order XXIX.—SAXIFRAGACEÆ (*Saxifrage Family*).

Herbs, or shrubs. Leaves alternate, or opposite, sometimes stipulate. Sepals 4 or 5, more or less cohering, persistent. Petals 4 or 5, inserted between the calyx-lobes, rarely wanting. Stamens 5-10, inserted on the calyx-tube. Ovary adherent to the calyx-tube, of 2, or sometimes 3-5 carpels, cohering below, distinct above. Styles 2, sometimes 3-5. Fruit a 1, or rarely 3-5-celled capsule.

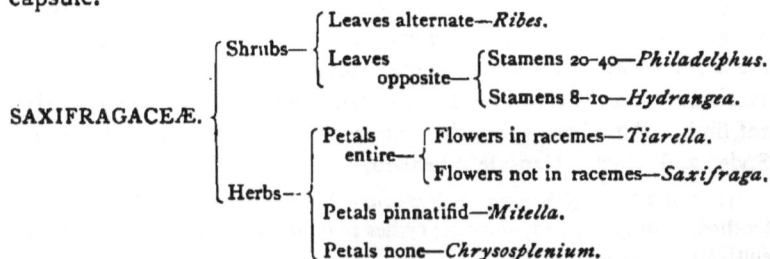

SAXIFRAGACEÆ.
- Shrubs—
  - Leaves alternate—*Ribes*.
  - Leaves opposite—
    - Stamens 20-40—*Philadelphus*.
    - Stamens 8-10—*Hydrangea*.
- Herbs—
  - Petals entire—
    - Flowers in racemes—*Tiarella*.
    - Flowers not in racemes—*Saxifraga*.
  - Petals pinnatifid—*Mitella*.
  - Petals none—*Chrysosplenium*.

1. Saxifraga.—Sepals 5, more or less united. Petals 5, inserted on the calyx-tube, entire. Stamens 10. Anthers 2-celled, opening longitudinally. Capsule of 2 carpels, 2-celled below, opening between the 2 divergent beaks. Seeds numerous. ♃

1. S. VIRGINIENSIS (*Early Saxifrage*).—Leaves mostly radical, ovate, spatulate, on broad petioles ; scape mostly naked ; flowers small, white, numerous, cymose. Early flowering plant, on rocks and dry hills, scape 3'-10' high. *April–May.*

2. S. PENNSYLVANICA (*Tall Saxifrage*).—Leaves radical, oval, rather acute, tapering at base, with short, margined petioles ; scape almost leafless, striate, viscid pubescent ; flowers yellowish-green, pedicellate ; petals linear-lanceolate. In swamps and meadows, with hollow scapes 1–3 feet high. *May.*

2. Mitella.—Calyx campanulate, 5-cleft. Petals 5, pinnatifid. Stamens 5-10, included. Styles 2, short, distinct. Capsule 1-celled, 2-valved. ♃

M. DIPHYLLA (*Common Mitella*).—Stem simple, pubescent ; leaves cordate, serrately toothed, pubescent ; flowers white, in long, terminal racemes on short pedicels, beautifully marked by the pectinate petals ; styles short. A plant 6'-12' high. *May–June.*

3. Tiarella.—Calyx 5-parted, with obtuse lobes, valvate in prefloration. Petals 5, entire. Stamens 10, inserted with the petals. Styles 2. Capsule 1-celled, 2-valved. ♃

T. CORDIFOLIA (*Mitrewort*).—Acaulescent ; leaves cordate, acutely 3-5-lobed, dentate, with mucronate teeth, hirsute above ; stolons creeping ; flowers white, in racemes ; bracts minute ; petals oblong. *May–June.*

4. Chrysosplenium.—Calyx coherent with the ovary, 4–5-lobed, colored within. Petals none. Stamens 8–10, with short filaments. Styles 2. Capsule obcordate, compressed, 1-celled, 2-valved. Seeds numerous. ♃

C. AMERICANUM (*Water-carpet*).—Stem slender, square, decumbent; leaves roundish-ovate, smooth; flowers remote, sessile; calyx usually 4-cleft, greenish yellow, marked with purple lines; stamens 8, very short, with orange-colored anthers. In shady springs and streams. *March–May.*

5. Hydrangea.—Flowers either all fertile, or more commonly the marginal ones are sterile. STERILE FLOWERS.—Calyx colored, membranaceous, veiny, 4–5-cleft. Petals, stamens, and styles none. FERTILE FLOWERS.—Calyx-tube hemispherical, with a 4–5-toothed, persistent limb. Petals ovate, sessile. Stamens twice as many as the petals. Styles 2, distinct. Capsule 2-beaked.

H. HORTENSIS (*Changeable Hydrangea*).—Leaves elliptical, serrated or toothed, strongly veined, smooth; cymes radiant; flowers mostly sterile. In cultivation, 1–2 feet high.

6. Philadelphus.—Calyx 4–5-parted, persistent, with the tube half adherent to the ovary. Petals 4–5, convolute in prefloration. Stamens 20–40. Capsule 4-celled, 4-valved. Seeds with an aril.

P. CORONARIUS (*False Syringa*).—Leaves ovate, smooth, petiolate; flowers numerous, white, very fragrant, in leafy clusters at the ends of the branches. A cultivated shrub, 5–7 feet high. *June.*

7. Ribes.—Leaves alternate and palmately lobed. Calyx 5-lobed. Petals 5. Stamens 5, alternate with petals. Low shrubs. ♃

1. R. RUBRUM (*Currant*). — Leaves subcordate, obliquely 3–5-lobed, mostly pubescent beneath, serrate; racemes nearly smooth, pendulous; flowers greenish, calyx rotate; fruit globose, smooth, red, or sometimes white. In gardens, it varying much in the size and color of fruit.

2. R. AUREUM (*Golden Currant*).—Glabrous; leaves 3-lobed; flowers numerous, golden yellow, very fragrant, in lax, many-flowered racemes; fruit smooth, globose, yellow, at length brown, pleasant. An ornamental shrub.

Order XXX.—CRASSULACEÆ.

Succulent herbs, or shrubby plants. Leaves simple, without stipules. Flowers usually in cymes. Sepals 3–20, more or less united at base, persistent. Petals as many as the sepals. Stamens as many as the sepals, and alternate with them, or twice as many, inserted on the calyx. Ovaries as many as the petals, and opposite to them. Follicles as many as the ovaries, many-seeded.

CRASSULACEÆ. {
   Leaves not fleshy—*Penthorum.*
   Leaves fleshy— {
     Pistils 6–12—*Sempervivum.*
     Pistils 4–5—*Sedum.*

1. Sedum.—Sepals 4–5, more or less united at base. Petals 4–5, distinct. Stamens 8-10. Carpels 4–5, distinct, many-seeded, with an entire scale at the base of each. ♃

1. S. TERNATUM (*Stone-crop*).—Leaves smooth, entire, lower ones ternately verticillate, obovate, tapering at base ; cyme of 3 spikes ; flowers several, white, sessile. In cultivation. *July–August.*

2. S. TELEPHIUM (*Orpine, Live-forever*).—Root tuberous, fleshy, white ; stem simple, erect, round, leafy ; leaves flattish, ovate, serrate, obtuse, scattered, sessile ; cymes corymbose, leafy ; flowers white and purple. In gardens. *August.*

3. S. ACRE (*Wall-pepper*).—Stems procumbent, branching at base ; leaves minute, somewhat ovate, fleshy, obtuse, alternate, crowded, sessile ; cymes few-flowered in 3 divisions, leafy ; flowers yellow. A little fleshy plant, rapidly spreading wherever it is planted, and thickly covering the surface. *June–July.*

2. Sempervivum.—Sepals 6–20, slightly united at base. Petals 6–20, acuminate. Stamens twice as many as the petals. Ovaries with lacerated scales at base. Carpels 6–20. ♃

S. TECTORUM (*House-leek*).— Herbaceous ; leaves thick and fleshy, fringed ; offsets spreading. A common plant in gardens, which sends out runners with offsets, and thus propagates itself, flowering only occasionally.

3. Penthorum.—Sepals 5, united at base. Petals 5, or none. Stamens 10. Capsules of 5 united carpels, 5-angled, 5-celled, and 5-beaked. ♃

P. SEDOIDES (*Virginia Stone-crop*).—Stem erect, somewhat branched, angular above ; leaves lanceolate, smooth, acute at both ends, serrate, almost sessile ; flowers yellowish-green, inodorous. In moist ground, 8'–15' high. *August–September.*

Order XXXI.—DROSERACEÆ (*Sundew Family*).

Marsh herbs with regular flowers in scapes, from a tuft of glandular leaves. Sepals and petals 5. Stamens 5–15. Pod 1-celled and many-seeded.

Drosera.—Sepals 5, united at base, equal, persistent ; petals 5 ; stamens 5 ; styles 3–5 ; capsule sub-globose, 3-valved, 1-celled, many-seeded.

D. ROTUNDIFOLIA (*Sundew*).—Leaves radical, in tufts, orbicular, on long petioles, lying flat on the ground, with long, reddish, glandular hairs ; scapes circinate, racemose, 1-sided ; flowers small, white. Marshes. *August.*

Order XXXII.—Hamamelaceæ (*Witch-Hazel Family*).

Shrubs. Leaves alternate. Stipules deciduous. Calyx 4-cleft. Petals 4, linear, sometimes none. Stamens 8, those opposite the petals barren, or else many, and all fertile ; inserted on the calyx. Ovary 2-celled. Styles 2, distinct. Capsule coriaceous, or woody, 2-beaked, 2-celled, free from the calyx at apex.

Hamamelis.—Calyx 4-parted, 2-3-bracted at base. Petals 4, very long, linear. Fertile stamens 4. Sterile ones 4, scale-like. Capsule 2-celled. ♃

H. Virginiana (*Witch Hazel*).—Leaves obovate or oval, toothed, on short petioles, nearly smooth ; flowers sessile, 3-4 together, axillary ; petals narrowly linear, curled or twisted, yellow ; calyx downy ; ovary hirsute. A shrub of irregular growth, 10-15 feet high. *December.*

Order XXXIII.—Haloragæ (*Water-Milfoil Family*).

Marshy plants with small flowers, sessile in the axils of leaves. Calyx-tube united with ovary. Stamens 1-8. Cotyledons small ; embryo in the axis of the albumen.

Floral parts in 4's—*Myriophyllum*.
Floral parts in 3's—*Proserpinaca*.

1. **Myriophyllum.**—Flowers monœcious, or frequently perfect. Calyx 4-toothed or 4-parted. Petals 4, often minute or wanting. Stamens 4-8. Fruit consisting of 4 nut-like, indehiscent carpels, cohering by their inner angles. ♃

1. M. spicatum (*Water Milfoil*).—Leaves in 3's, pinnately parted, with capillary segments ; flowers greenish, small, in terminal, nearly naked spikes ; petals broad-ovate. An aquatic plant in deep ponds. *July-August.*

2. M. ambiguum (*Milfoil*).—Leaves alternate ; submersed ones pinnately parted, with capillary segments ; upper ones linear, entire, or slightly toothed, petiolate ; flowers axillary, minute ; petals oblong, somewhat persistent. An aquatic, with floating stems, in ponds and ditches.

2. **Proserpinaca.**—Calyx-tube 3-sided, with a 3-parted limb. Petals none. Stamens 3. Stigmas 3, oblong. Fruit bony, 3-sided, 3-celled, crowned with the persistent calyx. ♃

P. palustris (*Mermaid-weed*).—Stem ascending at base, striate, smooth ; leaves alternate, lanceolate, sharply serrate ; those below the water pinnatifid ; flowers small, green, axillary, 1-3 together, followed by a hard, triangular nut. In shaded, shallow water, 6'-12' high. *June-July.*

Order XXXIV.—Melastomaceæ.

Trees, shrubs, or herbs, with square branches. Leaves opposite, ribbed, entire. Sepals 4-6, united, persistent, forming an urceolate tube, which coheres only with the angles of the ovary. Petals as many as the calyx-segments, twisted in prefloration. Stamens twice as many as the petals. Anthers 1-celled, before flowering contained in the cavities between the calyx and ovary. Fruit a capsule.

Rhexia.—Calyx-tube with a 4-cleft, persistent limb. Petals 4. Stamens 8. Style declined. Capsule 4-celled, with prominent placentæ. ♃

R. VIRGINICA (*Meadow Beauty*).—Stem somewhat hispid, with 4 slightly winged angles; leaves sessile, oval-lanceolate, strongly 3-nerved; flowers large, bright purple, showy and numerous, in corymbose cymes; petals obovate; anthers long, yellow, crooked. A very showy plant, 6'-12' high, in wet ground. *July-August.*

Order XXXV.—LYTHRACEÆ (*Loosestrife Family*).

Herbs with 4-sided branches and exstipulate, entire leaves. Calyx inclosing the many-seeded pod, and bearing petals and stamens on its throat. Style 1; stigma capitate. The flowers axillary or whorled. Seeds exalbuminous.

Nesæa.—Calyx broadly campanulate, 5 erect teeth and 5 elongated horns. Stamens 10, those opposite calyx-teeth very long. Style filiform. Stigma small. Capsule globose within calyx; many-seeded.

N. VERTICILLATA (*Swamp Loosestrife*).— Stem simple, woody, with recurved branches rooting at the summit; 4-6-angled; leaves opposite or whorled, entire, on short petioles; flowers purple, axillary, nearly sessile. Common in swamps. *August-September.*

Order XXXVI.—ONAGRACEÆ (*Evening Primrose Family*).

Herbs or shrubs. Flowers axillary, in spikes or racemes. Sepals united in a tubular 2-6-lobed calyx. Petals usually as many as the calyx-lobes, and alternate with them, sometimes none. Stamens as many or twice as many, inserted in the calyx-throat; filaments distinct. Ovary coherent with the calyx-tube, 2-4, or by abortion 1-2-celled. Style prolonged. Fruit baccate, or capsular.

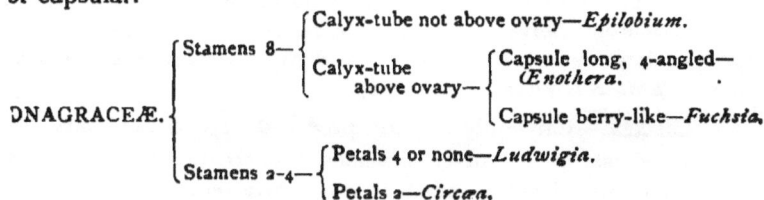

ONAGRACEÆ.
- Stamens 8—
  - Calyx-tube not above ovary—*Epilobium.*
  - Calyx-tube above ovary—
    - Capsule long, 4-angled—*Œnothera.*
    - Capsule berry-like—*Fuchsia.*
- Stamens 2-4—
  - Petals 4 or none—*Ludwigia.*
  - Petals 2—*Circæa.*

1. Epilobium.—Calyx-tube not prolonged beyond the ovary. Limb 4-cleft. with spreading and deciduous segments. Petals 4. Stamens 8. Anthers attached near the middle. Capsule linear, 4-sided, 4-celled, 4-valved. ♃

1. E. ANGUSTIFOLIUM (*Willow Herb*).—Stem erect, simple; leaves lanceolate, sessile; flowers numerous, large, of a purplish lilac color, in a long, terminal, spicate raceme; stamens and styles declined. A showy plant in low grounds; 2-6 feet high. *July–August.*

2. E. COLORATUM (*Colored Willow Herb*).—Stem erect, nearly terete, very branching; leaves mostly opposite, lanceolate, acute, very shortly petiolate; flowers numerous, rose-colored, small; petals cleft at apex, twice as long as the sepals; stigma clavate. In wet, swampy grounds; 1-3 feet high.

2. Œnothera.—Calyx-tube prolonged beyond the ovary, the segments 4, reflexed. Petals 4, equal. Stamens 8. Capsule 4-celled, 4-valved, many-seeded. Stigma 4-lobed. ②—♃

1. Œ. BIENNIS (*Evening Primrose*).—Stem erect, simple or branched; leaves ovate-lanceolate, pubescent; flowers large, yellow, in a terminal leafy spike, sessile. In fields. Plant 3-5 feet high. *June-August.*

2. Œ. PUMILA (*Dwarf Primrose*).—Stem slender, simple, reclined at base, ascending; leaves lanceolate, entire, obtuse, tapering at base, sessile; flowers rather small, yellow, in a terminal, leafy spike; petals obcordate; capsule oblong-clavate. In grassy fields, 8′-12′ high. *June–August.*

3. Fuchsia.—Calyx tubular, funnel-form, colored, deciduous, with a 4-lobed limb. Petals 4. Disk glandular, 8-furrowed. Capsule baccate, oblong, obtuse, 4-sided.

F. COCCINEA (*Lady's Eardrop*).—Stem shrubby, with smooth branches; ovate, acute, on short petioles; flowers axillary, nodding; sepals oblong, acute; petals convolute, half as long as the calyx. In cultivation, growing 1-6 feet high.

4. Ludwigia.—Calyx-tube not prolonged beyond the ovary. Petals 4, equal. Stamens 4, opposite the calyx-segments. Capsules short, 4-celled, 4-valved. ♃

1. L. ALTERNIFOLIA (*Seedbox*).—Stem erect, branching, slightly angled; nearly smooth; leaves lanceolate, or oblong-lanceolate, sessile; flowers yellow, on axillary, solitary peduncles; calyx-segments broadly ovate; capsule 4-winged, crowned with the calyx. In swamps, 18′-25′ high. *July–August.*

2. L. PALUSTRIS (*Water Purslane*).—Smooth and somewhat succulent; stems procumbent; leaves opposite, ovate, entire, petiolate, acute; flowers sessile, axillary; capsule oblong, 4-angled, short. Creeping plant in muddy places. *June–September.*

5. Circæa.—Calyx-tube slightly produced beyond the ovary, deciduous. Petals 2, obcordate. Stamens 2, alternate with the petals. Capsule obovate, hispid, 2-celled, at length 2-valved, 2-seeded. ♃

. C. Lutetiana (*Enchanter's Nightshade*).—Stem erect, branching, pubescent above ; leaves opposite, ovate ; flowers small, white, or pale rose-color, in elongated, naked racemes ; bracts none ; calyx reflexed ; fruit covered with hooked bristles. Plant 1–2 feet high.

Order XXXVII.—Cucurbitaceæ.

Succulent herbs, climbing, or creeping by tendrils. Leaves alternate, palmately lobed and veined, coarse and rough. Flowers monœcious, or polygamous. Calyx 5-lobed. Petals 5, united together, attached to the calyx, very cellular, and much reticulated in structure. Stamens 5. Anthers very long, variously wavy and contorted, 2-celled. Ovary adherent to the calyx-tube, 1-celled, with 3 parietal placentæ. Fruit a pepo, rarely membranous, and 1–4-seeded. Seeds flat.

CUCURBITACEÆ.
{
  Flowers large—
    { Both kinds of flowers solitary—*Cucurbita.*
    { Sterile flowers clustered—*Cucumis.*
  Flowers small, fruit 1-seeded—*Sicyos.*
}

BEGONIACEÆ.—*Begonia.*

1. Sicyos.—Flowers monœcious. Sterile flowers—Calyx 5-toothed ; teeth subulate or minute. Corolla rotate. Stamens 5, monadelphous, or in 3 parcels. Anthers contorted. Fertile flowers—Calyx campanulate, 5-toothed, contracted above. Petals 5. Fruit ovate, membranaceous, hispid or echinate with spiny bristles. Seed large. ①

S. angulatus (*Wild Cucumber*).—Stem climbing by tendrils, branching, hairy ; leaves roundish, cordate at base, 5-angled, 5-lobed ; lobes acuminate, denticulate ; tendrils 3–5-cleft ; flowers whitish ; *sterile* ones in crowded racemes, on long peduncles ; *fertile* ones on short peduncles, smaller ; both usually from the same axils ; fruit somewhat spiny, in crowded clusters, each containing one large seed. *July.*

2. Cucumis.—Flowers monœcious or perfect. Calyx tubular-campanulate. Corolla deeply 5-parted. Sterile flowers—Stamens 5, triadelphous. Fertile flowers—Style short. Stigmas thick, 2-parted. Pepo fleshy. Seeds ovate, without margins. ①

1. C. sativus (*Cucumber*).—Stem rough, prostrate and trailing ; tendrils simple ; leaves palmately 5-angled or lobed ; lobes nearly entire, acute ; the terminal one longest ; fruit oblong, obtusely angled, on a short peduncle, prickly when young ; flowers yellow, solitary, axillary. Cultivated for its green fruit. *June–September.*

2. C. melo (*Musk-melon*).—Stem prostrate, trailing, rough ; tendrils simple ; leaves roundish, palmately 5-angled or lobed ; lobes rounded, obtuse ; flowers sterile, fertile, and perfect, yellow ; fruit oval or sub-globose. Cultivated. *June–July.*

3. C. CITRULLUS (*Water-melon*).—Stem slender, prostrate, trailing, hairy; leaves palmately 5-lobed, very glaucous beneath; flowers yellow, solitary; fruit smooth, marked with various shades of green, very juicy. Cultivated. *June–August.*

3. **Cucurbita.**—Flowers monœcious. Corolla campanulate. Petals united and cohering with the calyx. STERILE FLOWERS—Calyx 5-toothed. Stamens 5, triadelphous, with united, straight anthers. FERTILE FLOWERS—Calyx 5-toothed. Stigmas 3, thick, 2-lobed. Pepo fleshy or woody, 3–5-celled. Seeds obovate, smooth, with thickened margins. ①

C. PEPO (*Pumpkin*).—Plant rough and hispid; stem procumbent; tendrils branched; leaves very large, cordate, palmately 5-lobed; flowers large, axillary, yellow; sterile ones on long peduncles; fruit very large, roundish, and yellow when ripe. Common in cultivation. *July.*

Order XXXVIII.—CACTACEÆ (*Cactus Family*).

Succulent, shrubby plants, almost always destitute of leaves, and producing spinose buds. Stems usually angular, or flattened. Flowers sessile, showy. Sepals numerous. forming a tube, which adheres to the ovary, completely inclosing it. Petals indefinite, often passing into the sepals, inserted into the calyx-tube over the summit of the ovary. Stamens indefinite, attached to the petals, with long filaments and versatile anthers. Ovary 1-celled, fleshy. Style single, forming a stellate cluster with several anthers. Fruit a many-seeded berry.

Opuntia.—Sepals and petals numerous, united in a tube which adheres to the ovary. Stamens numerous, shorter than the petals. Style cylindrical, with numerous, thick, erect stigmas. Berry prickly. ♃

O. VULGARIS (*Prickly Pear*).— Prostrate, creeping, with articulated branches and broad and flattened joints, with fascicles of prickles regularly arranged; prickles short and numerous, each fascicle usually consisting of several strong subulate spines; flowers yellow; fruit crimson, nearly smooth, eatable. In sandy fields. *June–July.*

Order XXXIX.—UMBELLIFERÆ.

Herbs, rarely suffrutescent. Stems usually hollow and furrowed. Leaves alternate, usually compound, the petioles becoming dilated, and sheathing at base. Flowers in umbels, usually with an involucre. Calyx adherent to the ovary, the very small border 5-toothed, or entire. Petals 5, usually with an inflexed point, inserted between the calyx-teeth in a disk which crowns

the ovary. Stamens 5, alternate with the petals. Ovary of 2
united carpels, 2-celled, with 1 ovule in each cell. Styles 2, distinct, or united and thickened at the base. Fruit consisting of 2
dry carpels, which adhere by their opposite faces (*commissure*)
to a common axis (*carpophore*), at length separating, and suspended from the forked summit of the carpophore. Each carpel
is indehiscent, marked with 5 longitudinal primary ribs, and often
with secondary ones alternate with the first. In the substance of
the pericarp, little oil-tubes (*vittæ*) are usually imbedded opposite
the intervals between the ribs, or opposite the ribs themselves.

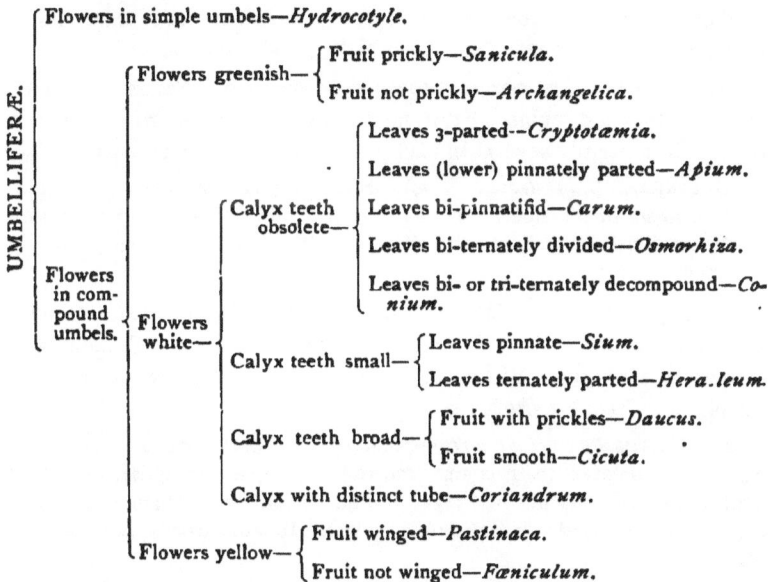

Flowers in simple umbels—*Hydrocotyle.*

UMBELLIFERÆ.

Flowers in compound umbels.

Flowers greenish—
- Fruit prickly—*Sanicula.*
- Fruit not prickly—*Archangelica.*

Flowers white—
- Calyx teeth obsolete—
  - Leaves 3-parted--*Cryptotæmia.*
  - Leaves (lower) pinnately parted—*Apium.*
  - Leaves bi-pinnatifid—*Carum.*
  - Leaves bi-ternately divided—*Osmorhiza.*
  - Leaves bi- or tri-ternately decompound—*Conium.*
- Calyx teeth small—
  - Leaves pinnate—*Sium.*
  - Leaves ternately parted—*Hera.leum.*
- Calyx teeth broad—
  - Fruit with prickles—*Daucus.*
  - Fruit smooth—*Cicuta.*
- Calyx with distinct tube—*Coriandrum.*

Flowers yellow—
- Fruit winged—*Pastinaca.*
- Fruit not winged—*Fœniculum.*

1. Hydrocotyle.—Calyx obsolete. Petals ovate, entire, acute,
spreading, with the point not inflexed. Fruit flattened laterally, with
a narrow commissure. Carpels without vittæ. 2⊥

H. AMERICANA (*Penny-wort*).—Smooth and shining; stem branching,
often decumbent; leaves orbicular-reniform, somewhat lobed, doubly crenate; flowers minute, greenish, in axillary, sessile, few-flowered, greenish,
capitate umbels; fruit very minute, orbicular. In wet places. 2′-4′ long.
June-August.

2. Sanicula.—Flowers polygamous. Calyx-tube echinate. Petals
obovate, erect, with a long inflexed point. Fruit sub-globose, armed with
hooked prickles. Carpels not ribbed, furnished with numerous vittæ. 2⊥

S. MARILANDICA (*Sanicle*).—Leaves digitately 5-7-parted; segments oblong, incisely serrate; cauline leaves few, nearly sessile; flowers greenish,

small ; umbels with few rays ; umbellets with numerous rays ; involucre 6-leaved, serrate ; sterile flowers numerous, on pedicels as long as the fertile, sessile flowers ; style elongated, conspicuous and recurved. In thickets, 2-3-feet high. *June-July.*

3. **Cicuta.**—Calyx-margin with 5 broad teeth. Petals obcordate, with an inflexed point. Fruit roundish. Carpels with 5 flattish, equal ribs. Intervals filled with single vittæ. Commissure with 2 vittæ. Carpophore 2-parted. Seeds terete. Involucels many-leaved.

C. MACULATA (*Water Hemlock*).—Stem smooth, hollow, often streaked with purple ; lower leaves triternately divided ; upper ones biternately divided ; leaflets lanceolate, serrate, smooth ; flowers white, in umbels ; involucel of 5-6 linear divisions ; calyx and style persistent ; root thick, fleshy, poisonous. In wet meadows, 4-8 feet high. *July-August.*

4. **Sium.**—Calyx-margin 5-toothed or obsolete. Petals obcordate, with an inflexed point. Fruit nearly oval. Carpels with 5 obtuse ribs. Vittæ usually several in each interval. Carpophore 2-parted. ♃

S. LINEARE (*Long-leaved Sium*).—Stem angular, sulcate ; leaves pinnate ; leaflets linear, finely serrate ; flowers white, small ; involucres with 5-6 linear divisions ; calyx-teeth minute ; fruit obovate. A stout plant, 3-5 feet high. *July.*

5. **Cryptotænia.**—Calyx-margin obsolete. Petals obcordate, with an inflexed point. Fruit linear-oblong or ovate-oblong. Carpels with 5 equal, obtuse ribs. Vittæ very narrow, twice as many as the ribs. Carpohore free, 2-parted. ♃

C. CANADENSIS (*Honewort*).—Stem smooth, branching above ; leaves 3-parted, petiolate ; teeth coarse, mucronate ; umbels irregular, somewhat paniculate, with very unequal rays ; flowers small, white ; involucres none ; involucels few-leaved ; fruit oblong-elliptical. In moist woods, 1-2 feet high. *July.*

6. **Carum.**—Calyx-margin obsolete. Petals obovate, emarginate, with an inflexed point. Styles spreading, dilated at base. Fruit oval, compressed laterally. Carpels 5-ribbed, lateral ribs marginal. Intervals with single vittæ ; commissure with 2. ♃

C. CARUI (*Caraway*).—Leaves bi-pinnatifidly divided ; segments numerous, linear ; involucre 1-leaved, or none ; involucels none ; flowers white. Cultivated for its aromatic fruit. *June.*

7. **Apium.**—Calyx-margin obsolete. Petals roundish, with an inflexed point. Fruit roundish, laterally compressed. Carpels 5-ribbed ; the lateral ribs marginal. Intervals with single vittæ. Carpophore undivided. ②

A. GRAVEOLENS (*Celery*).—Stem branching, furrowed ; lower leaves pinnately dissected, on very long petioles ; segments incised ; upper leaves 3-

parted ; segments lobed and dentate at apex ; flowers white, in umbels, with unequal, spreading rays. Cultivated in gardens. *July–August.*

8. Fœniculum.—Calyx-margin obsolete. Petals revolute, with a broad, retuse apex. Fruit oblong, laterally compressed. Carpels with 5 obtuse ribs ; marginal ones a little broader. Intervals with single vittæ. ①

F. VULGARE (*Fennel*).—Stem round, smoothed, branched ; leaves biternately dissected, with linear-subulate, elongated segments ; umbels with numerous unequal, spreading rays ; involucre and involucels none ; carpels turgid, oblong-ovate ; flowers yellow. Common in gardens. *July.*

9. Archangelica.—Calyx-teeth short. Petals equal, entire, acuminate, with the point inflexed. Fruit dorsally compressed. Carpels with 3 carinate dorsal ribs, with the 2 lateral ones dilated into wings. Vittæ very numerous. ♃

A. ATROPURPUREA (*Angelica*).—Stem mostly dark purple, furrowed ; leaves 3-parted, on large, inflated petioles ; divisions of the leaves bipinnately divided, with 5–7 segments ; flowers greenish, in very large umbels, on nearly smooth peduncles ; involucels many-leaved ; fruit smooth. A rank plant in meadows, 4–6 feet high. *June.*

10. Pastinaca.—Calyx-teeth obsolete, or minute. Petals roundish, entire, involute, with an inflexed point. Fruit much compressed, with a broad, flat margin. Carpels with 5 nearly obsolete ribs. Intervals with single vittæ ; commissure with 2 or none. Carpophore 2-parted. Seeds flat. Involucre and involucels few-leaved ; or none. ♃

P. SATIVA (*Parsnip*).—Roots fleshy, stem smooth ; leaves pinnately divided, slightly pubescent, especially beneath ; leaflets ovate, or oblong ; umbels large, on long peduncles ; flowers yellow ; fruit oval. Common in cultivation, and also naturalized in waste places. Stem 3–5 feet high. *July–September.* ②

11. Heracleum.—Calyx with 5 small, distinct teeth. Petals obcordate, with an inflexed point, in the exterior flowers deeply 2-cleft. Fruit compressed, flat, with broad, flat margins. Carpels with 3 obtuse dorsal ribs. Seeds flat. Involucre caducous, mostly few-leaved. Involucels many-leaved. ♃

H. LANATUM.—Stem branching, hollow, pubescent ; leaves very large and broad ; leaflets petiolate, cordate ; lobes acuminate ; flowers white, in very large umbels ; segments of the involucre lanceolate, deciduous, those of the involucels lanceolate, acuminate ; fruit nearly orbicular. A rank plant, 4–8 feet high, in meadows. *June.*

12. Daucus.—Calyx-margin 5-toothed. Petals obovate, emarginate, with an inflexed point. Fruit ovoid, or oblong. Carpels with 5 primary ribs, 3 dorsal and 2 on the flat commissure, and 4 secondary ribs, the latter more prominent, winged, and each bearing a single row of prickles, with single vittæ beneath. Carpophore entirely free. ②

D. CAROTA (*Carrot*).—Stem hispid, branching; leaves bi- or tri-pinnatifid; segments pinnatifid; leaflets lanceolate or linear; leaflets of the involucre pinnatifid; flowers white, sometimes yellowish; the central flower in each umbellet abortive, rose-colored. Root conical. Common in cultivation. *July.*

13. **Osmorhiza.**—Calyx-margin obsolete. Petals oblong, entire; the cuspidate point inflexed. Styles conical at base. Fruit very long, linear, clavate, attenuate at base. Carpels with 5 acute, bristly ribs. Intervals without vittæ. Commissure with a deep bristly channel. ♃

O. BREVISTYLIS (*Hairy Cicely*).—Stem erect, branching; leaves biternately divided; segments pinnatifid, hairy; flowers white; fruit somewhat tapering, with the persistent styles at length converging. In woods, 1–3 feet high. *May–June.*

14. **Conium.**—Calyx-margin obsolete. Petals obcordate, with a short inflexed point. Fruit ovate, with compressed sides. Carpels with 5 prominent, equal, undulate-crenulate ribs; the lateral ones marginal. Intervals without vittæ. Seeds with a deep, narrow groove in the face. ②

C. MACULATUM (*Poison Hemlock*).—Stem smooth, branching, hollow, spotted; leaves decompound, bipinnately divided; leaflets lanceolate, pinnatifid, with acute lobes; involucel of 3–5 unilateral leaflets; flowers small, white, in terminal umbels; fruit smooth. A poisonous weed, 3–8 feet high, in waste places. *July–August.*

15. **Coriandrum.**—Calyx with 5 conspicuous teeth. Petals obcordate, inflexed at the point; outer ones much larger, bifid. Fruit globose. Carpels cohering together, with 5 depressed, primary ribs, and 4 secondary, more prominent ones. Seeds concave on the face. ①

C. SATIVUM (*Coriander*).—Glabrous; leaves bipinnately divided; lower ones with broad, cuneate segments; upper ones with linear segments; involucel 3-leaved, unilateral; flowers white; carpels hemispherical. A garden plant, 2–3 feet high. *July.*

Order XL.—ARALIACEÆ (*Ginseng Family*).

Herbs, shrubs, or trees. Leaves compound or simple, exstipulate. Flowers in umbels, which are often arranged in racemes or panicles. Calyx adherent to the ovary, with a small, entire, or 5-toothed limb. Petals 5–10, very rarely wanting, inserted in a disk which crowns the ovary. Stamens as many as the petals, alternate with them, Ovary 2-15-celled, with 1 ovule in each cell. Styles erect, connivent, as many as the cells. Fruit drupaceous, or baccate.

ARALIACEÆ. { Leaves compound—*Aralia*.
{ Leaves simple—*Hedera*.

1. **Aralia.**—Calyx-limb 5-toothed or entire, short.   Petals 5, spreading. Stamens 5, alternating with the petals.   Styles 5, at length diverging. Fruit baccate, 5-lobed, 5-celled, 5-seeded. ♃

1. A. RACEMOSA (*Spikenard*).—Stem smooth, herbaceous; leaves decompound, 3-5-parted; each division with 3-5 ovate leaflets; umbels small, numerous, arranged in branching, compound racemes, forming panicles on axillary peduncles; flowers small, greenish white; fruit small, dark purple. In rich, rocky woodlands, 3-6 feet high. *July.*

2. A. NUDICAULIS (*Sarsaparilla*).—Nearly acaulescent; leaf radical, solitary, on a long, 3-cleft petiole; each division pinnately 3-5 foliate; leaflets oval or obovate, sharply serrate; scape naked, bearing 3 simple, pedunculate umbels; flowers small, greenish; root long, creeping, aromatic. In rich woods, with a scape 1 foot high. *May-June.*

3. A. TRIFOLIUM (*Dwarf Ginseng*).—Root globose; leaves 3, verticillate, 3-5-foliate; leaflets oblong-lanceolate, serrate, subsessile; peduncle nearly as long as the leaves; flowers white, on short pedicels; styles 3; berries 3-seeded. In low woods, 3'-6' high. *May.*

2. **Hedera.**—Calyx 5-toothed. Petals 5, dilated at base.   Berry 5-seeded, surrounded by the persistent calyx. *Evergreen.* ♃

H. HELIX (*English Ivy*).—Stem and branches long and flexible, attaching themselves to the earth, walls, or trees, by numerous rootlets; leaves dark green, smooth, petiolate, with white veins; lower ones 5-lobed; upper ovate; flowers green, in numerous umbels, arranged in corymbs; berry black. A climbing, shrubby plant, in cultivation.

Order XLI.—CORNACEÆ (*Cornel Family*).

Shrubs or trees. Leaves simple. Flowers small; calyx united to the 1-2-celled ovary. Petals valvate in bud. Style 1. Fruit a drupe or berry.

1. **Cornus.**—Calyx-limb 4-toothed, with minute segments.   Petals 4, oblong, spreading. Stamens 4, with filiform filaments.   Style 1. Drupes baccate. *Trees, shrubs, and perennial herbs.*

1. C. FLORIDA (*Boxwood*).—Leaves ovate, acuminate, entire; flowers small, greenish-yellow, surrounded by a large 4-leaved involucre, the segments of which are obcordate, with a callous point at apex, white and showy, often tinged with red; drupes oval, bright red. A tree 15-30 feet high. *May-June.*

2. C. Canadensis (*Low Cornel*).—Herbaceous; flowering stems low, simple, erect; rhizoma creeping, somewhat woody; upper leaves about 6, somewhat verticillate, oval, acute; involucre 4-leaved, much larger than the flowers; leaflets broad-ovate, greenish-white, petaloid, inclosing the umbel of greenish-yellow flowers; drupes red, baccate, rather large, and of a sweetish taste. In damp woods, 4'-6' high. *May-June.*

SUPERIOR MONOPETALOUS EXOGENS.

Order XLII.—Caprifoliaceæ (*Honeysuckle Family*).

Shrubs, often climbing, rarely herbs. Leaves opposite. Stipules none. Calyx-tube adherent to the ovary; limb 4–5-cleft. Corolla regular, or irregular; limb 4–5-lobed. Stamens 4–5, alternate with the corolla-segments when equaling them in number. Ovary 3–5-celled. Style 1. Fruit always crowned with the persistent calyx-teeth. Seeds pendulous.

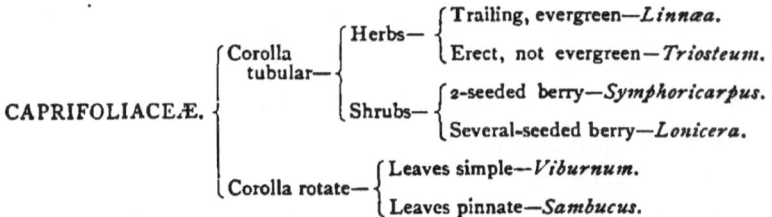

CAPRIFOLIACEÆ.
- Corolla tubular—
  - Herbs—
    - Trailing, evergreen—*Linnæa.*
    - Erect, not evergreen—*Triosteum.*
  - Shrubs—
    - 2-seeded berry—*Symphoricarpus.*
    - Several-seeded berry—*Lonicera.*
- Corolla rotate—
  - Leaves simple—*Viburnum.*
  - Leaves pinnate—*Sambucus.*

1. **Lonicera.**—Calyx-limb with 5 short teeth. Corolla tubular or funnel-form, with a 5-cleft, usually quite irregular limb. Stamens 5. Ovary 2–3-celled. Berry few-seeded. ♃

1. L. sempervirens (*Trumpet Honeysuckle*).—Leaves oblong, evergreen, pale beneath, upper pairs connate; flowers in whorls, almost regular, ventricose above, scarlet without and yellow within, nearly 2' long, inodorous. A climbing evergreen, in cultivation. *May-August.*

2. L. caprifolium (*Italian Honeysuckle*).—Leaves deciduous, the upper pair connate; flowers in a terminal whorl; corolla ringent, varying through red, yellow, and white, very fragrant. Cultivated species. *June-August.*

2. **Triosteum.**—Calyx-limb with 5 linear, foliaceous, persistent teeth. Corolla tubular; limb with 5 subequal lobes. Stigma capitate. Fruit dry, drupaceous. Seeds 3-angled, bony. ♃

T. perfoliatum (*Feverwort*).—Herbaceous; stem hollow, pubescent, simple; leaves ovate, entire, connate, pubescent; flowers sessile, in verticils of 5–8; corolla viscid-pubescent, dull purple, with a curved tube; fruit somewhat 3-sided, orange-colored when ripe. A coarse, hairy plant, 2–3 feet high. *June.*

3. **Symphoricarpus.**—Calyx-tube globose ; limb with 4–5 persistent teeth. Corolla bell-shaped. Stamens 4–5, inserted on the corolla. Berry globose, 4-celled, 2-seeded ; 2 cells abortive. ♉

S. RACEMOSUS (*Snow-berry*).—Leaves oval, often undulate at the margin, mostly smooth, paler beneath, on short petioles ; flowers in terminal, loose, interrupted, and somewhat leafy racemes ; corolla densely bearded within, rose-colored ; stamens and style included ; berries large, roundish. In cultivation. *June–September.*

4. **Linnæa.**—Calyx-limb deciduous, with 5 subulate teeth. Corolla campanulate ; limb with 5 nearly equal lobes. Stamens 4, 2 longer than the other 2. Berry 3-celled, dry, indehiscent, 1-seeded, with 2 abortive cells. ♉

L. BOREALIS (*Twin-flower*). — Evergreen ; stems filiform, creeping, branching and rooting through their whole length ; leaves small, roundish, crenate, with short petioles and a few scattered hairs ; peduncles erect, filiform ; corolla rose-colored, of a deeper hue inside. In moist woods. *June.*

5. **Sambucus.**—Calyx with 5 minute or obsolete teeth. Corolla with 5 spreading segments. Stamens ·5. Stigmas 3. Berry globose, pulpy, 3-seeded. ♉

1. S. CANADENSIS (*Elder*).—Stem shrubby ; leaves pinnate ; leaflets 5–11, oblong or oval, serrate, smooth ; cyme flat, 5-parted ; flowers white, very numerous, with a rather oppressive odor ; berry dark purple, juicy. Shrub, in waste grounds, 6–10 feet high. *May–July.*

2. S. PUBENS (*Red-berried Elder*).—Stem shrubby, with a warted bark ; leaves pinnate ; leaflets 5–7, oval-lanceolate, acuminate, and with the petiole pubescent beneath ; cymes densely panicled, or pyramidal ; flowers white ; fruit scarlet, small. *May–June.*

6. **Viburnum.**—Calyx persistent, 5-toothed. Corolla with 5 obtuse, spreading segments. Stamens 5. Stigmas 3. Fruit a 1-celled, 1-seeded drupe.

1. V. OPULUS (*Cranberry-Tree*).—Leaves chiefly 3-lobed, rounded, rarely tapering at base ; dentate ; cymes pedunculate ; fruit ovoid, red, acid. A handsome shrub of erect growth.

2. V. ACERIFOLIUM (*Maple-leaved Viburnum*).—Leaves 3-veined, 3-lobed, somewhat cordate at base, sharply serrate, pubescent beneath ; petioles and young branches pubescent ; cymes on long peduncles, flat ; fruit oval, compressed ; flowers dull white. A shrub 4–6 feet high. *June.*

3. V. LENTAGO (*Sweet Viburnum*).—Leaves ovate, acuminate, finely and sharply serrate ; petioles long ; flowers white, in broad cymes ; berries oval, finally black and edible, with a sweetish taste. A handsome shrub or small tree, 10–15 feet high. *May–June.*

· 12

4. V. DENTATUM (*Arrow-wood*).—Smooth ; leaves roundish-ovate, sharply and coarsely serrate, with very prominent veins ; flowers small, white, in smooth cymes ; fruit small, roundish, dark blue. A common shrub, 6-10 feet high. *June–July.*

Order XLIII.—Rubiaceæ (*Madder Family*).

Trees, shrubs, or herbs. Leaves opposite, sometimes verticillate, entire. Stipules present, interpetiolar, sometimes taking the size and appearance of leaves. Calyx-tube more or less adherent to the ovary. Limb 3–5-cleft, sometimes obsolete. Corolla regular, 3–5-lobed. Stamens equaling the number of the corolla-lobes, alternate with them, inserted in the tube. Ovary 2–5-celled. Style entire, or partially divided. Fruit of various forms.

RUBIACEÆ.
{
Leaves whorled, without stipules—*Galium.*

Leaves opposite, with stipules—
{
Low herbs—
{
Ovary-cells one-seeded—*Mitchella.*

Ovary-cells many-seeded—*Houstonia.*
}

Shrubs or trees—*Cephalanthus.*
}
}

1. Galium.—Calyx minute, with 3-4 teeth. Corolla rotate, 3-4-cleft. Stamens 3-4, short. Styles 2. Fruit of 2 united 1-seeded, indehiscent capsules. Stem 4-angled. ♃

1. G. ASPRELLUM (*Rough Cleavers*).—Stem weak, very branching, prickly backward, supporting itself by its prickles ; leaves in verticels of about 6 on the main stems, and 4 on the branches, oblong-lanceolate ; flowers numerous, minute, white ; fruit mostly smooth. In low grounds, 4-6 feet high. *July.*

2. G. TRIFIDUM (*Goose-grass*).—Stem slender, decumbent, or nearly erect, weak, rough backward ; leaves in whorls of 4-6, oblong-linear, with rough margins ; peduncles 1-3-flowered ; pedicels slender ; flowers white, the parts mostly in 3's, minute ; fruit smooth. In wet grounds, 4'-18' high. *June–August.*

3. G. TRIFLORUM (*Three-flowered Cleavers*).—Stem weak, procumbent, rough backward on the angles ; leaves mostly in 6's, oval-lanceolate, mucronate ; peduncles 3-flowered ; flowers pedicellate, greenish ; fruit hispid. In moist woods, 1-3 feet long. *July.*

2. Mitchella.—Flowers in pairs, with united ovaries. Calyx 4-parted. Corolla funnel-shaped, bearded within, 4-lobed. Stamens 4. Stigmas 4. Fruit a baccate drupe. ♃

M. REPENS (*Partridge-berry*).—Evergreen ; stem creeping ; leaves dark green, roundish-ovate, opposite ; flowers white, or tinged with rose, fragrant, 2 together, on a double ovary ; berries small, bright red, edible but dry. Creeping plant in woods. *June–July.*

3. Houstonia.—Calyx 4-parted, persistent, inserted on the corolla. Stigmas 2. Capsule 2-celled, many-seeded. ♃

H. CŒRULEA (*Bluets, Innocence*).—Smooth; stem slender, dichotomous; radical leaves spatulate; peduncles long, filiform, 1-2-flowered; corolla pale blue, fading to white, with a yellow base, somewhat salver-form. Little plant, 2'-8' high, in moist grounds. *April–September.*

4. Cephalanthus.—Calyx-limb 4-toothed. Corolla tubular, slender, 4-toothed. Stamens 4. Style filiform, much exserted. Stigma capitate. ♃

C. OCCIDENTALIS (*Button-bush*).—Leaves oval, entire, smooth, acute, petiolate; flowers in large, globose heads, white. Common in wet grounds. *July.*

Order XLIV.—Dipsaceæ.

Herbs. Leaves opposite, or verticillate, sessile. Stipules none. Flowers in dense involucrate heads. Calyx-tube adherent to the ovary. Limb somewhat campanulate, sometimes taking the form of a pappus. Corolla tubular, with a 4–5-lobed, slightly irregular limb. Stamens 4, distinct, rarely united in pairs, often unequal, inserted on the corolla. Ovary 1-celled, containing 1 ovule. Fruit a bony achenium.

Dipsacus.—Flowers in heads. Involucre many-leaved. Calyx-tube adhering to the ovary. Corolla tubular, 4-cleft. Stamens 4. Fruit 1-seeded, crowned with the calyx. ♃

D. SYLVESTRIS (*Teazel*).—Prickly; leaves lanceolate-oblong, opposite; heads cylindrical; bent inward; bracts terminating in a long, straight awn; flowers blue. A prickly plant, 2-4 feet high. *July.*

Order XLV.—Compositæ.

Herbs, or shrubs. Leaves alternate, or opposite, without stipules. Flowers arranged in dense heads, on a common receptacle, and surrounded by an involucre of bracts; the separate flowers often with chaffy bracteoles somewhat like a calyx. Calyx-tube adherent to the ovary; limb obsolete, or present, and assuming the various forms of bristles, hairs, scales, and is termed pappus. Corolla ligulate, or tubular, often 5-cleft, and rarely wanting. Stamens 5, their anthers united in a tube. Ovary 1-celled, 1-ovuled. Style 2-cleft. Fruit a dry, indehiscent achenium crowned with the pappus.

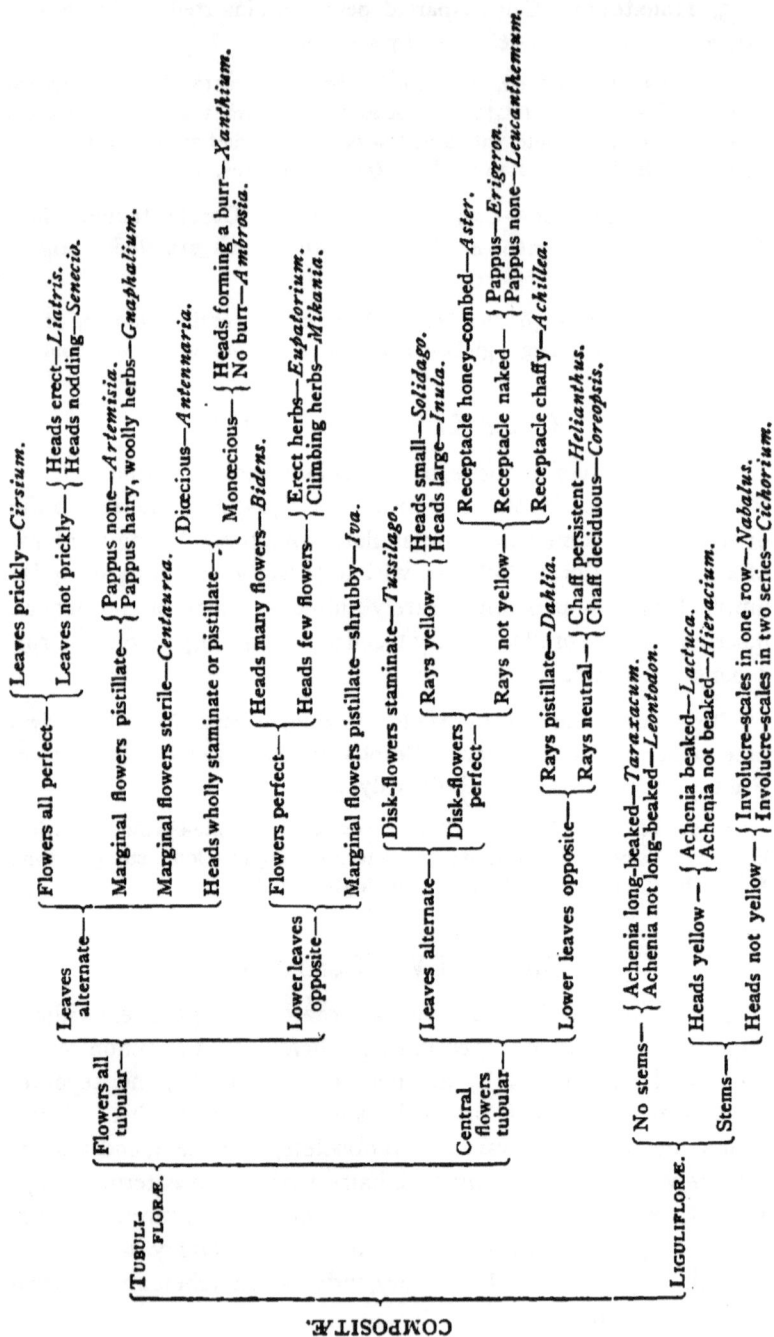

COMPOSITÆ.

- **TUBULIFLORÆ.**
  - **Flowers all tubular—**
    - **Leaves alternate—**
      - Flowers all perfect—
        - Leaves prickly—*Cirsium.*
        - Leaves not prickly—
          - Heads erect—*Liatris.*
          - Heads nodding—*Senecio.*
      - Marginal flowers pistillate—
        - Pappus none—*Artemisia.*
        - Pappus hairy, woolly herbs—*Gnaphalium.*
      - Marginal flowers sterile—*Centaurea.*
      - Heads wholly staminate or pistillate—
        - Diœcious—*Antennaria.*
        - Monœcious—
          - Heads forming a burr—*Xanthium.*
          - No burr—*Ambrosia.*
    - **Lower leaves opposite—**
      - Flowers perfect—
        - Heads many flowers—*Bidens.*
        - Heads few flowers—
          - Erect herbs—*Eupatorium.*
          - Climbing herbs—*Mikania.*
      - Marginal flowers pistillate—shrubby—*Iva.*
  - **Central flowers tubular—**
    - **Leaves alternate—**
      - Disk-flowers staminate—*Tussilago.*
      - Disk-flowers perfect—
        - Rays yellow—
          - Heads small—*Solidago.*
          - Heads large—*Inula.*
        - Rays not yellow—
          - Receptacle honey-combed—*Aster.*
          - Receptacle naked—
            - Pappus—*Erigeron.*
            - Pappus none—*Leucanthemum.*
          - Receptacle chaffy—*Achillea.*
    - **Lower leaves opposite—**
      - Rays pistillate—*Dahlia.*
      - Rays neutral—
        - Chaff persistent—*Helianthus.*
        - Chaff deciduous—*Coreopsis.*
- **LIGULIFLORÆ.**
  - **No stems—**
    - Achenia long-beaked—*Taraxacum.*
    - Achenia not long-beaked—*Leontodon.*
  - **Stems—**
    - Heads yellow—
      - Achenia beaked—*Lactuca.*
      - Achenia not beaked—*Hieracium.*
    - Heads not yellow—
      - Involucre-scales in one row—*Nabalus.*
      - Involucre-scales in two series—*Cichorium.*

1. Eupatorium.—Heads 3 to many-flowered. Involucre cylindrical, imbricate. Receptacle flat. Style much exserted. Pappus simple, roughish. Achenia 5-angled. ♃

1. E. PURPUREUM (*Trumpet-weed*).—Stem tall, simple; leaves broad-ovate to lanceolate, veiny, rough, toothed, 3–6 in a whorl; heads 5–10-flowered; scales of the involucre purplish, numerous, closely imbricated in several rows of unequal length; flowers light purple, in dense compound corymbs. Low ground, 2–10 feet high. *August–September.*

2. E. PERFOLIATUM (*Boneset*).—Stem erect, stout, rough, hairy; leaves lanceolate, connate-perfoliate; heads white, about 12-flowered, in a flat-topped, pubescent corymb; scales of the involucre linear-lanceolate. In low grounds. *August–September.*

2. Mikania.—Involucre about 4-leaved. Heads with about 4 flowers. Receptacle naked. Pappus simple, capillary, roughish. Achenia 5-angled. ♃

M. SCANDENS (*Climbing Mikania*).—Smooth; stem climbing; leaves opposite, cordate; heads in numerous, axillary, pedunculate corymbs; flowers pale pink or flesh-color. In wet thickets. *August–September.*

3. Liatris.—Heads few to many-flowered. Involucres imbricate, with appressed scales. Pappus abundant, more or less plumose. Achenia tapering at base. Styles much exserted. ♃

L. SCARIOSA (*Blazing Star*).—Stem erect, simple, rough; lower leaves lanceolate, on long petioles; upper ones linear, and smaller; heads large, purple, 20–40-flowered, in a long, terminal raceme; involucre somewhat hemispherical; scales obovate, very obtuse, with scarious and often purplish tips. In dry, sandy soils, 2–4 feet high. *August–September.*

4. Tussilago.—Heads many-flowered. Ray-flowers narrow, in many rows, pistillate. Disk-flowers few, staminate. Involucre mostly simple. Receptacle naked, flat. Pappus capillary. ♃

T. FARFARA (*Colt's-foot*).—Acaulescent; rhizoma creeping; leaves large, cordate, angular, toothed; scapes simple, 1-flowered, scaly, preceding the leaves; flowers yellow, with numerous, very narrow rays. *April.*

5. Aster.—Heads many-flowered. Involucre-scales generally imbricated, often with herbaceous tips. Ray-flowers pistillate, fertile, in a single row. Disk-flowers tubular, perfect. Receptacle flat, alveolate. Pappus simple, capillary. Achenia usually compressed. ♃

1. A. CORYMBOSUS (*Corymbed Aster*).—Stem slender, flexuous, smooth, with pubescent branches; leaves ovate, sharply and irregularly serrate, nearly smooth; lower and radical ones cordate; uppermost ovate-lanceolate, sessile; petioles naked; involucre of closely appressed, obtuse scales; rays 6–9, narrow. Dry woods and thickets. *August.*

2. A. MACROPHYLLUS (*Large-leaved Aster*).—Stem stout, branched, not flexuous; leaves rough, finely serrate; lower ones large, cordate, on long

petioles; upper ones ovate or oblong, sessile, or on winged petioles; heads in large, flat corymbs; rays 12–15, white or bluish; involucre with oblong, acute scales. In woods, 1–2 feet high. *August–September.*

3. A. CORDIFOLIUS (*Heart-leaved Aster*).—Stem erect, mostly smooth, with many divaricate branches above; leaves cordate, sharply serrate; heads racemed on the branches; rays 10–15, pale blue; involucre-scales appressed, with short green tips. In rocky woods, 2 feet high.

4. A. UNDULATUS (*Variable Aster*).—Pubescent and somewhat hairy; stem erect, panicled above; lower leaves cordate, on winged petioles; upper ones ovate or ovate-lanceolate, undulate or slightly serrate, on short, broadly margined petioles; all somewhat rough above, pubescent beneath; heads solitary, in somewhat unilateral racemes, arranged in a terminal panicle; rays pale blue. In dry woods and thickets. *August–September.*

5. A. PATENS (*Spreading Aster*).—Pubescent and somewhat rough; stem branching; leaves oblong-ovate, or oblong, sessile, cordate, and clasping the stem at base, rough above and on the margin, entire; heads large, with very showy violet-purple rays, solitary, on leafy branchlets, forming a loose, terminal panicle; involucre-scales lanceolate, with spreading, green tips; achenia silky. In dry fields, 2–3 feet high. *August–October.*

6. A. LÆVIS (*Smooth Aster*).—Very smooth and often glaucous; stem angular; leaves lanceolate, or ovate-lanceolate, somewhat fleshy, mostly entire; the upper ones somewhat cordate, or auriculate at base; the lower and radical ones tapering to a winged petiole; involucre-scales with broad-linear, appressed, green tips; heads large at the ends of the branchlets, with bright blue, showy rays, forming a terminal panicle. In low grounds, 2–3 feet high. *September–November.*

7. A. PUNICEUS (*Rough Aster*).—Stem erect, very branching, pubescent, rough, paniculate above; leaves lanceolate, auriculate, and clasping at base, slightly serrate, pubescent; heads large, with very numerous and narrow, pale-purple rays, showy, forming a very large and leafy panicle; involucre-scales narrow-linear, long and revolute in 2 rows. Swamps and low grounds. *September–October.*

8. A. NOVÆ ANGLIÆ (*New England Aster*).—Stem stout, hispid, paniculate above; leaves lanceolate, entire, acute, auriculate and clasping at base, somewhat pubescent, thickly clothing the stem; heads large, with numerous, deep-purple rays, somewhat paniculately corymbose. In moist grounds. *September.*

9. A. TRADESCANTI (*Narrow-leaved Aster*).—Nearly or quite smooth; stem terete, with virgate, erect, spreading, or diverging branches; leaves linear-lanceolate, the lower ones commonly serrate in the middle, the others entire; heads very numerous, rather small, mostly with rays, densely racemose on the branches; involucre-scales narrow-linear. In moist fields. *August–October.*

10. A. MULTIFLORUS (*Many-flowered Aster*). — Pubescent and somewhat rough; stem very branching; leaves linear, crowded on the stem, entire, sessile, those of the branches much smaller; heads small, with white rays, very numerous, densely racemose on the spreading branches; involucre-

scales linear-spatulate, with spreading, green tips. Dry soils, 2 feet high. *September.*

6. Erigeron.— Heads many-flowered, somewhat hemispherical. Rays narrow, very numerous, pistillate. Disk-flowers perfect. Receptacle flat, naked.

1. E. PHILADELPHICUM (*Purple Fleabane*).—Hairy ; stem slender, leafy ; leaves thin, oblong, clasping at base, mostly entire ; heads with exceedingly numerous and narrow, reddish-purple, or flesh-colored rays, broadly corymbed at the summit of the stem. In fields, 1–3 feet high. *June–August.*

2. E. ANNUUM (*Daisy Fleabane*).—Stem tall, furrowed, rough, pubescent, branching ; leaves hairy, closely serrate, the lowest ovate ; upper ones ovate-lanceolate, crowded, acute, tapering at base, sessile, the uppermost lanceolate ; heads large, with very numerous, narrow, and short white rays tinged with purple, corymbose at the summit of the stem. A tall plant, growing as a weed in fields and waste places. Very common. *August.*

3. E. CANADENSE (*Fleabane*). — Hairy ; stem erect, furrowed, with numerous short branches ; leaves linear-lanceolate, radical ones incised ; heads very numerous, small, with numerous white rays scarcely longer than the involucre, racemose on the branches. In waste places, 6'–6 feet high. *July–October.*

7. Dahlia.—Heads many-flowered. Disk-flowers pistillate. Involucre double. Outer scales, double series. Receptacle chaffy. No pappus. ♃

D. VARIABILIS. (*Dahlia*).—Stem smooth, green ; leaves pinnate, opposite ; leaflets about 5, ovate ; outer involucre reflexed. Very common in cultivation.

8. Solidago.— Heads few or many-flowered. Disk-flowers perfect. Involucre with imbricated, appressed scales. Receptacle small. Pappus simple, capillary. Achenia nearly round. Heads, with 1 exception, yellow. ♃

1. S. BICOLOR (*White-rayed Golden-rod*).—Pubescent ; stem mostly simple ; leaves oblong, or oblong-lanceolate, acute at each end ; lower ones oval and petiolate, slightly serrate ; heads in numerous, erect, densely flowered, axillary, racemose clusters, forming a long, terminal, interrupted spike ; involucre-scales ovate, obtuse ; rays short, pale cream-color, or white, about 8 in number. Dry fields and woods. *August–September.*

2. S. GIGANTEA (*Large Golden-rod*).—Stem stout, smooth ; leaves lanceolate, nearly or quite smooth on both sides, acuminate, sharply serrate, tapering and entire at base, and ciliate on the margin ; panicles large, with pubescent branches. In low grounds, 4–6 feet high. *August–October.*

3. S. ODORA (*Spicy Golden-rod*).—Nearly or quite smooth ; stem slender, erect, or reclined ; leaves linear-lanceolate, entire, shining above, very smooth, fragrant, with pellucid dots ; heads small, with 3–4 rays, in short, spreading racemes, forming rather small, unilateral panicles. *July–September.*

4. S. ARGUTA (*Sharp-toothed Golden-rod*).—Smooth ; stem erect, thick, furrowed ; leaves sharply serrate, with diverging teeth, acuminate, tapering at base ; lower and radical ones oval-lanceolate or lanceolate, attenuate to marginal and ciliate petioles ; cauline ones lanceolate, or oblong, the highest entire.

9. **Inula.**—Heads many-flowered, involucre imbricated. Rays numerous in one row. Disk-flowers perfect. Receptacle naked. Pappus capillary. ♃

HELENIUM. — Stem stout ; leaves ovate, serrate ; heads very large, solitary, terminating the branches ; rays linear, light yellow, 2–3-toothed at apex ; involucre-scales ovate, foliaceous. Common by roadsides, 4–6 feet high. *July–August.*

10. **Helianthus.**—Heads many-flowered. Rays neutral. Disk-flowers perfect. Involucre-scales imbricated in several rows. Chaff persistent with the 4-sided, laterally compressed achenia. Pappus of 2 deciduous, chaffy awns. ♃

1. H. ANNUUS (*Sunflower*).—Leaves cordate, petiolate, 3-veined, the lowest opposite the others, alternate ; heads very large, on nodding peduncles ; rays numerous, broad, bright yellow. In cultivation, 8–10 feet high. *July–September.*

2. H. DECAPETALUS (*Ten-rayed Sunflower*).— Stem tall, branching, rough above, smooth below ; leaves opposite, ovate, acuminate, coarsely serrate, 3-veined, of the same color on both sides, abrupt at base, with winged petioles ; heads rather large, with about 10 pale-yellow rays ; involucre-scales linear-lanceolate, spreading. Along river-banks, 2–5 feet high. *August–September.*

3. H. DIVARICATUS (*Slender Sunflower*).— Stem smooth, simple, or sparingly branched ; leaves opposite, ovate-lanceolate, 3-veined, sessile, serrate, rough above ; heads small, few, somewhat corymbose ; involucre-scales lanceolate, acuminate. In dry grounds, 2–5 feet high. *August–September.*

11. **Achillea.**—Heads many-flowered. Rays 5–10, pistillate. Involucre with imbricate, unequal scales. Receptacle flat, chaffy. Pappus none. ♃

A. MILLEFOLIUM (*Yarrow*).—Stem erect, furrowed, branching above ; leaves alternate, bi-pinnately divided, segments linear, toothed or lobed ; heads small, numerous, in dense, flat, terminal corymbs ; rays about 5, short, white (sometimes rose-color) ; involucre furrowed, oblong. In fields, 1 foot high. *June–August.*

12. **Leucanthemum.**—Heads many-flowered. Rays numerous, pistillate. Involucre depressed, flattish ; scales imbricated, with scarious margins. Receptacle naked, flat. Achenia striate. Pappus none. ♃

L. VULGARE (*Ox-eye Daisy*).—Stem erect, simple, or sparingly branched, furrowed. Leaves few and rather small, cut pinnatifid, incised at base ;

heads large, solitary, on long, naked, furrowed peduncles ; rays white ; disk yellow ; involucre-scales with brownish margins. *July–September.*

13. Coreopsis.—Heads many-flowered. Rays about 8 ; neutral. Involucre double. Each series 6–10-leaved. Receptacle flat and chaffy. Achenia compressed, emarginate, 2-awned.

C. TINCTORIA (*Coreopsis*).—Smooth ; stem erect, branching ; radical leaves somewhat bi-pinnate, segments oval, entire ; cauline ones somewhat pinnate, with linear segments ; heads large, numerous, brilliant ; rays bright yellow, brownish purple at base ; achenia smooth. In cultivation, 1–3 feet high.

14. Bidens.—Heads many-flowered. Rays neutral. Involucre double. Outer series large and foliaceous. Receptacle chaffy, flat. Achenia compressed, or slender and 4-sided, armed with 2–4 rigid, persistent awns. ①

B. FRONDOSA (*Beggar-Ticks*).— Smooth, or slightly pubescent ; stem erect, tall, with spreading branches ; leaves pinnately 3–5 parted, segments lanceolate, acuminate, serrate, mostly petiolate ; leaflets of the outer involucre much longer than the flower, ciliate at base ; heads discoid ; achenia flat, cuneate-obovate, 2-awned, with rough margins. A troublesome weed in moist grounds. *July–September.*

15. Senecio.—Heads many-flowered, either discoid, with tubular, perfect flowers, or radiate, with pistillate rays. Involucre-scales mostly in a single row. Receptacle flat, naked. Pappus simple.

1. S. VULGARIS (*Common Groundsel*).—Stem erect, branching, angular, mostly smooth ; leaves pinnatifid, toothed, clasping ; radical ones petiolate ; heads discoid, terminal, yellow, in loose corymbs, nodding. In waste grounds, 15′ high.

2. S. AUREUS (*Golden Senecio*).—Mostly smooth ; stem furrowed, erect, nearly simple ; radical leaves undivided and roundish, mostly cordate, crenate, on long petioles ; lower cauline ones lyrate ; upper lanceolate, pinnatifid, sessile ; heads large, showy, with golden-yellow rays, somewhat umbellate, in flat, terminal corymbs ; involucre-scales linear, acute. In meadows, 10′–20′ high.

16. Artemisia.—Heads discoid, few, or many-flowered. Flowers all tubular. Involucre-scales imbricate, dry and scarious on the margins. Receptacle flat, naked, or slightly hairy. Achenia with a small disk at summit. Pappus none.

1. A. VULGARIS (*Mugwort*).—Stem erect ; leaves whitish tomentose beneath ; cauline ones pinnatifid, with linear-lanceolate, entire, or incised lobes ; heads few, erect, nearly sessile, purplish, racemose, forming a loose, leafy, terminal panicle ; involucre tomentose. Plant 2–3 feet high. *July–August.*

2. A. ABSINTHIUM (*Wormwood*).—Stem erect, furrowed, very branch-

ing, somewhat shrubby, covered with white, silky down; leaves bi- or tripinnately parted; segments lanceolate, obtuse, often incised; heads very
numerous, yellowish, nodding, racemose on the branches, forming a large,
leafy panicle. *August.*

17. Gnaphalium.—Heads many-flowered, discoid; outer flowers
pistillate and slender; central ones perfect. Involucre-scales imbricated, scarious, white or colored. Receptacle flat, naked. Pappus
simple, rough, capillary.

1. G. POLYCEPHALUM (*Life Everlasting*). — Stem erect, branching,
covered with cottony down; leaves linear-lanceolate, tapering at base, sessile, white downy beneath, nearly smooth above; heads in dense clusters at
the summit of the branches, corymbose, fragrant; involucre-scales ovate,
acute, whitish; flowers yellowish. In fields, 1-2 feet high. *August-September.*

2. G. ULIGINOSUM (*Cudweed*).—Woolly; stem low, diffusely branched;
leaves lanceolate or linear; heads small, in sessile, terminal, crowded, leafy
clusters; involucre-scales oblong, yellowish. In low grounds, 3'-6' high.
August-September.

18. Antennaria.—Heads many-flowered, diœcious; pistillate heads
with filiform corollas. Involucre-scales imbricated, scarious, white or
colored. Receptacle not chaffy. Pappus simple, bristly. ♃

1. A. MARGARITACEA (*Pearl Everlasting*).—Stem erect, leafy, white-
downy, corymbose above; leaves linear-lanceolate, sessile; heads in a terminal, flat corymb; involucre-scales elliptic, obtuse, pearly white; flowers
yellowish. In fields and pastures, 1-2 feet high. *August.*

2. A. PLANTAGINIFOLIA (*Mouse-ear Everlasting*).—Stoloniferous; stem
simple, downy; leaves white and silky when young, at length green above
and hoary beneath; heads small, aggregate in a dense, terminal corymb;
involucre-scales mostly white; outer ones more or less obtuse. In old fields
and pastures, 3'-6' high. *April-July.*

19. Xanthium.—Sterile and fertile flowers in different heads upon
the same plant. Sterile involucre imbricated; receptacle chaffy. Fertile involucre closed, 2-leaved, covered with hooked prickles, 2-flow-
ered. ①

X. STRUMARIUM (*Clot-weed*).—Stem erect, unarmed, branching; leaves
cordate, 3-5-lobed, dentate, rough; fruit oval, with 2 straight beaks. Stem
2-4 feet high. *August-September.*

20. Ambrosia.—Sterile and fertile flowers in different heads upon
the same plant. Sterile involucre hemispherical. Staminate flowers
5 or more, funnel-form. Fertile involucre 1-leaved, closed, 1-flow-
ered. ①

1. A. TRIFIDA (*Great Rag-weed*).—Hairy, rough; stem tall, square,
stout, usually branching; leaves large, opposite, usually 3-lobed, serrate;

sterile flowers in long, naked racemes ; fertile flowers sessile below, each with a 6-ribbed involucre, terminating in 6 tubercles. A tall, rank, herbaceous plant. In low grounds. *August.*

2. A. ARTEMISIÆFOLIA (*Roman Wormwood*).—Stem erect, slender, branching, more or less hairy ; leaves bipinnatifid, nearly smooth above ; lower ones opposite, upper ones alternate ; sterile racemes naked, terminal, loosely panicled ; fertile flowers sessile in the axils of the upper leaves. Homely weed, 2–4 feet high. *August–September.*

21. **Iva.**—Heads discoid ; marginal flowers 1–5, pistillate, with a tubular corolla. Involucre-scales few, mostly in one row. Receptacle hairy. Achenia obovoid, obtuse. Pappus none. ♃

1. FRUTESCENS (*Marsh Elder. Highwater Shrub*).—Nearly or quite smooth ; stem shrubby, with opposite branches ; leaves oval or lanceolate, coarsely serrate, with 3 prominent veins, petiolate, the lower ones opposite, the upper alternate, narrow ; heads greenish-white, on short, recurved pedicels, in long, axillary racemes, arranged in a long, leafy, terminal panicle. Along the sea-shore, 3–6 feet high. *August.*

22. **Centaurea.**—Heads many-flowered. Flowers all tubular ; marginal ones mostly enlarged and sterile, resembling ray-flowers. Involucre imbricate. Receptacle bristly. ①

C. CYANUS (*Bachelor's Button*).—Stem erect, branching, downy ; leaves linear, sessile, downy ; heads solitary, ovoid ; sterile flowers longer than the disk ; involucre-scales fringed. A garden annual. *July.*

23. **Cirsium.**—Heads discoid, many-flowered. Flowers all perfect. Involucre-scales imbricated in many rows. Receptacle bristly. Pappus of capillary, plumose bristles in a ring. Achenia oblong, compressed, smooth.

1. C. ARVENSE (*Canada Thistle*).—Root creeping ; stem erect, branching ; leaves pinnatifid, with spiny teeth, smooth or somewhat downy beneath ; heads small, light purple, numerous, in a terminal, loose panicle ; involucre-scales closely appressed, tipped with minute spines. A weed, spreading rapidly by its creeping roots. *July.*

2. C. LANCEOLATUM (*Common Thistle*).—Stem very branching, leafy ; leaves deeply pinnatifid, decurrent, hispid above, white and woolly beneath ; heads large, numerous, purple ; involucre-scales spreading, lanceolate, tipped with long and formidable spines. *July–September.*

3. C. PUMILUM (*Pasture Thistle*).—Stem very stout, hairy, nearly or quite simple ; leaves pinnatifid, clasping at base, green on both sides ; segments variously lobed and cut, ciliate, spinose ; heads very large, 1'–3' in diameter, fragrant, about 1–3 in number, purple, rarely whitish ; involucre-scales spinous. *August.*

24. **Cichorium.**—Heads many-flowered. Involucre double ; outer of 5 short, leafy scales ; the inner of 8–10. Receptacle chaffy. Pappus short, chaffy. Achenia striate. ♃

C. INTYBUS (*Succory*).—Stem terete, with several long, nearly simple branches; upper ones oblong or lanceolate, more or less clasping at base, slightly dentate or entire, small and inconspicuous; heads 2–3 together, axillary, sessile, light blue, showy; corolla 5-toothed. *July–September.*

25. Leontodon.—Heads many-flowered. Involucre scarcely imbricated, with several small scales at base. Receptacle naked. Pappus plumose, persistent. Achenia striate, somewhat rostrate. ♃

L. AUTUMNALIS (*Autumn Dandelion*).—Acaulescent; scape smooth, branching; leaves radical, lanceolate, pinnatifid; peduncles scaly, thickened upward; heads yellow, on separate peduncles, resembling somewhat those of the dandelion. *July–November.*

26. Lactuca.—Heads several-flowered. Involucre-scales in 2 or more rows. Pappus copious, fugacious, soft and capillary. Achenia compressed, with long, filiform beaks.

1. L. ELONGATA (*Wild Lettuce*).—Stem tall, stout, usually leafy, branching above; leaves smooth, paler beneath, entire, sessile; heads small, numerous, racemose on the branches, forming a long, naked panicle; corolla yellow, rarely purple; achenia oval. Plant, abounding in a milky juice, 2–6 feet high. In rich, moist soils.

2. L. SATIVA (*Garden Lettuce*).—Stem smooth, branching, corymbose above; leaves more or less orbicular, very smooth; cauline ones cordate; heads numerous, small, yellow, in terminal corymbs. Universally cultivated. *June–July.*

27. Hieracium.—Heads many-flowered. Involucre-scales usually more or less imbricate. Achenia striate, oblong, more or less rostrate. Pappus of tawny, fragile, capillary bristles, in a single row. ♃

1. H. CANADENSE (*Canadian Hawkweed*).—Stem leafy, somewhat pubescent; leaves lanceolate, or oblong-ovate, acute, dentate with coarse and acute teeth, sessile; heads large, on hairy peduncles, forming a terminal, paniculate corymb; involucre-scales linear, imbricated. In dry woods, 1–2 feet high. *August.*

2. H. VENOSUM (*Veiny Hawkweed*).—Mostly acaulescent; stem or scape slender, smooth, often with 1 or 2 leaves, branching, loosely corymbose above; radical leaves obovate, sometimes oblong, nearly entire, on very short petioles, hairy above, ciliate, marked with purplish veins; heads rather small, solitary, bright yellow. In dry woods, 1–2 feet high. *July–August.*

3. H. SCABRUM (*Rough Hawkweed*).—Stem leafy, nearly simple, rough, hairy; leaves hairy, nearly entire; lower ones obovate, slightly petiolate, upper ones oval, sessile; peduncle thick, glandular-hispid; heads 40–50-flowered. In dry soils, 15′–24′ high.

28. Nabalus.—Involucre cylindrical, with 5–10 linear scales in one row, and a few scales at base. Receptacle naked. Pappus-bristles copious, capillary, brownish or straw-color, in two series. Achenia striate, linear-oblong, without beaks. ♃

1. N. ALBUS (*White Lettuce*).—Stem tall, smooth, somewhat glaucous, corymbosely paniculate above; leaves angular-hastate, often 3–5-lobed; upper ones ovate, dentate, or oblong, entire; involucre purplish; heads 8–12-flowered in nodding racemes; pappus brownish. In woods, 2–4 feet high. *August.*

2. N. ALTISSIMUS (*Tall White Lettuce*).—Smooth; stem slender, tall; leaves triangular-ovate, cordate, petiolate, variously lobed and toothed; petioles naked or margined; involucre of 5 greenish scales; heads nodding, 5–6-flowered in racemes, forming a long, leafy, virgate panicle; pappus dirty white or pale straw-color. In woods, 3–6 feet high. *August.*

29. Taraxacum.—Heads many-flowered. Involucre double; outer series of short scales. Receptacle naked. Achenia oblong, with a long, filiform beak crowned with the white, copious, capillary pappus. ♃

T. DENS-LEONIS (*Dandelion*).—Acaulescent; smooth or nearly so; scapes several, hollow, naked, 1-flowered; leaves runcinately toothed; heads large, erect, yellow; outer involucre reflexed. A common plant in fields and pastures. *April-October.*

Order XLVI.—CAMPANULACEÆ (*Campanula Family*).

Herbs with a somewhat milky juice. Leaves alternate. Stipules none. Flowers usually blue and showy. Calyx-tube adherent to the ovary; limb usually 5-cleft, persistent. Corolla regular, campanulate, usually 5-lobed, withering. Stamens 5, distinct, inserted on the calyx, alternate with the 5 lobes of the corolla. Anthers 2-celled. Ovary 2–5-celled. Capsule crowned with the persistent calyx-tube, opening with loculicidal dehiscence, many-seeded.

CAMPANULACEÆ. { Corolla regular— { Campanulate—*Campanula.* / Rotate—*Specularia.* / Corolla irregular—*Lobelia.*

1. Campanula.—Calyx 5-cleft. Corolla mostly campanulate, 5-lobed. Stamens 5, broad at base. Stigmas 3–5. Capsule 3–5-celled, opening laterally by pores.

1. C. ROTUNDIFOLIA (*Harebell*).—Stem slender, branching, weak; radical leaves ovate, or roundish, cordate, crenate, on long petioles, soon withering and disappearing; cauline narrow-linear, entire, smooth; flowers bright blue, nodding; corolla twice as long as the subulate calyx-teeth. A foot high, growing on rocky banks. *July-September.* ♃

2. C. APARINOIDES (*Prickly Bell-flower*).—Stem slender, weak, branching, 3-angled, the angles rough backward; leaves linear-lanceolate, denticulate, rough backward on the margin and veins; flowers small, nearly white, solitary; corolla much longer than the triangular calyx-teeth. In low grounds. *June-August.*

3. C. AMERICANA (*American Bell-flower*).—Stem erect, virgate, nearly simple ; leaves ovate-lanceolate, tapering at both ends, serrate, slightly hairy, with ciliate petioles ; the lowest sometimes cordate ; flowers large, blue, nearly rotate, deeply cleft, axillary, sessile, solitary, or several together ; calyx-teeth subulate, shorter than the corolla. In cultivation. *July–August.*

4. C. MEDIUM (*Canterbury Bell*).—Stem erect, simple, hispid ; leaves lanceolate, obtusely serrate, sessile, with 3 veins at base ; flowers very large, broad at base, with a reflexed limb, deep blue, erect. In gardens.

2. **Specularia.**—Calyx 5-lobed. Corolla rotate, 5-lobed. Stamens with hairy filaments. Style hairy, included. Stigmas 3. Capsule prismatic, 3-celled, opening by 3 lateral valves. ①

S. PERFOLIATA (*Clasping Bell-flower*).—Stem erect, simple, somewhat pubescent ; leaves nearly orbicular, clasping and cordate at base, crenate ; flowers sessile, deep blue ; the upper ones only opening ; corolla with spreading segments ; calyx-segments acute, lanceolate. In dry, sandy fields, 8'–12' high. *June–July.*

3. **Lobelia.**—Calyx 5-cleft, with a short tube. Corolla tubular, irregular, deeply cleft on the upper side ; upper lip nearly erect, 2-cleft ; lower lip spreading, 3-cleft. Capsule 2-celled, opening at summit. Seeds minute, many.

1. L. CARDINALIS (*Cardinal Flower*).—Stem erect, simple ; leaves ovate-lanceolate, acute, serrate ; flowers deep scarlet, large, in a long, terminal raceme ; bracts linear, leaflike ; pedicels short ; corolla much longer than the calyx. In low grounds, 2–3 feet high. *July–September.*

2. L. SPICATA (*Slender Lobelia*).—Slightly pubescent ; stem erect, simple, slender ; radical leaves oblong, or spatulate, all but the uppermost dentate ; flowers small, pale blue, in long, spicate racemes ; bracts narrow-linear, nearly as long as the pedicels ; calyx-teeth as long as the corolla, subulate. In fields, 1–2 feet high. *July–August.*

INFERIOR MONOPETALOUS EXOGENS WITH
REGULAR FLOWERS.

Order XLVII.—ERICACEÆ (*Heath Family*).

Shrubs ; or evergreen or leafless herbs. Leaves simple, alternate, rarely opposite, often evergreen. Stipules none. Calyx-tube usually free from the ovary, sometimes adherent ; limb 4–6, usually 5-cleft, rarely entire. Corolla regular, or sometimes irregular, 4–6, usually 5-cleft, rarely with 5 distinct petals. Stamens inserted with the corolla. Anthers 2-celled, opening by pores, often appendaged at top. Ovary 2–10-celled. Style 1. Stigma 1. Fruit a berry, drupe, or capsule.

```
          ┌                      ┌ Ovaries 8-10-celled, anthers awnless—Gaylussacia.
          │ Calyx united to ovary—│
          │                      └ Ovaries 4-5-celled, anthers often awned—Vaccinium.
          │
          │                              ┌ Corolla bell-shaped, 5-lobed—Andromeda.
E         │                    ┌ Upright │
R         │                    │ shrubs—┤ Corolla not  ┌ Flower-buds scaly—Aza-
I         │          ┌ Shrubs or│         │ bell-shaped. │  lea.
C         │          │  trees— ┤         └ 5-toothed—  │
A         │          │         │                       └ No scaly buds—Kalmia.
C  ERICACEÆ │        │         │
E         │          │         │          ┌ Fruit a berry, corolla urn-shaped—Arc-
Æ         │          │         │ Trailing │   tostaphylos.
.         │  Calyx   │         └ shrubs— ┤          ┌ Calyx berry-like—Gaulthe-
          │  free— ─┤                    │ Fruit a  │  ria.
          │          │                    └ dry pod—┤
          │          │                              └ Calyx dry—Epigæa.
          │          │
          │          │                   ┌ Flowers racemed—Pyrola.
          │          │ Evergreen herbs—┤
          │          │                   └ Flowers corymbed—Chimaphila.
          │          │
          └          └ Fleshy herbs—Monotropa.
```

1. Gaylussacia.—Calyx 5-toothed. Corolla with a 5-cleft, reflexed limb. Stamens 10. Anthers awnless. Fruit a drupe resembling a berry, with 8–10 seeds. ♃

    1. G. FRONDOSA (*Dangleberry*).—Smooth, with terete, slender branches; leaves oblong-obovate, obtuse, entire, covered with minute, resinous dots; flowers in loose, bracteate racemes; corolla ovoid-campanulate, nearly globose, small, of a reddish-white color; berries large, blue, ripening late, covered when mature with a glaucous bloom, sweet and edible. In low woodlands, 3–5 feet high. *June.*

    2. G. RESINOSA (*Huckleberry. Whortleberry*).—Very branching; leaves oval, or oblong, entire, clammy with resinous dots when young, petiolate; flowers small, greenish, striped with red, covered with resinous dots, in short, clustered, drooping racemes; corolla ovoid-conic, contracted at apex; berries black, destitute of bloom, ripe in July and August. In woods and pastures, 1–4 feet high. *May–June.*

    2. Vaccinium.—Calyx 5-toothed. Corolla campanulate, or cylindrical. Limb 4–5-cleft, revolute. Stamens 8–10. Anthers often 2-awned. Berry 4–5-celled, many-seeded, sometimes apparently 8–10-celled. ♃

    1. V. MACROCARPON (*Cranberry*).—Evergreen; stem trailing, with erect branches; leaves oblong, obtuse, glaucous beneath, with slightly revolute edges; flowers rather large, on long pedicels; corolla deeply 4-parted, flesh-colored; berries on drooping pedicels, globular, bright scarlet, smooth, juicy, of a keen, acid taste, ripe in October. In boggy meadows. *June.*

    2. V. VACILLANS (*Blueberry*).—Shrub, with angular, green branches; leaves oval, or obovate, of a pale, dull green, smooth on both sides, glaucous beneath; flowers in dense, sessile racemes, on nearly naked branchlets; corolla yellowish or reddish-white; berries blue, large and sweet, ripe in July and August. · In open woods, 1–2 feet high. *May–June.*

3. V. CORYMBOSUM (*Swamp Huckleberry*).—Tall; leaves oblong, or oval-obovate, smooth on both sides, slightly pubescent beneath when young; flowers in short, sessile racemes; corolla large, white, or slightly tinged with red, cylindric, slightly contracted at the mouth; stamens included; berries large, deep blue, ripe in August and September. In swamps, 4–8 feet high. *May–June.*

3. **Arctostaphylos.**—Calyx 5-parted, persistent. Corolla ovoid; limb short, revolute, 5-toothed. Stamens 10. Drupe 5-seeded. ♃

A. UVA-URSI (*Bearberry*).—Stem woody, trailing; leaves evergreen, thick and leathery, obovate, entire, smooth and shining; flowers white, tinged with rose, in short, drooping racemes, terminating the branches; corolla bell-form, hairy inside; berry red, insipid. Mountains and hilly woods. *May–June.*

4. **Gaultheria.**—Calyx 5-cleft, with 2 bracts at base. Corolla ovoid-cylindric, with 5 short, revolute teeth. Stamens 10, hairy, included. Fruit 5-celled, 5-valved, inclosed in the fleshy lobes of the calyx.

G. PROCUMBENS (*Checker-berry, Partridge-berry*).—Stem creeping and throwing up simple, erect branches; leaves evergreen, obovate or oval, shining above, in tufts; flowers few, axillary, nodding; corolla white, contracted at the mouth; berry bright red, and together with the leaves of a pleasant, spicy flavor. In woods, 2'–4' high. *June–July.*

5. **Epigæa.**—Calyx 5-parted, with 3 bracts at base. Corolla salver form; tube hairy within. Stamens 10, with filiform filaments. Capsule 5-celled, 5-valved, many-seeded. ♃

E. REPENS (*Trailing Arbutus. May-flower*).—Stem trailing, clothed with long, rusty hairs; leaves evergreen, ovate, entire, with a bristly, reddish pubescence; flowers erect, in small clusters, very fragrant; corolla white, often tinged with rose-color. *April–May.*

6. **Andromeda.** — Calyx 5-parted, minute, persistent. Corolla ovoid-cylindric; limb with 5 reflexed teeth. Stamens 8–10, included. Capsule 5-celled, 5-valved, many-seeded.

A. LIGUSTRINA (*Panicled Andromeda*).—Leaves deciduous, obovate, or oblong-obovate, pubescent beneath, nearly entire; flowers small, dull white, in dense racemes; corolla subglobose, pubescent without; filaments pubescent. In low grounds, 3–5 feet high. *June.*

7. **Azalea.** — Calyx 5-parted, persistent. Corolla funnel-form, 5-lobed, with spreading, unequal lobes. Stamens 5–10. Capsule 5-celled, 5-valved.

1. A. NUDIFLORA (*Swamp Pink*).—Branchlets slightly hairy; leaves obovate, downy beneath; flowers large, pale pink or purple; calyx-teeth minute; stamens and style much exserted. In low grounds, 4–8 feet high. *May.*

2. A. VISCOSA (*White Swamp Pink*).—Branchlets hispid; leaves obovate, or oblong; flowers white, sometimes tinged with rose-color, in large

clusters, fragrant; corolla clammy, with viscid hairs; stamens slightly exserted; style much exserted. In low grounds. Stem 4-8 feet high. *June-July.*

8. Kalmia.—Calyx 5-parted. Corolla rotate-campanulate, 5-lobed, with 10 cavities inside, in which the anthers are lodged. Capsule globose, 5-celled, many-seeded. Evergreen shrubs.

1. K. LATIFOLIA (*High Laurel*).— Leaves oval-lanceolate, smooth; flowers in terminal corymbs, white, shaded with pink; peduncles clammy-pubescent; pedicels bracted. In dry thickets, 4-8 feet high. *June.*

2. K. ANGUSTIFOLIA (*Low Laurel. Sheep Laurel*).—Leaves opposite, or in threes, narrow-oblong, light green, paler beneath; corymbs lateral and axillary; flowers deep red; bracts minute, linear lanceolate, 3 at the base of each pedicel. In damp grounds, 2-4 feet high. *June-July.*

9. Pyrola.—Calyx 5-parted, persistent. Petals 5, concave, deciduous. Stamens 10. Filaments subulate. Anthers large, pendulous, opening by 2 pores at apex. Style long. Stigma 5-rayed, 5-tubercled at apex. Capsule 5-celled, 5-valved, many-seeded.

1. P. ROTUNDIFOLIA (*Round-leaved Pyrola*).—Leaves orbicular, thick and shining, entire; scapes 3-angled; flowers white, large, drooping, fragrant, in a long, terminal raceme; petals round-obovate. Woods, 6'-12' high. *July.*

2. P. ELLIPTICA (*Oval-leaved Pyrola*).—Leaves thin, elliptical, smooth, mostly larger than the marginal petioles; racemes many-flowered; flowers white, nodding, fragrant; calyx-teeth ovate, acute. In woods, 5'-10' high. *July.*

10. Chimaphila.—Petals 5, concave, spreading. Stamens 10. Style short and thick. Stigma broad, orbicular, obscurely 5-toothed on the margin. Capsule 5-celled, opening downward.

C. UMBELLATA (*Prince's Pine*).—Leaves wedge-lanceolate, tapering at base, serrate, coriaceous; flowers large, light purple or whitish, fragrant, 3-7 in a terminal corymb; anthers violet. *July.*

11. Monotropa.—Sepals 4-5, bractlike, deciduous. Petals 4-5, distinct, fleshy, gibbous at base. Stamens 8-10. Anthers 2-celled. Style columnar, hollow. Stigma disk-like, bearded at the margin. Capsule 4-5-celled, 4-5-valved. ①

M. UNIFLORA (*Indian Pipe*).—Plant smooth, fleshy, white throughout, scentless; stem low, simple, furnished with lanceolate scales instead of leaves, one-flowered; flower large, smooth inside and out, nodding at first, finally erect. In rich woods. *June-July.*

Order XLVIII.—PLUMBAGINACEÆ (*Leadwort Family*).

Herbs, or somewhat suffruticose. Leaves simple, alternate, or all radical. Flowers often on simple or branching scapes. Calyx tubular, 5-toothed, plaited, persistent. Corolla with the 5 stamens inserted opposite its lobes. Styles 5. Ovary 1-celled, free from the calyx. Fruit a 1-seeded utricle; or else opening by 5 valves.

Statice. — Flowers scattered, or loosely spicate in a compound corymb, 1-sided, 2–3-bracted. Calyx funnel-form, dry and membranous, persistent. Petals 5. Stamens 5, attached at base. Styles 5, distinct. Fruit indehiscent. ♃

S. LIMONIUM (*Marsh Rosemary*).—Leaves radical, lanceolate, or oblong-obovate, 1-veined, entire, mucronate below the tip, thick and fleshy, dull green, on long petioles; scape with withering sheaths, very branching, forming a large, flat-topped, compound corymb, of small, pale-blue flowers, which are sessile in secund spikes upon the branchlets. In salt marshes, a foot high. *August–October.*

Order XLIX.—PRIMULACEÆ (*Primrose Family*).

Herbs. Leaves opposite, verticillate, or alternate, or all radical. Stipules none. Calyx 4–5-cleft, usually persistent, nearly or quite free from the ovary. Corolla regular, 4–5-cleft. Stamens as many as the lobes of the corolla, and inserted opposite them. Ovary 1-celled, with a free, central placenta. Style 1. Stigma 1. Capsule many-seeded; the placenta attached only to the base of the cell.

PRIMULACEÆ. { Flower parts 7—*Trientalis.*
{ Flower parts 5—*Lysimachia.*

1. **Trientalis.**—Calyx and corolla mostly 7-parted. Stamens mostly 7. Filaments united in a ring at base. Capsule many-seeded. ♃

T. AMERICANA (*Chick Wintergreen*).—Stem low, slender, crowned by a whorl of leaves; flowers few, on very slender peduncles, projecting from among the leaves, white and starlike; sepals linear. In damp, rich woods, 6'–10' high. *May.*

2. **Lysimachia.**—Calyx 5-parted. Corolla 5-parted, rotate, with a very short tube. Limb 5-parted, spreading. Stamens 5. Capsule globose, 5–10-valved, opening at apex. ♃

L. STRICTA (*Upright Loosestrife*).—Stem erect, simple, or branching; leaves opposite, or in threes, lanceolate, tapering at both ends, smooth, punctate, sessile ; flowers numerous, on slender pedicels, whorled, the numerous whorls forming a long, cylindrical raceme ; pedicels nearly horizontal ; corolla yellow, spotted with purple ; capsule 5-seeded. In swamps, 1–2 feet high. *July.*

Order L.—Oleaceæ (*Olive Family*).

Trees, or shrubs. Leaves opposite, simple, or pinnate. Flowers perfect, or polygamous. Sepals united at base, persistent, sometimes none. Petals 4, united below, sometimes distinct, valvate in prefloration, sometimes none. Stamens 2. Anthers 2-celled. Ovary free, 2-celled. Ovules pendulous. Style 1. Stigma 1, or bifid. Fruit drupaceous, baccate, or a samara, usually 1-celled, 1–2-seeded, by abortion.

OLEACEÆ.
- Flowers perfect—
  - Fruit a pod—*Syringa.*
  - Fruit a berry—*Ligustrum.*
  - Fruit a drupe—*Chionanthus.*
- Flowers imperfect—*Fraxinus.*

1. **Syringa.**—Calyx small, with 4 erect lobes. Corolla salver-form; tube much longer than the calyx-limb, 4-cleft, with obtuse, spreading segments. Stamens short, included in the tube. Capsule 2-celled, 2-valved. Shrubs.

1. S. VULGARIS (*Lilac*).—Leaves cordate, entire, smooth, green on both sides; flowers light purple, large, fragrant, in dense thyrses; corolla-limb somewhat concave. Universally cultivated, 5–8 feet high. *April–May.*

2. S. PERSICA (*Persian Lilac*).—Leaves smooth, lanceolate or pinnatifid, green on both sides; limb of the corolla flat. A smaller and more delicate shrub than the last, frequent in cultivation, 3–6 feet high. *April–May.*

2. **Ligustrum.**—Calyx tubular, short, deciduous; with 4 minute teeth. Corolla funnel-form, 4-lobed; lobes spreading, obtuse. Stamens inserted on the corolla-tube. Stigma 2-cleft. Berry 2-celled, 2–4-seeded.

L. VULGARE (*Prim*).—Shrubby; leaves oblong-lanceolate, varying to obovate, acute, or obtuse, entire, smooth, dark green, on short petioles; flowers small, white, in dense panicles; anthers large, exserted; berries black, bitter. Used for hedges, 4–6 feet high. *May–June.*

3. **Chionanthus.**—Calyx small, persistent, 4-parted. Corolla in 4 long and linear divisions. Stamens very short, inserted at the base of the corolla. Style very short. Drupe fleshy, 1-celled, 1-seeded. Trees.

C. VIRGINICA (*Fringe-tree*).—Leaves oval-oblong, smooth or somewhat downy, petiolate, entire; flowers snow-white, on long pedicels, in racemes, forming drooping panicles; calyx smooth; segments of the corolla linear; drupes purple, covered with a bloom. *May–June.*

4. **Fraxinus.**—Flowers polygamous or diœcious, often perfect. Staminate flowers—calyx small, 4-cleft, or wanting; stamens usually 2.

Pistillate flowers—calyx and corolla as in the staminate ; style single ; stigma 2-cleft. Fruit a 1-2-celled samara, flattened, winged at apex. Trees.

1. F. AMERICANA (*White Ash*).—Leaflets 7–9, petiolate, oblong or oblong-ovate, acuminate, glaucous beneath, mostly smooth ; calyx present ; corolla wanting ; fertile flowers in loose panicles ; the barren in dense, contracted ones ; samara obtuse, narrow, spatulate, with a long, tapering base. *April–May.*

2. F. PUBESCENS (*Red Ash*).—Leaflets 7–9, petiolate, lanceolate or lance-ovate, soft-downy ; calyx present ; corolla wanting ; samara obtuse, abruptly tapering at base. *April–May.*

3. F. SAMBUCIFOLIA (*Black Ash*).—Leaflets 7–11, sessile, ovate-lanceolate, serrate, hairy on the veins beneath ; calyx and corolla both wanting ; samara oblong, extremely obtuse at both ends. In moist woods and swamps. *May.*

Order LI.—APOCYNACEÆ (*Dogbane Family*).

Trees, shrubs, or herbs, with a milky juice. Leaves opposite, or verticillate, rarely alternate, without stipules. Flowers regular. Sepals 5, united, persistent. Corolla 5-lobed, twisted in prefloration. Stamens 5, alternate with the segments of the corolla. Filaments distinct. Anthers 2-celled, sometimes slightly connected. Ovaries 2, distinct, rarely united, but with 2 united styles or stigmas. Fruit usually a pair of follicles, 1 sometimes abortive. Seeds often with a coma, or tuft of hairs.

APOCYNACEÆ. { Shrubs—*Nerium.*
Herbs— { Upright herbs—*Apocynum.*
Trailing or creeping—*Vinca.*

1. **Apocynum.**—Calyx very small, 5-parted. Corolla campanulate, with 5 short lobes. Stamens 5, inserted at the base of the corolla. Anthers sagittate, converging, much longer than the very short filaments. Stigma ovoid, obscurely 2-lobed. Fruit 2 long, slender follicles.

1. A. ANDROSÆMIFOLIUM (*Dog's-bane*).—Smooth ; stem erect, branching above, reddened by the sun on one side, with diverging, forked branches ; leaves ovate, entire ; flowers in loose cymes ; corolla white, striped with rose-color, with 5 acute, revolute segments ; follicles 2'–3' long, nodding. In thickets, 2–3 feet high. *June–July.* ♃

2. A. CANNABINUM (*Indian Hemp*).—Stem erect, dividing above into long, ascending branches ; leaves oblong ; flowers very small, greenish-white, in dense, erect, many-flowered cymes, shorter than the leaves ; corolla-lobes nearly erect, the tube scarcely longer than the lanceolate calyx-teeth. In thickets, 2–3 feet high. *June–July.*

2. Vinca.—Corolla salver-form, contorted ; limb 5-cleft ; lobes oblique ; throat 5-angled. Ovary with 2 glands at base. Capsule follicular, erect.

V. MINOR (*Small Periwinkle*).—Evergreen ; stems procumbent, shrubby, terete, smooth, leafy ; leaves smooth and shining, elliptic-lanceolate ; flowers solitary, alternate, pedunculate, violet ; sepals lanceolate. In cultivation. *May.*

3. Nerium.—Calyx with 5 teeth at base. Corolla salver-form. Filaments inserted into the middle of the corolla-tube. Anthers sagittate, adhering to the stigma by the middle. ♃

N. OLEANDER (*Oleander*).—Evergreen, shrubby ; leaves linear-lanceolate, smooth, entire, 3 together, prominently veined beneath ; flowers large, in terminal clusters, rose-colored. In house cultivation, 4–6 feet high.

Order LII.—Asclepiadaceæ.

Herbs, or shrubs, usually with a milky juice. Leaves usually opposite, sometimes alternate or verticillate. Flowers generally in umbels, sometimes in racemes or corymbs. Sepals 5, slightly united at base. Corolla regular, consisting of 5 petals. Stamens 5, inserted at the base of the corolla, alternate with its segments united into a tube. Anthers 2-celled. Pollen cohering in masses. Ovaries 2. Styles 2, often very short. Stigmas united into 1 column for both ovaries. Fruit consisting of 2 follicles, 1 sometimes abortive. Seeds usually with a coma.

Asclepias.—Calyx 5-parted ; lobes small, spreading. Corolla deeply 5-cleft ; segments valvate in prefloration, reflexed when open, deciduous. Crown consisting of 5 hooded lobes, resting on the united mass of the stamens, and furnished with an incurved, horn-like process. Filaments united into a tube, inclosing the style. Anthers adhering to the stigma, with 2 cells opening longitudinally, each containing pollen-masses. Seeds flat, furnished with a long tuft of silky hairs. ♃

1. A. CORNUTI (*Common Milkweed*).—Stem erect, simple, rarely branching ; leaves nearly oval, tapering at both extremities, petiolate ; flowers in large, dense, simple, globose umbels, odorous ; calyx-segments lanceolate ; petals reflexed, dull purple ; horn short and stout. In rich soils, 3–5 feet high. *July.*

2. A. INCARNATA (*Swamp Milkweed*).—Nearly smooth ; stem erect, branching above, marked with 2 pubescent lines ; leaves oblong-lanceolate, obtuse at base, with distinct petioles ; umbels numerous, many-flowered, erect, often opposite ; peduncles half as long as the leaves ; segments of the corolla reddish-purple ; hoods of the crown flesh-colored, entire. In wet grounds, 2–3 feet high. *July–August.*

3. A. QUADRIFOLIA (*Four-leaved Milkweed*).—Smooth; stem erect, slender, simple; leaves ovate, smooth and thin, mostly in whorls of 4; umbels few, loose; pedicels filiform, marked with a pubescent line; segments of the corolla white, tinged with pink; hoods of the crown white, 2-toothed; horn stout and thick. In dry woods, 1-2 feet high. *July.*

Order LIII.—Gentianaceæ (*Gentian Family*).

Herbs, usually smooth, with a watery juice. Leaves usually opposite, rarely alternate, radical or single. Flowers regular, usually terminal or axillary, often showy. Calyx of 4-12 sepals, united at base. Corolla convolute, sometimes induplicate in prefloration, 4-12-parted, regular. Stamens as many as the segments of the corolla, inserted on the tube, alternately with them. Ovary 1-celled, free, sometimes apparently 2-celled, on account of the 2 introflexed placentæ. Style 1, or wanting. Stigmas usually 2, sometimes 1.

GENTIANACEÆ. { Leaves opposite, sessile— { Leafy plant—*Gentiana.* Leaves reduced to scales—*Bartonia.* } Leaves alternate, petioled— { Leaves simple—*Limnanthemum.* Leaves trifoliate—*Menyanthes.* } }

1. Gentiana.—Calyx 4-5-cleft. Corolla marcescent, regular, tubular at base; limb 4-5-cleft. Stamens 4-5, inserted on the corolla-tube. Stigmas 2, persistent. Capsule 1-celled, 2-valved, many-seeded.

1. G. CRINITA (*Fringed Gentian*).—Stem round, erect, branching; branches spreading at base; leaves lanceolate, sessile, cordate, or rounded at base; flowers of a rich blue, solitary, showy, terminating the branches; calyx 4-angled, 4-parted; corolla campanulate at base, open at summit, expanding when the sun shines; segments fringed on the margin. In low, grassy meadows, 10'-15' high. *October.*

2. G. ANDREWSII (*Soap-wort Gentian*).—Smooth; stem erect, simple; leaves lanceolate, acute or narrowed at base, 3-veined; flowers large, purplish-blue, in sessile heads; corolla inflated, club-shaped, closed at top. Stem 1-2 feet high. *September-October.*

2. Bartonia.—Calyx 4-parted. Corolla deeply 4-cleft; segments but slightly united, erect. Stamens short. Stigma large, persistent, at length 2-lobed. Capsule oblong, 1-celled, 2-valved. ①

B. TENELLA (*Screw-stem*).—Stem slender, erect, square, branching above; leaves minute, scale-like; flowers small, yellowish-white, 1-3 on the opposite branches; style none. Damp grounds, 3'-8' high. *August.*

3. Limnanthemum.—Calyx 5-parted. Corolla-tube short; limb 5-lobed; lobes deciduous, fringed merely at the base or margin. Style short, or none. Stigma 2-lobed, persistent. Capsule 1-celled, valve-less. ♃

L. LACUNOSUM (*Lake-flower*).—Floating ; stem filiform, bearing at top a single leaf, an umbel of flowers, and a tuft of short radicles ; leaves reni-form, floating at top, somewhat peltate, rough above ; flowers 5-6, in an umbel beneath the water ; corolla white ; lobes oval. 1-3 feet long. *July.*

4. **Menyanthes.**—Calyx 5-parted. Corolla funnel-form, 5-parted, deciduous. Stamens 5. Styles slender, persistent. Stigma 2-lobed, capitate, 1-celled. ♃

M. TRIFOLIATA (*Buckbean*).—Scape round, erect ; leaves radical, tri-foliate, on long petioles, with sheathing, membranous bases ; leaflets oval, varying to obovate, entire, sessile ; flowers in long, naked racemes ; corolla white or flesh-colored. Bogs, 1 foot high. *May.*

Order LIV.—POLEMONIACEÆ.

Herbs. Leaves opposite, rarely alternate, simple or compound. Calyx free from the ovary, 5-cleft, persistent. Corolla regular, with a 5-lobed limb, convolute in prefloration. Stamens 5, inserted on the corolla, alternately with its lobes, often unequal in length. Ovary 3-celled. Style 1. Stigma trifid. Capsule 3-celled, 3-valved, loculicidal, the valves separating from the 3-angled axis, which bears the seeds.

POLEMONIACEÆ. { Leaves entire—*Phlox.*
{ Leaves not entire— { Stamens declined—*Gilia.*
{ Stamens not declined—*Polemonium.*

1. **Phlox.**—Calyx somewhat prismatic, deeply 5-cleft. Corolla salver-form, with the slender tube more or less curved. Stamens very unequal, inserted in the corolla-tube above the middle. Capsule ovoid, 3-celled ; cells 1-seeded.

1. P. DIVARICATA (*Early Phlox*).—Low, diffuse, covered with minute down ; stems branching at base into a few, weak, ascending flowering-branches ; leaves oblong-lanceolate ; floral leaves narrow-linear ; flowers in terminal, loose corymbs ; corolla bright bluish-purple. In damp woods, 1-2 feet long. *May.*

2. P. DRUMMONDII (*Drummond's Phlox*).—Plant clothed with rough, glandular hairs ; stem erect, dichotomously branching ; leaves oblong or lan-ceolate ; flowers very showy, in dense, terminal cymes ; calyx hairy ; corolla varying from white to dark purple. In gardens, 8'-12' high.

3. P. SUBULATA (*Dwarf Phlox*)—Stems procumbent, tufted, clothed with minute down, very branching ; leaves rigid, or very narrowly linear, small, crowded, with fascicles of smaller ones in their axils ; cymes few-flow-ered ; corolla pink or rose-colored, rarely white. In gardens. *May.*

2. Gilia.—Calyx 5-cleft ; segments acute. Corolla-tube long or short ; limb regularly 5-lobed. Stamens 5, equal, inserted at the top of the tube. Capsule oblong or ovoid, few to many-seeded.

G. TRICOLOR (*Three-colored Gilia*).—Stem erect, nearly smooth ; leaves alternate, twice and thrice pinnatifid ; segments narrowly-linear ; flowers 3-6 together, in cymes, arranged in panicles, bractless. A garden annual, one foot high.

3. Polemonium.—Calyx campanulate, 5-cleft. Corolla rotate-campanulate, 5-lobed, erect ; tube very short. Filaments furnished with hairy appendages at base. Cells of the capsule few, many-seeded. ♃

P. CŒRULEUM (*Greek Valerian*).—Stems stout, clustered, smooth, simple, erect, hollow ; leaves mostly radical, alternate, in long, channeled petioles, pinnately parted ; flowers erect, in a terminal, corymbose panicle ; corolla blue, rather large. In cultivation, 1-2 feet high. *June.*

Order LV.—BORRAGINACEÆ (*Borage Family*).

Herbs ; sometimes shrubby plants. Stems round. Leaves alternate, usually rough. Flowers often in 1-sided clusters, unfolding spirally. Calyx free from the ovary, persistent, regular, consisting of 5 sepals, more or less united at base. Corolla regular, rarely irregular, the limb 5-toothed, often with a row of scales in the throat. Stamens 5, inserted on the corolla alternately with its lobes. Ovary deeply 4-lobed. Style 1, usually central, proceeding from base of the ovary, sometimes terminal. Fruit consisting of 4 achenia.

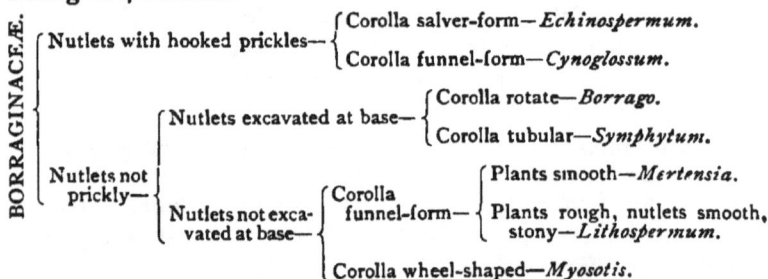

BORRAGINACEÆ.

Nutlets with hooked prickles—
- Corolla salver-form—*Echinospermum.*
- Corolla funnel-form—*Cynoglossum.*

Nutlets not prickly—
- Nutlets excavated at base—
  - Corolla rotate—*Borrago.*
  - Corolla tubular—*Symphytum.*
- Nutlets not excavated at base—
  - Corolla funnel-form—
    - Plants smooth—*Mertensia.*
    - Plants rough, nutlets smooth, stony—*Lithospermum.*
  - Corolla wheel-shaped—*Myosotis.*

1. Borrago.—Calyx 5-parted. Corolla rotate, 5-cleft ; segments acute ; tube with a crown at throat. Achenia rounded, with a perforation at base. ①

B. OFFICINALIS (*Borage*).—Rough with scattered bristles ; stem erect, branching ; leaves ovate ; lower with short petioles ; upper sessile ; flowers large, in nodding racemes, sky-blue ; calyx spreading. Common. In gardens, 2 feet high. *June-September.*

2. Symphytum.—Calyx 5-parted. Corolla tubular-campanulate, inflated above, 5-parted ; segments short, spreading. Stamens included in the corolla. Style filiform. Achenia smooth, perforated. ①

S. OFFICINALE (*Comfrey*).—Hairy ; stem erect, branching above ; lower leaves ovate-lanceolate ; upper ones decurrent ; flowers in 1-sided, nodding racemes ; corolla yellowish-white, occasionally pink or red ; sepals lanceolate. In low grounds, 3–5 feet high. *June–August.*

3. Lithospermum.—Calyx 5-parted, persistent. Corolla funnelform ; limb 5-lobed ; throat open at the orifice. Stamens included in the corolla. Anthers oblong. Achenia smooth or rugose. ①

L. ARVENSE (*Gromwell*).—Stem erect, slender ; leaves lanceolate, sessile, entire ; flowers in nodding racemes, which become erect and elongated ; lower flowers remote. In dry grounds, 6′–12′ high. *June–July.*

4. Mertensia.—Calyx 5-parted. Corolla-tube cylindric, expanding ; limb 5-lobed. Stamens inserted. Style long, filiform. Achenia smooth or somewhat wrinkled. ♃

M. VIRGINICA (*Virginian Lungwort*).—Smooth, stem erect, simple ; leaves obovate, ovate ; entire, pale green ; flowers large, in racemes, destitute of bracts, corolla brilliant purplish blue, rarely white. In rich woods, 10′–20′ high. *May.*

5. Myosotis.—Calyx 5-cleft. Corolla salver-form ; tube as long as calyx, 5-lobed ; throat with 5 short, concave scales. Achenia ovate, smooth, flattened.

M. PALUSTRIS (*Forget-me-not*).—Nearly smooth ; stem ascending, rooting near the base ; leaves linear-oblong, obtuse ; flowers small, in long, bractless, 1-sided racemes ; calyx in 5 short, spreading segments, open in fruit. In wet grounds, 6′–12′ high. *June–September.*

6. Echinospermum.—Calyx 5-parted. Corolla salver-form, short, closed at the throat by 5 short, concave scales. Achenia compressed, armed on the back with barbed prickles. ①

E. LAPPULA (*Burr-seed*). — Rough-hairy ; stem erect, very branching above ; leaves lanceolate ; flowers small, blue, in bracted racemes ; corolla longer than the calyx. In waste places, 1–2 feet high. *July.*

7. Cynoglossum.—Calyx 5-parted. Corolla funnel-form. Achenia depressed, affixed laterally to the base of the style, covered with short, hooked prickles.

1. C. OFFICINALE (*Hound's-tongue*).—Plant with a soft, silky pubescence ; stem erect, branching, leafy ; leaves lanceolate, acute ; upper ones clasping with a rounded base ; entire ; flowers large, in nearly bractless racemes ; corolla dull red or purplish. Road-sides, 1–2 feet high. *July.*

2. C. MORISONI (*Beggar's Lice*).—Hairy ; stem erect, very branching above, leafy ; leaves oblong-ovate, acuminate, remote, entire, thin, rough

13

above ; flowers very small, in leafy, bracteate, forking racemes ; pedicels nodding in fruit ; corolla minute ; achenia convex with hooked prickles. In thickets, 2–4 feet high. *July.*

Order LVI.—Convolvulaceæ (*Convolvulus Family*).

Herbs, or shrubs ; often with a milky juice. Stems trailing or climbing, rarely erect. Leaves alternate, sometimes none. Stipules none. Flowers often showy. Sepals 5, usually more or less united at base, persistent. Corolla regular, limb 5-cleft or entire, twisted and plaited in prefloration. Stamens 5, inserted at the base of the corolla, alternate with its segments, when lobed. Ovary 2–4, rarely 1-celled, free from the calyx. Style 1, rarely more. Fruit a capsule, 2–4-celled, opening by septifragal dehiscence. Seeds few, large.

CONVOLVULACEÆ. { With green herbage— { Calyx naked— { Stamens included, stigmas 2, linear—*Ipomœa.* Stamens protruded—*Quamoclit.* } Calyx inclosed in bracts—*Calystegia.* } Without green herbage—*Cuscuta.*

1. **Ipomœa.**—Calyx 5-parted, naked. Corolla campanulate, funnel-form ; limb with 5 plaits, and the border entire, or 5-lobed. Style 1, often 2-cleft at apex. Capsule 2–4-celled, 4–6-seeded.

I. purpurea (*Common Morning-glory*).—Stem twining, rough with reflexed hairs ; leaves cordate, entire ; peduncles ·elongated, 2–5-flowered ; pedicels thickened ; sepals hispid, ovate-lanceolate, acute ; corolla funnelform, large, 2′ long, with a spreading, entire border. *June–September.*

2. **Calystegia.**—Calyx 5-parted, included in 2 large, leafy bracts. Corolla funnel-form. Stamens nearly equal, shorter than the limb. Style 1. Stigmas 2. Ovary imperfectly 2–4-celled. Capsule 1-celled, 4-seeded. ♃

C. sepium (*Wild Morning-glory*).—Stem twining, mostly smooth ; leaves sagittate ; peduncles sharply 4-angled, 1-flowered ; bracts cordate, much longer than the concealed calyx ; flowers large, 2′ long, white, varying to pale rose-color. In low thickets, 5–10 feet long. *June–July.* ♃

3. **Quamoclit.**—Sepals 5, mostly mucronate. Corolla tubular cylindric. Stamens exserted. Style 1. Stigma capitate, 2-lobed. Ovary 4-celled ; cells 1-seeded. ①

Q. vulgaris (*Cypress Vine*).—Smooth ; stem very slender, twining ; leaves deeply pinnatifid ; segments linear, parallel, acute ; peduncles 1-flowered ; flowers small, brilliant, scarlet. In cultivation. *July–August.*

4. Cuscuta.—Calyx 5, rarely 4-cleft. Corolla globose-campanu-late ; border spreading, 5, rarely 4-cleft. Stamens 5, rarely 4. Stig-mas 2. Capsule 2-celled ; cells 2-seeded. ①

C. GRONOVII (*Dodder*).—Plant leafless, parasitic, destitute of all verdure ; stem filiform, orange-yellow ; flowers sessile, in dense clusters, white ; corolla campanulate, withering at the base of the globose capsule. *July-Sep-tember.*

Order LVII.—SOLANACEÆ (*Nightshade Family*).

Herbs, or shrubby plants with a colorless juice. Leaves alter-nate. Calyx free from the ovary, consisting of 4–5 persistent sepals, more or less united at base. Corolla regular, rarely slight-ly irregular, limb 4–5-cleft, plaited in prefloration. Stamens as many as the corolla-lobes, alternate with its segments. Ovary 2, and rarely 4 or 6-celled, with a central placenta. Fruit a many-seeded capsule or berry.

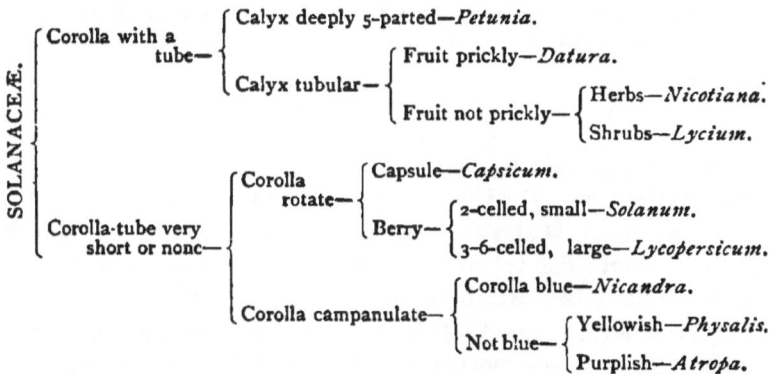

SOLANACEÆ.

{ Corolla with a tube—
  { Calyx deeply 5-parted—*Petunia.*
    Calyx tubular—
      { Fruit prickly—*Datura.*
        Fruit not prickly—
          { Herbs—*Nicotiana.*
            Shrubs—*Lycium.*

Corolla-tube very short or none—
  { Corolla rotate—
      { Capsule—*Capsicum.*
        Berry—
          { 2-celled, small—*Solanum.*
            3-6-celled, large—*Lycopersicum.*
    Corolla campanulate—
      { Corolla blue—*Nicandra.*
        Not blue—
          { Yellowish—*Physalis.*
            Purplish—*Atropa.*

1. Petunia.—Calyx with a short tube and a 5-cleft, leafy limb. Corolla salver-form ; tube cylindric ; limb in 5 unequal, flat, folded lobes. Stamens 5, unequal. Capsule 2-valved.

P. VIOLACEA (*Purple Petunia*).—Stem weak, hairy, viscid ; leaves broad-ovate, acute, on short, winged petioles, entire ; peduncles axillary ; sepals obtuse ; corolla-limb bright purple, divided into 5 unequal, rounded, acute lobes. In cultivation, 2-4 feet long. *July.*

2. Nicotiana.—Calyx urn-shaped, 5-cleft. Corolla funnel-form, or salver-form, regular ; limb plaited, 5-lobed. Stigma capitate. Cap-sule 2-celled, 2-4-valved. Seeds minute. ①

N. TABACUM (*Tobacco*).—Viscid-pubescent ; stem erect, paniculate above ; leaves very long, lanceolate, sessile, decurrent ; flowers dull rose-color. Stem 4-6 feet high. *July.*

3. Datura.—Calyx tubular, ventricose, 5-angled, 5-toothed. Corolla funnel-form, with a long-cylindrical tube ; limb plaited, 5-parted. Stigma 2-lipped. Capsule globular, prickly, 2-celled, 2-valved. Seeds large. ①

D. STRAMONIUM (*Thorn Apple*).—Smooth ; stem erect, fleshy, hollow, sometimes spotted with purple ; leaves large, ovate, irregularly dentate ; flowers large, 2'-3' long, dull white ; calyx-teeth acuminate ; fruit of the size and shape of a hen's egg, covered with short, sharp spines. Poisonous weed, in waste grounds, 1-3 feet high.

4. Nicandra.—Calyx 5-cleft, 5-angled ; angles compressed. Segments sagittate, enlarged in fruit. Corolla campanulate ; border open, plaited, nearly entire. Stamens 5, converging. ①

N. PHYSALOIDES (*Apple of Peru*).—Smooth, herbaceous ; stem erect, branching ; leaves large, broad-ovate ; flowers axillary, terminal, solitary, pale blue, white in the center, with 5 blue spots ; calyx closed, with the angles very acute. In gardens, 2-5 feet high. *August.* ①

5. Physalis.—Calyx 5-cleft, persistent, reticulated, inflated after flowering. Corolla spreading, campanulate, with a very short tube ; limb obscurely 5-lobed. Stamens 5, converging. ①

P. VISCOSA (*Yellow Henbane*).—Viscid-pubescent, branching, herbaceous ; leaves ovate, or lance-ovate, cordate or tapering at base, repandly toothed, or entire ; flowers nodding ; corolla greenish-yellow, with 5 brownish spots at the base inside ; fruit yellow or orange-color, inclosed in the inflated, angular calyx. Dry hills, a foot high. *July–August.*

6. Capsicum.—Calyx erect, 5-parted, persistent. Corolla rotate ; tube very short ; limb plaited, 5-lobed. Stamens converging. Capsule dry, inflated, 2-3-celled. Seeds flat, extremely acrid. ①

C. ANNUUM (*Red Pepper*).—Smooth ; stem herbaceous, angular, branching above ; leaves ovate-acuminate, petiolate, entire ; flowers nodding ; calyx angular, with 5 short, acute lobes ; corolla white, lobes spreading ; fruit oblong, red to yellow. Cultivated, 1-2 feet high.

7. Solanum.—Calyx mostly 5-parted, spreading, persistent. Corolla usually rotate ; tube very short ; limb mostly 5-cleft, plaited in the bud. Filaments very short. Anthers opening at top by 2 pores.

1. S. DULCAMARA (*Bitter-sweet*).—Stem shrubby toward the base, climbing, more or less smooth ; leaves ovate-cordate, hastate ; flowers in corymbose clusters ; corolla dull purple, the segments reflexed ; berries oval, scarlet, poisonous. In moist thickets, 4-6 feet long. *July.*

2. S. NIGRUM (*Nightshade*).—Smooth, herbaceous ; stem very branching, with rough angles ; leaves ovate, toothed and undulate ; flowers small, white, in drooping, lateral umbels ; anthers yellow ; berry globular, black. In waste grounds. *July–August.* ①

3. S. TUBEROSUM (*Potato*).—Rhizoma producing tubers ; stem ascend-

ing, herbaceous, nearly simple, with winged angles ; leaves interruptedly pinnate ; alternate leaflets much the smallest, all entire ; flowers dull-white, sometimes purplish, nodding, in terminal umbels, pedicellate. *June-July.*

4. S. PSEUDO-CAPSICUM (*Jerusalem Cherry*).—Evergreen ; stem shrubby, branching above ; leaves oblong-lanceolate, dark green, smooth and shining ; flowers solitary, nodding ; corolla white ; anthers orange ; berries globose, scarlet, as large as small cherries. Cultivated, 2–4 feet high.

5. S. MELONGENA (*Egg-plant*).—Stem prickly, herbaceous, branching ; leaves ovate, downy, prickly ; flowers small, whitish ; fruit large, ovate, varying from 2'–8' in length, smooth, glossy, purple. Cultivated, 2–3 feet high. *July-September.*

8. Lycopersicum.—Calyx mostly 5-parted, persistent. Corolla rotate ; tube very short ; limb mostly 5-lobed, plicate. Anthers converging, opening at top by 2 pores. Berry 3–6-celled.

L. ESCULENTUM (*Tomato*).—Hairy ; stem herbaceous ; leaves unequally pinnatifid ; segments incised, glaucous beneath ; peduncles bearing clusters of greenish-yellow flowers ; fruit torulose, furrowed, smooth, green at first, but bright red and juicy when mature. Stem 3–5 feet long.

9. Atropa.—Calyx persistent, 5-cleft. Corolla campanulate. Stamens distant. Berry globose, sitting on the calyx, 2-celled. ①

A. BELLADONNA (*Deadly Nightshade*).—Smooth, herbaceous ; stem branching below ; leaves large, ovate, entire ; flowers dull, lurid purple ; berries large, green at first, black when mature, full of purple juice ; stem 4 feet high. A poisonous plant. Gardens. *July-August.*

10. Lycium.—Calyx 2–5-cleft, short. Corolla tubular, limb mostly 5-lobed, spreading. Stamens 4–5. Filaments bearded, closing the throat of the corolla. Berry 2-celled. Seeds several, reniform.

L. VULGARE (*Matrimony Vine*).—Shrubby ; stem branching ; branches long, pendulous, ending in a spiny point, often furnished with axillary spines ; leaves lanceolate, often in clusters, smooth, acute or obtuse, tapering to a petiole ; flowers axillary, greenish-purple ; berries orange-red. In cultivation. *July.*

INFERIOR MONOPETALOUS EXOGENS WITH IRREGULAR FLOWERS.

Order LVIII.—SCROPHULARIACEÆ (*Figwort Family*).

Herbs, or sometimes shrubby. Leaves opposite, or alternate, sometimes verticillate or radical. Sepals 4–5, persistent, more or less united. Corolla bilabiate, personate, sometimes nearly regular, with 4–5 more or less unequal segments, the lobes imbricated in prefloration. Stamens didynamous, often with the rudiments of a 5th, which is sometimes perfect ; oftener still only 2. Ovary free, 2-celled. Style 1. Stigma 2-lobed. Capsule 2-valved, many-seeded.

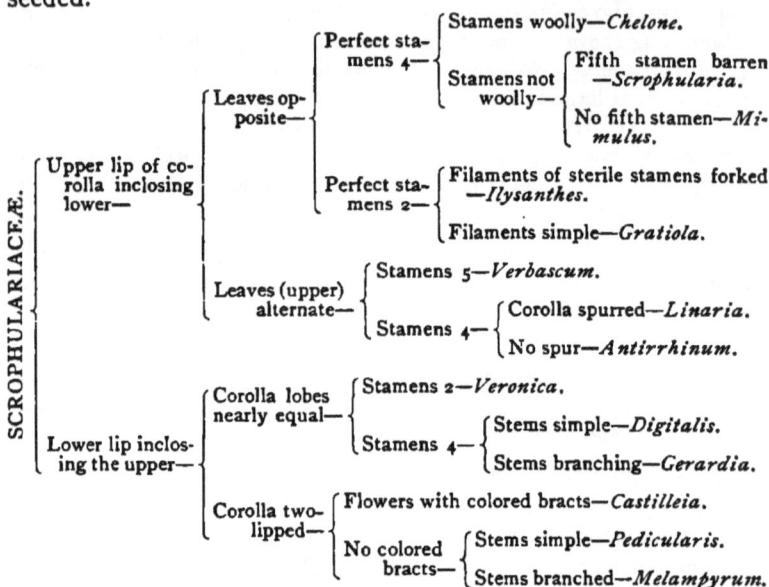

SCROPHULARIACEÆ.

Upper lip of corolla inclosing lower—

  Leaves opposite—
    Perfect stamens 4—
      Stamens woolly—*Chelone.*
      Stamens not woolly—
        Fifth stamen barren —*Scrophularia.*
        No fifth stamen—*Mimulus.*

    Perfect stamens 2—
      Filaments of sterile stamens forked —*Ilysanthes.*
      Filaments simple—*Gratiola.*

  Leaves (upper) alternate—
    Stamens 5—*Verbascum.*
    Stamens 4—
      Corolla spurred—*Linaria.*
      No spur—*Antirrhinum.*

Lower lip inclosing the upper—

  Corolla lobes nearly equal—
    Stamens 2—*Veronica.*
    Stamens 4—
      Stems simple—*Digitalis.*
      Stems branching—*Gerardia.*

  Corolla two-lipped—
    Flowers with colored bracts—*Castilleia.*
    No colored bracts—
      Stems simple—*Pedicularis.*
      Stems branched—*Melampyrum.*

1. Verbascum.—Calyx 5-parted. Corolla rotate, with 5 nearly equal lobes. Stamens 5, all perfect, declinate. Capsule globose or ovoid, many-seeded. ②

    1. V. THAPSUS (*Mullein*).—Plant densely woolly throughout ; stem tall, rigidly erect, usually simple ; leaves decurrent, oblong, acute ; flowers yellow, in a long, dense, terminal, cylindrical spike ; two lower stamens usually beardless. In neglected fields, 4–7 feet high. *June–August.*

    2. V. BLATTARIA (*Moth Mullein*).—Stem simple or branching above, leafy, erect ; leaves smooth, clasping, oblong, coarsely serrate, the lower petiolate ; flowers in a long, leafy raceme, yellow or white, usually tinged with purple ; stamens unequal, with purple, woolly filaments. Road-sides, 2–3 feet high. *June–July.*

2. Linaria.—Calyx 5-parted. Corolla personate ; upper lip bifid, reflexed ; lower lip 3-cleft ; palate prominent ; tube inflated and spurred. Stamens 4. Capsule opening at the summit by 1–2 pores.

    1. L. VULGARIS (*Toad-flax Snapdragon*).—Smooth and glaucous ; stem erect, with short, leafy branches ; leaves alternate, crowded, linear-lanceolate ; flowers yellow, in dense, terminal spikes ; corolla with a long spur, the throat completely closed by the orange-colored palate. Common along road-sides. *July–August.*

    2. L. CANADENSIS (*Canadian Snapdragon*).—Smooth ; stem slender,

erect, nearly simple ; leaves linear, erect, smooth, scattered, obtuse ; flowers small, blue, in an elongated, slender, terminal raceme, on short pedicels ; spur filiform, curved, as long as the corolla.   In sandy soils, 6'-15' high. *June-October.*

3. **Antirrhinum.**—Calyx 5-sepaled.   Corolla gibbous at base ; upper lip bifid, reflexed ; lower lip trifid, closed.   Capsule without valves, opening by 3 pores.   ♃

A. MAJUS (*Snapdragon*).—Stem erect ; leaves lanceolate, opposite ; upper ones alternate ; flowers in terminal racemes, pink, with the lip white, and the mouth yellow ; sepals lanceolate, acute, covered with glandular hairs. Gardens. *July-August.*

4. **Scrophularia.**—Calyx-segments 5, acute.   Corolla-tube subglobose ; limb contracted ; upper lip with 4 erect lobes ; lower lip spreading.   Stamens 4, declinate ; a 5th stamen as a scale on the corolla-tube.   Capsule 2-celled, many-seeded. ♃

S. NODOSA (*Figwort*).—Smooth ; stem tall, angular, branching ; leaves ovate or ovate-oblong, the upper ones varying to lanceolate, all acute, dentate or serrate, petiolate, mostly cordate or rounded at base ; flowers of a dull purple, in loose cymes ; calyx-teeth broad, obtuse, somewhat margined. In low grounds, 4–6 feet high. *July.*

5. **Chelone.**—Calyx-sepals distinct, with 3 bracts at base.   Corolla tubular, inflated, 2-lipped ; upper lip arched, emarginate ; lower lip bearded at the throat, 3-lobed.   Stamens with woolly filaments and anthers ; 5th filament sterile and smaller.   Seeds with broad, membranous margins. ♃

C. GLABRA (*Snake-head*).—Smooth ; stem erect, simple or branching ; leaves opposite, lanceolate, acuminate, serrate, on very short petioles ; flowers large, white, varying to rose-color, in dense, short spikes ; corolla with an open throat and contracted mouth ; style long, exserted.   In wet grounds, 2–3 feet high. *July-September.*

6. **Mimulus.**—Calyx prismatic, 5-toothed.   Corolla tubular, ringent ; upper lip erect, and reflexed at the sides, 2-lobed ; lower lip with a prominent palate, 3-lobed.   Stamens 4.   Stigma thick, 2-lipped. Capsule 2-celled, many-seeded. ♃

M. RINGENS (*Monkey-flower*).—Smooth ; stem erect, square, branching ; leaves sessile, oblong-lanceolate or lanceolate, acuminate, serrate ; flowers large, on solitary, square peduncles, curved upward ; corolla pale blue, with a yellow throat.   In wet places, 1–2 feet high. *July-August.*

7. **Gratiola.**—Calyx-segments nearly equal.   Corolla 2-lipped ; upper lip entire or 2-cleft ; lower lip without a prominent palate, 3-cleft. Fertile stamens 2.   Style dilated or 2-lipped at apex.   Capsule 2-celled, 4-valved, many-seeded. ♃

1. G. VIRGINIANA (*Virginian Hedge Hyssop*).—Nearly or quite smooth ; stem low, erect, simple or branching ; leaves lanceolate or oblong-lanceolate, sessile, opposite,'slightly serrate, tapering at base ; flowers small, on axillary peduncles ; corolla whitish, generally with a pale yellow tube ; sterile filaments none. In muddy grounds, 3'-8' high. *July–August.*

2. G. AUREA (*Golden Hedge Hyssop*).—Smooth ; stem decumbent at base, erect above, square, simple, or with ascending branches ; leaves oblong-lanceolate, nearly entire, sessile ; flowers solitary, golden yellow ; sterile filaments 2, minute. Common on the borders of ponds, 3'-8' high. *August–September.*

8. Ilysanthes.—Calyx 5-parted. Corolla 2-lipped ; upper lip short, erect, 2-cleft ; lower large, spreading, 3-cleft. Fertile stamens 2, included, posterior. Sterile stamens 2, anterior, forked. Style 2-lipped at apex. Capsule many-seeded. ♃

I. GRATIOLOIDES (*False Pimpernel*).—Smooth ; stem ascending, branching, low ; leaves opposite, sessile, ovate or oblong, sparingly serrate, more or less obtuse, the lower ones sometimes obovate and tapering at base ; flowers small, pale blue, solitary, on axillary, bractless peduncles ; corolla erect. In wet grounds, 2'-4' high. *July–August.* ♃

9. Digitalis.—Calyx 5-parted. Corolla campanulate, ventricose. Limb of 5 nearly equal lobes. Capsule ovate, 2-celled, 2-valved, with a double dissepiment. ②

D. PURPUREA (*Foxglove*).—Stem erect ; leaves oblong, rugose, downy, crenate, lower ones crowded, petiolate ; flowers large, crimson, beautifully spotted within, in a long, showy, 1-sided raceme ; calyx-segments ovate-oblong ; corolla obtuse, upper lip entire. *July.*

10. Veronica.—Calyx 4-parted. Corolla rotate or tubular, deeply 4-cleft. Stamens 2, exserted, one on each side of the upper lobe of the corolla. Style entire. Stigma single. Capsule compressed, 2-furrowed.

1. V. VIRGINICA (*Culver's Physic*).—Mostly smooth ; stem erect, simple, straight, tall ; leaves lanceolate, petiolate, acute, or acuminate, finely serrate, in whorls of 4-7 ; flowers white, in panicled spikes ; corolla tubular, pubescent within ; stamens and style twice as long as the corolla. In rich, low grounds, 2–6 feet high. *July.*

2. V. AMERICANA (*Brooklime*).—Smooth and rather fleshy ; stem decumbent at base, and then erect ; leaves ovate, or oblong, serrate ; flowers small, in opposite, loose racemes, on slender, spreading pedicels twice longer than the bracts ; corolla pale blue, marked with brownish lines ; capsule turgid, emarginate. In wet grounds, 6'-12' high. *June–August.*

3. V. SERPYLLIFOLIA (*Common Speedwell*).—Nearly or quite smooth ; stem low, prostrate, much branched at base, with ascending, simple branches ; leaves ovate or oblong, obtuse ; lowest roundish, petiolate ; upper sessile, entire bracts ; flowers in loose, bracted, terminal racemes, elongated in fruit ;

corolla blue and white, marked with purple lines ; capsule obtusely emarginate. In grassy fields, 2'–6' high. *May–September.* ①

11. Gerardia.—Calyx 5-parted. Corolla tubular, swelling above, with 5 spreading, more or less unequal lobes ; 2 upper ones usually the smallest. Stamens 4, included, hairy. Style elongated. Capsule ovate, acuminate, many-seeded.

1. G. PURPUREA (*Purple Gerardia*).—Smooth ; stem erect, angular, with long, spreading branches ; leaves linear, acute, rough on the margin ; flowers axillary, solitary ; calyx-teeth subulate ; corolla bright purple, showy, smooth or slightly downy. In wet, grassy grounds, 8'–20' high. *August.*

2. G. FLAVA (*Yellow Gerardia*).—Pubescent ; stem erect, simple, or branching toward the summit ; leaves opposite, sessile, ovate-lanceolate, or oblong, entire, obtuse ; flowers large, opposite, axillary, on very short peduncles ; calyx-segments oblong, obtuse. In dry woods, 2–3 feet high. *August.*

3. G. QUERCIFOLIA (*Oak-leaved Gerardia*).—Smooth and glaucous ; stem tall, simple, or somewhat branching ; lower leaves twice pinnatifid ; upper oblong-lanceolate, pinnatifid or entire ; flowers pedunculate, axillary, opposite, of a brilliant yellow, large and showy ; segments of the calyx linear-lanceolate, equaling the tube. In rich woods, 4–6 feet high. *August.*

4. G. PEDICULARIA (*Bushy Gerardia*).—Stem erect, very branching ; leaves opposite, ovate, pinnatifid ; the lobes variously cut and toothed ; petioles short, hairy ; flowers large, yellow ; segments of the calyx usually toothed, as long as the hairy tube ; corolla 1' long, with rounded, spreading segments. In dry woods, 3–4 feet high. *August.*

12. Castilleia.—Calyx tubular, flattened, 2–4-cleft, included in colored bracts. Corolla-tube included in the calyx ; upper lip long and narrow, arched, inclosing the stamens ; lower lip short, 3-lobed. Stamens 4. Anthers oblong-linear, 2-lobed, with unequal lobes. Capsule many-seeded. ♃

C. COCCINEA (*Painted Cup*).—Pubescent ; stem erect, angular, simple ; leaves alternate, sessile, pinnatifid, radical ones clustered at base ; bracts 3-cleft, colored with bright scarlet at apex, rarely yellow, longer than the corolla ; flowers in short, dense, terminal spikes ; calyx and corolla greenish yellow, the former tinged with scarlet at tip. In meadows, 10'–20' high. *May–June.*

13. Pedicularis.—Calyx campanulate or tubular, 2–5-cleft ; the segments leafy. Corolla strongly bilabiate ; upper lip arched, emarginate ; lower lip spreading, 3-lobed. Stamens 4, included in the upper lip. Capsule oblique. ♃

P. CANADENSIS (*Lousewort*).—Stems low, erect, simple, clustered ; leaves petiolate, alternate ; lowest pinnately dissected ; lobes oblong-ovate, crenately toothed ; flowers in short, dense, hairy, terminal heads ; calyx 2-toothed ;

corolla greenish yellow, or dull red ; upper lip vaulted, terminating in 2 teeth.
turned downward. In fields, 6'-15' high. *May-July.*

14. Melampyrum.—Calyx campanulate, 4-cleft ; the lobes with
long, bristly points. Corolla-tube cylindrical, larger above ; upper lip
arching. Stamens 4, included in the upper lip. Capsule usually 4-
seeded, oblique, compressed. ①

M. PRATENSE (*Cow-wheat*).—Smooth ; stem erect, branching ; leaves
opposite, lanceolate, or linear, petiolate ; upper ones larger, with a few long
teeth ; flowers solitary, remote ; calyx smooth ; corolla yellowish. In dry
woods, 6'-10' high. *July-August.*

Order LIX.—LENTIBULACEÆ (*Bladderwort Family*).

Herbs, growing in the water or mud. Leaves radical ; when
floating in the water much dissected, and furnished with air-
bladders ; when growing on land, entire and fleshy. Flowers
showy, very irregular. Calyx of 2-5 sepals, distinct, or partially
united. Corolla bilabiate, personate, tube very short, spurred.
Stamens 2, inserted on the upper lip. Anthers 1-celled. Ovary
free from the çalyx, 1-celled. Style 1. Fruit a many-seeded
capsule,

Utricularia.—Calyx 2-parted, with nearly equal lips. Corolla
irregularly 2-lipped, personate ; the lower lip projecting, and sometimes
closing the throat. ♃

1. U. INFLATA (*Inflated Bladderwort*).—Upper leaves floating in a
whorl of 5 or 6, which are inflated into oblong bladders, but dissected at
apex into capillary segments ; lower leaves submerged, very finely dissected,
and bearing many little bladders ; scape projecting above the water, 4-6-
flowered ; flowers large, yellow, very irregular, spurred, striate, emarginate,
upper lip of the corolla broad-ovate, entire ; lower 3-lobed. Common in
ponds. *July-August.*

2. U. VULGARIS (*Common Bladderwort*).—Leaves all submerged,
crowded, dissected into very numerous, capillary segments, furnished with
little bladders ; flowers 5-12, pedicellate, yellow, very showy, alternate ; spur
conical, obtuse, much shorter than the corolla. Common in ponds. *June-
August.*

Order LX.—BIGNONIACEÆ.

Trees, or shrubby, climbing, or twining plants. Flowers usually large and showy. Leaves simple, or pinnately parted. Corolla broad at the throat, with a bilabiate or irregularly 5-lobed
limb. Stamens 5, 1 or 3 sterile, when 4, often didynamous.
Ovary 2-celled, free from the calyx, surrounded by a fleshy disk at

base. Style 1. Fruit a woody or coriaceous 2-valved, many-seeded pod. Seeds winged.

BIGNONIACEÆ. $\begin{cases} \text{Climbers—}Tecoma. \\ \text{Trees—}Catalpa. \end{cases}$

1. Tecoma.—Calyx campanulate, 5-toothed. Corolla funnel-form, with a 5-lobed limb, somewhat bilabiate. Stamens didynamous. Capsule long and narrow, 2-celled, 2-valved. Seeds winged.

T. RADICANS (*Trumpet Creeper*).—Stem woody, climbing by means of rootlets; leaves pinnate; leaflets 5-11, ovate, acuminate, dentate; flowers corymbed; corolla large, 2'-3' long, orange and scarlet, very showy; stamens included. *July-August.*

2. Catalpa.—Calyx 2-lipped. Corolla campanulate; tube inflated; limb irregular. Stamens 5, 2 only usually have anthers. Stigma 2-lipped. Capsule long and slender, 2-celled.

C. BIGNONIOIDES (*Catalpa*).—Leaves cordate, or ovate-cordate, acuminate, entire; flowers in compound panicles, white, tinged and spotted with purple and yellow, large and showy; calyx-teeth mucronate; capsule cylindric, pendent, 6'-12' long. *June-July.*

Order LXI.—Verbenaceæ (*Vervain Family*).

Herbs, shrubs; or trees in the tropics. Leaves usually opposite, without stipules. Calyx tubular, free from the ovary, 4-5-cleft, persistent. Corolla tubular; the limb bilabiate, or with 4-5 more or less unequal lobes. Stamens 4, didynamous, sometimes only 2. Ovary entire, 2-4-celled, rarely 1-celled, each cell with 1 ovule. Style 1. Fruit separating into 2 or more indehiscent 1-seeded portions, rarely a single achenium.

A large, chiefly tropical order, represented here chiefly by Verbena, of which we have several native and exotic species.

VERBENACEÆ. $\begin{cases} \text{Fruit 4 achenia—}Verbena. \\ \text{Fruit 1-seeded—}Phryma. \end{cases}$

1. Verbena.—Calyx tubular, 5-toothed. Corolla funnel-form; limb with 5 slightly unequal lobes. Stamens 4, rarely 2, included. Style slender. Stigma capitate. Fruit splitting into 4 achenia.

1. V. HASTATA (*Blue Vervain*).—Stem tall, erect, with a few opposite branches above; leaves lanceolate, sharply serrate; flowers sessile, in dense, slender, erect spikes, usually arranged in terminal panicles; corolla purplish blue; stamens 4. In low grounds, 3-5 feet high. *July-September.*

2. V. URTICIFOLIA (*Nettle-leaved Vervain*).—Stem tall, erect, branching; leaves ovate, or oblong-ovate, acute, coarsely serrate, strongly nerved;

flowers minute, remote, white, sessile in elongated, very slender spikes. Common in waste places. 2–3 feet high. *July–August.*

3. V. AUBLETIA (*Garden Verbena*).—Stem weak, decumbent at base, erect above, square, viscid-pubescent, with opposite branches ; leaves oval, deeply cut and toothed ; flowers-large, in solitary, dense, corymbose clusters, on long peduncles ; bracts downy, nearly as long as the downy calyx, narrow, persistent ; corolla rose-red or scarlet, with emarginate lobes. In house cultivation, also in the open air. 1–2 feet high. *May.*

2. **Phryma.**—Calyx cylindrical, bilabiate ; upper lip longer, with 3 bristly teeth ; lower lip 2-toothed. Corolla bilabiate. Style slender. Stigma 2-lobed. Fruit oblong, 1-celled, 1-seeded. ♃

P. LEPTOSTACHYA (*Lopseed*).—Stem erect, slender, square, branching, pubescent ; leaves ovate, coarsely serrate, thin and large ; flowers opposite, rather small, in very long, slender spikes ; corolla light purple ; pedicels of the fruit deflexed ; seed solitary, inclosed in a thin pericarp, all invested by the closed calyx. In rich woodlands. 1–3 feet high. *July.*

Order LXII.—LABIATÆ (*Mint Family*).

Herbs, or slightly shrubby plants. Stems square, usually with opposite branches. Leaves opposite or verticillate, usually containing receptacles of volatile oil. Flowers axillary or terminal, in whorls, which sometimes take the form of dense heads ; rarely solitary. Calyx tubular, free from the ovary, persistent, 4–5-cleft, or bilabiate. Corolla bilabiate, rarely almost regular, with 4–5 subequal lobes. Stamens didynamous, sometimes only 2, the upper pair being abortive, or wanting. Ovary deeply 4-lobed. Style 1, central, arising from the base of the 4 lobes of the ovary, which in fruit consists of 4, rarely fewer, little separate nuts or achenia, contained in the tube of the persistent calyx.

1. **Lavandula.**— Calyx ovoid-cylindric, with 5 short teeth ; the upper ones often the largest. Upper lip of the corolla 2-lobed ; lower lip 3-lobed ; lobes nearly equal. Stamens included. ♃

L. VERA (*Lavender*).—Stem suffruticose and branching at base ; leaves linear-lanceolate, sessile, revolute on the margin, white-downy ; flowers bluish, in interrupted spikes ; corolla much exsert. *July.*

2. **Mentha.**—Calyx somewhat campanulate, with 5 equal teeth. Corolla with a short tube, nearly regular, 4-cleft ; upper lobe broadest, entire or emarginate. Stamens 4, nearly equal, straight, distant. Achenia smooth. ♃

·1. M. CANADENSIS (*Horsemint*).—Stem low, ascending, simple or branching, pubescent with reversed hairs on the angles ; leaves oblong, or oblong-lanceolate, serrate ; flowers small, pale purple, in dense, axillary, globular

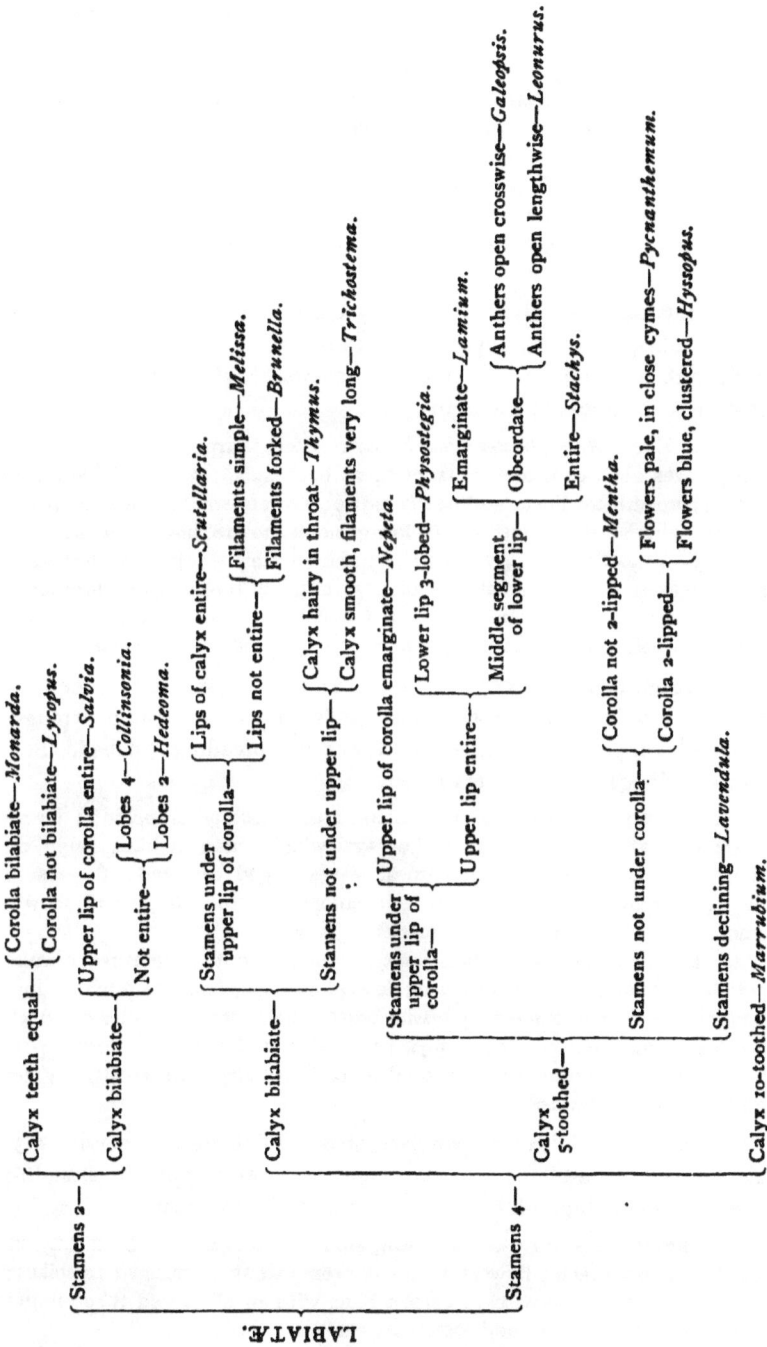

LABIATÆ.

- **Stamens 2—**
  - Calyx teeth equal—
    - Corolla bilabiate—*Monarda.*
    - Corolla not bilabiate—*Lycopus.*
  - Calyx bilabiate—
    - Upper lip of corolla entire—*Salvia.*
    - Not entire—
      - Lobes 4—*Collinsonia.*
      - Lobes 2—*Hedeoma.*

- **Stamens 4—**
  - Calyx bilabiate—
    - Stamens under upper lip of corolla—
      - Lips of calyx entire—*Scutellaria.*
      - Lips not entire—
        - Filaments simple—*Melissa.*
        - Filaments forked—*Brunella.*
    - Stamens not under upper lip—
      - Calyx hairy in throat—*Thymus.*
      - Calyx smooth, filaments very long—*Trichostema.*
  - Calyx 5-toothed—
    - Stamens under upper lip of corolla—
      - Upper lip of corolla emarginate—*Nepeta.*
      - Upper lip entire—
        - Lower lip 3-lobed—*Physostegia.*
        - Middle segment of lower lip—
          - Emarginate—*Lamium.*
          - Obcordate—
            - Anthers open crosswise—*Galeopsis.*
            - Anthers open lengthwise—*Leonurus.*
          - Entire—*Stachys.*
    - Stamens not under corolla—
      - Corolla not 2-lipped—*Mentha.*
      - Corolla 2-lipped—
        - Flowers pale, in close cymes—*Pycnanthemum.*
        - Flowers blue, clustered—*Hyssopus.*
    - Stamens declining—*Lavendula.*
  - Calyx 10-toothed—*Marrubium.*

whorls; calyx hairy; stamens exserted. In wet grounds, 1-2 feet high. *July–September.*

2. M. VIRIDIS (*Spearmint*).—Stem smooth, upright, with erect branches; leaves ovate-lanceolate, acute, unequally serrate, smooth; flowers pale purple, in distinct, axillary whorls, arranged in slender, interrupted spikes; calyx-teeth somewhat hairy. Wet grounds, 1-2 feet high. *July.*

3. M. PIPERITA (*Peppermint*).—Smooth; stem purplish, erect, branching; leaves ovate, acute, serrate, petiolate, dark green; flowers purplish, in dense whorls, forming compact spikes; bracts lanceolate. In wet grounds, 2-3 feet high. *July.*

3. Lycopus.—Calyx tubular, 4–5-toothed. Corolla campanulate, with 4 subequal lobes; upper segment broadest, emarginate. Stamens 2, distant, sometimes with 2 sterile rudiments. Style straight. Achenia smooth, with thickened margins. ♃

1. L. EUROPÆUS (*Water Hoarhound*).—Stem sharply 4-angled, branching; leaves oblong-lanceolate, tapering at both ends, petiolate; lowest pinnatifid; uppermost linear and nearly entire; flowers small, white, in dense, axillary whorls; calyx-teeth 5. In low grounds, 10′-18′ high. *August.*

2. L. VIRGINICUS (*Bugle Weed*).—Stem obtusely 4-angled, with concave sides; leaves ovate-lanceolate, or oblong, coarsely toothed, on short petioles; flowers small, purplish white, in few-flowered, axillary whorls; calyx-teeth 4, ovate, obtuse. In low grounds, 10′-15′ high. *July–August.*

4. Monarda.—Calyx elongated, tubular, 15-nerved, 5-toothed. Corolla tubular, with a somewhat inflated throat. Limb strongly 2-lipped; upper lip linear, erect; lower lip spreading, 3-lobed. Stamens 2, elongated. Anthers linear. ♃

1. M. DIDYMA (*Bee Balm*). — Stem erect, usually branching; leaves ovate, or ovate-lanceolate, rounded or somewhat cordate at base, rough, on short petioles; bracts and uppermost leaves tinged with red; flowers in dense, terminal heads; calyx smooth, colored; corolla large and showy, smooth, very long, bright red or scarlet. *July.*

2. M. FISTULOSA (*Wild Bergamot*).—Nearly smooth; stem erect; leaves petiolate, ovate-lanceolate, rounded; flowers in a few, dense, terminal, many-flowered heads, surrounded by sessile bracts, the upper and outer of which are leafy and often colored; calyx very hairy at the throat; corolla pale purple, greenish white or rose-colored, smooth or hairy. In woods, 2-4 feet high. *July–September.*

5. Salvia.—Calyx campanulate, striate, bilabiate; upper lip 2–3-cleft; lower lip 2-cleft. Corolla deeply 2-lipped; upper lip straight; lower lip spreading, 3-lobed. Stamens 2, with short filaments. ♃

S. OFFICINALIS (*Sage*).—Stem low, shrubby; leaves ovate-lanceolate, of a dull, grayish green; flowers in few-flowered whorls, arranged in spikes; calyx-teeth mucronate, viscid; corolla blue, with an elongated tube; upper lip as long as the lower, and somewhat vaulted. *July.*

6. Pycnanthemum.—Calyx more or less tubular, about 13-nerved, 5-toothed. Corolla bilabiate, with a short tube ; upper lip nearly flat ; lower lip with 3 ovate, obtuse lobes. Stamens distant. Anther-cells parallel. ♃

1. P. INCANUM (*Mountain Mint*).—Stem erect, obtusely 4-angled, white-downy, leaves oblong-ovate, remotely toothed, acute, hoary beneath ; uppermost whitish on both sides ; flowers in dense whorls, forming loose, compound cymes or heads ; corolla flesh-colored or pale purple ; the lower lip spotted with dark purple ; calyx bilabiate, the 3 upper segments being united and bearded at the extremity. In rocky woods, 2–3 feet high. *July.*

2. P. LANCEOLATUM (*Narrow-leaved Wild Basil*).—Smooth or slightly downy ; stem erect, pubescent on the angles, very branching, leafy ; leaves lanceolate ; flowers in dense heads, forming a terminal corymb ; bracts ovate-lanceolate, acuminate, white-downy ; calyx-teeth short, triangular, white-downy ; corolla pale purplish white, spotted with brownish purple. Thickets and fields, 1–3 feet high. *July–August.*

7. Thymus.—Calyx bilabiate, about 13-nerved, hairy in the throat ; upper lip 3-toothed ; lower lip 2-cleft. Corolla short, somewhat 2-lipped ; upper lip flattish, lower lip spreading, with 3 nearly equal teeth. Stamens 4, distant.

1. T. VULGARIS (*Thyme*).—Stems decumbent at base, ascending, branching, tufted ; leaves oblong-ovate, or lanceolate, entire, veiny, revolute on the margins ; flowers in whorls, arranged so as to form terminal, leafy spikes ; corolla purplish. *June–August.*

2. T. SERPYLLUM (*Creeping Thyme*).—Stems decumbent at base, ascending, wiry, branching, slender, leafy, downy above, tufted ; leaves ovate, obtuse, entire ; flowers in dense, oblong heads ; corolla purple, spotted. *June–July.*

8. Hyssopus.—Corolla bilabiate ; upper lip erect, flat, emarginate ; lower lip 3-parted ; tube about as long as the calyx. ♃

H. OFFICINALIS (*Hyssop*).—Stems erect, tufted ; leaves linear-lanceolate, acute, entire, sessile ; flowers in racemose, one-sided verticils, bright blue ; calyx-teeth erect. In gardens, 2 feet high. *July.*

9. Collinsonia.—Calyx bilabiate ; upper lip truncate, 3-toothed ; lower lip 2-toothed. Corolla somewhat bilabiate ; upper lip with 4 nearly equal lobes ; lower lip longer, fringed. Stamens 2, rarely wanting, much exserted, diverging. ♃

C. CANADENSIS (*Horse-Balm*).—Stem erect, square, branching ; leaves ovate, coarsely serrate ; flowers rather large, in racemes, forming a terminal panicle ; corolla pale yellow, with a conspicuously fringed lower lip ; calyx-teeth subulate ; style and stamens very long. In damp, rich soils, 3–5 feet high. *July–September.*

10. Hedeoma.—Calyx nearly tubular, 13-ribbed, bilabiate ; upper lip 3-toothed ; lower lip 2-toothed ; throat hairy. Corolla bilabiate ; upper lip flat, erect ; lower lip spreading, 3-lobed ; lobes nearly equal. Stamens 2, sometimes accompanied by 2 sterile filaments. ①

H. PULEGIOIDES *(Pennyroyal).*—Stem erect, branching ; leaves ovate, or oblong-ovate, few-toothed, on short petioles, smooth above ; flowers small, pale purple, in whorls ; corolla pubescent.   In dry, barren fields, 3′-8′ high. *July–September.*

11. **Melissa.**—Calyx 13-ribbed, flattish above ; upper lip 3-toothed ; lower lip 2-toothed.   Corolla bilabiate ; upper lip erect, flattish ; lower lip spreading, 3-lobed ; middle lobe broadest.   Stamens ascending. ♃

M. OFFICINALIS *(Balm).*—Stem erect, branching ; leaves ovate, acute, coarsely toothed ; flowers in half whorls, white or yellow ; bracts few, ovate-lanceolate, petiolate. *June–August.*

12. **Scutellaria.**—Calyx campanulate, bilabiate ; lips entire ; upper sepal arched.   Corolla bilabiate, with the tube elongated ; upper lip arched, entire ; lower lip with its middle lobe dilated and convex. Stamens 4, ascending beneath the upper lip.   Anthers approximate in pairs, ciliate. ♃

1. S. GALERICULATA *(Common Skullcap).*—Stem simple or slightly branched ; leaves all alike, ovate, or ovate-lanceolate, acute, serrate, more or less cordate at base, on very short petioles, almost sessile ; flowers large, axillary, usually solitary ; corolla blue, greatly expanded above.   In swamps and meadows, 1-2 feet high. *August.*

2. S. LATERIFLORA *(Side-flowering Skullcap).*—Stem erect, with opposite branches ; leaves ovate-lanceolate ; lower floral leaves resembling the others ; upper floral leaves small, resembling bracts ; flowers small, in lateral, axillary, leafy, long-peduncled, somewhat one-sided racemes ; corolla blue. In meadows and low grounds, 10′-18′ high. *July–August.*

13. **Brunella.**—Calyx tubular-campanulate, bilabiate, closed in fruit ; upper lip broad and flat, with 3 short teeth ; lower with 2 lanceolate teeth.   Corolla bilabiate ; upper lip vaulted, erect, entire ; lower lip spreading, reflexed, 3-cleft.   Stamens 4.   Filaments forked, the lower division bearing the anther. ♃

P. VULGARIS *(Self-heal).*—Stem low, simple, or slightly branched, marked with pubescent lines ; leaves oblong-ovate, entire, or slightly toothed, petiolate ; flowers in dense, sessile, bracted, 5-6-flowered verticils, forming a dense, terminal spike ; bracts reniform, 2 to each verticil, membranous, ciliate ; corolla blue or violet of various shades.   In meadows, 6′-12′ high. *June–August.*

14. **Nepeta.**—Calyx tubular, obliquely 5-toothed.   Corolla naked and expanded at the throat, bilabiate ; upper lip erect, emarginate ; lower lip spreading, 3-cleft.   Stamens 4, approximate in pairs. ♃

1. N. CATARIA *(Catnip).*—Stem tall, erect, branching ; leaves cordate, crenate, soft and velvety ; flowers in slightly pedunculate whorls, arranged

in interrupted spikes ; corolla purplish white, twice as long as the calyx ; lower lip dotted with purple. In waste places, 3-4 feet high. *July.*

2. N. GLECHOMA (*Gill. Ground Ivy*). — Stem creeping, rooting at base ; leaves reniform, crenate, glaucous green ; flowers 3-5 together, in loose clusters ; corolla light blue, variegated at the throat ; anther-cells diverging at a right angle. In waste grounds, 1-2 feet long. *May-August.*

15. Physostegia.—Calyx campanulate, with 5 nearly equal teeth, inflated after flowering. Corolla with the throat inflated ; upper lip nearly erect, entire ; lower lip spreading, 3-parted ; middle lobe broad and rounded, emarginate. Stamens 4. ♃

P. VIRGINIANA (*False Dragon's Head*).—Stem erect, thick, and rigid ; leaves lanceolate-ovate ; flowers large, showy, in dense, terminal, 4-rowed spikes ; corolla pale purple or flesh-color, spotted inside. Stem 1-3 feet high. *July-September.*

16. Lamium.—Calyx tubular-campanulate, 5-ribbed, with 5 nearly equal teeth. Corolla dilated at throat, bilabiate ; upper lip vaulted, narrowed at base ; lower lip 3-parted ; lateral lobes small, attached to the margin of the throat. Stamens 4. ①

L. AMPLEXICAULE (*Dead Nettle*).—Stems decumbent at base; leaves broad, nearly round, hairy ; lower ones small ; cauline ones cordate ; floral leaves similar, but nearly or quite sessile ; flowers in dense verticils, sessile in the axils of the upper leaves ; calyx hairy ; corolla light purple, elongated ; upper lip downy ; lower lip spotted. In waste grounds, 5'-18' high. *May-October.*

17. Leonurus.—Calyx turbinate, 5-ribbed, with 5 subequal teeth. Corolla bilabiate ; upper lip erect, oblong, entire, hairy ; lower lip 3-lobed, spreading ; middle lobe obcordate. Stamens 4. ♃

L. CARDIACA (*Motherwort*).—Stem erect, branching, often purplish ; leaves palmately lobed ; floral leaves trifid, variously toothed and arranged in 4 rows on the stem ; flowers in dense, axillary whorls ; corolla purplish, hairy outside, variegated inside. In waste places, 3-5 feet high. *July-September.*

18. Galeopsis.—Calyx tubular-campanulate, 5-ribbed, with 5 subequal teeth. Corolla bilabiate, dilated at the throat ; upper lip ovate, arched ; lower lip 3-cleft, spreading ; middle lobe obcordate, toothed ; the palate with 2 teeth on the upper side. Stamens 4. ①

G. TETRAHIT (*Hemp Nettle*).—Stem hispid, swollen below the joints ; leaves ovate, coarsely serrate, hispid, acute ; flowers in dense, axillary verticils ; corolla purple, variegated with white. In waste places, 1-2 feet high. *June-July.*

19. Stachys.—Calyx tubular-campanulate, angular, 5-10-ribbed, 5-toothed. Corolla bilabiate ; upper lip erect, spreading, or arched, entire ; lower lip spreading, 3-lobed ; middle lobe nearly entire. Stamens 4, ascending beneath the lower lip. Anthers approximate in pairs. ♃

S. PALUSTRIS (*Hedge Nettle. Woundwort*).—Stem erect, nearly simple, clothed with stiff, deflexed bristles, especially on the angles ; leaves ovate-lanceolate or oblong-lanceolate, acute, serrate, bristly, especially on the mid-rib and veins, rounded at base, on short petioles ; flowers arranged in spikes ; calyx with bristly teeth ; corolla pale purple. In swamps and meadows, 1–3 feet high. *July.*

20. **Marrubium.**—Calyx tubular, 5–10-ribbed, with 5–10 nearly equal teeth. Corolla bilabiate ; upper lip erect, flattish, emarginate ; lower lip spreading, 3-cleft. Stamens 4, included in the corolla-tube. ♃

M. VULGARE (*Hoarhound*).—Stem ascending, white-downy ; leaves round-ish-ovate, crenately toothed, petiolate, white-downy beneath ; flowers in dense, hairy, axillary whorls ; calyx-teeth 10, recurved ; alternate ones short-er ; corolla small, white. A bitter, aromatic herb, rather frequent in waste grounds. Introduced. Stem 1–2 feet high. *August.*

21. **Trichostema.**—Calyx campanulate, oblique, deeply and un-equally 5-toothed ; upper lip with 3 nearly equal teeth. Corolla with a very short tube, unequally 5-lobed ; lobes oblong, declined. Sta-mens 4. ①

T. DICHOTOMUM (*Blue Curls*).—Stem erect, dichotomously branching, hairy ; leaves oblong-lanceolate ; flowers axillary and terminal, on slender, 1-flowered pedicels, becoming inverted by the twisting of their stalks ; corolla small, bright blue ; stamens very long, much exserted, bright blue, very con-spicuous, curving from the lower lip to the upper. In dry pastures, 6′–8′ high. *August.*

Order LXIII.—PLANTAGINACEÆ (*Plantain Family*).

Herbs usually acaulescent. Leaves in a radical tuft, often ribbed. Calyx 4-cleft, free from the ovary, persistent. Corolla more or less tubular, 4-cleft, scarious, persistent. Stamens 4, alternate with the corolla-lobes. Filaments long and weak. An-thers versatile. Ovary 2-celled. Style 1. Capsule a membrana-ceous pyxis, the cells 1 or several-seeded.

Plantago.—Calyx-teeth 4, persistent, dry. Corolla tubular, with-ering, with a 4-cleft, reflexed border. Capsule an ovoid, 2-celled pyxis. Acaulescent.

1. P. MAJOR (*Common Plantain*).—Leaves large, ovate, tapering abruptly at base, very strongly ribbed ; spike long, cylindrical ; flowers densely im-bricated, whitish, inconspicuous. In damp soils, 8′–2 feet high. *June–Sep-tember.*

2. P. LANCEOLATA (*Ribwort*).—Leaves lanceolate, strongly 3-5-nerved, nearly entire, tapering at both ends ; petioles slender ; scape slender, angular, more or less pubescent ; spike ovate, densely flowered ; stamens whitish. In dry, grassy fields, 6'-20' high. *May-October.*

APETALOUS EXOGENS WITH PERFECT FLOWERS.

Order LXIV.—NYCTAGINACEÆ.

Herbs, or shrubs. Leaves opposite, one of each pair smaller than the other. Calyx colored, at length separating from the lower part, which hardens and incloses the achenium. Stamens hypogynous, 1-20. Ovary free, 1-celled. Style 1. Stigma 1. Fruit an achenium.

Mirabilis.—Calyx funnel-form, with 2 bracts at base ; tube contracted, free from the ovary ; limb plaited, entire, deciduous. Stamens 5. Style 1. Stigma globose. ①

M. JALAPA (*Four-o'clock*).—Stem erect, dichotomous, smooth ; leaves opposite, acuminate, smooth ; flowers pedicellate, in axillary and terminal clusters, large and fragrant ; root large and tuberous. A showy plant, in cultivation, 2-3 feet high. *June-September.*

Order LXV.—AMARANTACEÆ (*Amaranth Family*).

Herbs, or shrubs. Leaves alternate, without stipules. Flowers in heads or spikes, or dense clusters, furnished with dry and scarious, usually colored bracts. Calyx consisting of 3-5, dry and scarious, persistent sepals. Stamens 3-5, or more, hypogynous, distinct or monadelphous. Ovary free, 1-celled, with 1 or more ovules. Fruit a utricle, rarely a many-seeded capsule.

AMARANTACEÆ. { Leaves opposite—*Gomphrena.*
{ Leaves alternate— { Flowers perfect—*Celosia.*
{ Flowers imperfect—*Amarantus.*

1. **Amarantus.**—Flowers monœcious or polygamous, sometimes diœcious, with 3 bracts. Sepals 3-5, mostly colored, persistent. Stamens 3-5. Styles 2-3, sometimes 4, filiform. Utricle indehiscent, 1-seeded. ①

A. ALBUS (*White Pigweed*).—Smooth ; stem erect or ascending, angular ; leaves obovate, entire, retuse, with a mucronate point, light green ; flowers greenish, inconspicuous, in axillary clusters. Common weed, 1-2 feet high. *July.*

2. Gomphrena.—Flowers perfect. Bracts 5, colored ; 3 outer converging. Sepals 5, hairy. Disk cylindric, 5-toothed. Stamens 5. Stigma 1. Utricle 1-celled. ①

G. GLOBOSA (*Globe Amaranth*).—Stems erect, hairy, with opposite branches ; leaves opposite, obtuse, pubescent ; flowers purple, in dense, globose, solitary heads. Cultivated for its fadeless heads of flowers. 10′-18′ high. *July–September.*

3. Celosia.—Sepals 3–5, colored. Stamens 5, united at base by a plicate disk. Style 2–3-cleft. Utricle circumscissile. ①

C. CRISTATA (*Cockscomb*).—Stem erect, mostly simple ; leaves ovate, acuminate, mostly alternate ; stipules falcate, striate ; flowers small, densely crowded, in large, compressed, thin clusters, bright, purplish red. The crests of flowers vary 2′-8′ in breadth. Common in cultivation, 1-2 feet high. *June–September.*

Order LXVI.—CHENOPODIACEÆ (*Goosefoot·Family*).

Herbs, rarely shrubby. Leaves alternate, often more or less succulent or fleshy. Sometimes none. Flowers greenish, inconspicuous, usually perfect, sometimes diœcious or polygamous. Calyx sometimes tubular at base, persistent. Stamens as many as the calyx-segments, or fewer, inserted at their base. Ovary free, 1-celled, 1-ovuled. Styles 2-4, rarely 1. Fruit a utricle.

CHENOPODIACEÆ.
{ Flowers perfect— { Root large—*Beta*.
{ Root small ; weeds—*Chenopodium*.
{ Flowers imperfect—*Spinacia*.

1. Chenopodium.—Flowers perfect, bractless. Calyx 5-parted, at length dry, partially enveloping the fruit. Stamens 5. Styles 2. Utricle membranaceous. Seed lenticular. ①

1. C. ALBUM (*Lamb's Quarters*).—Smooth ; stem erect, branching ; leaves rhomboid-ovate, coarsely toothed, entire at base, pale green, petiolate, white and mealy beneath ; flowers greenish, mealy, sessile, forming large, terminal panicles. Homely weed, 2-5 feet high, in waste grounds. *July–September.*

2. C. HYBRIDUM (*Tall Lamb's Quarters*).—Smooth ; stem erect, slender, very branching ; leaves ovate, cordate at base, angular, with a few large, remote teeth, light green on both sides ; flowers greenish, sessile, in racemes. A common weed in waste grounds, stem 2-3 feet high. *July–August.*

2. Beta.—Flowers perfect. Calyx of 5 sepals. Stamens 5. Styles 2, very short, erect. Stigmas acute. Seeds reniform, inclosed in the fleshy calyx. ②

B. VULGARIS (*Common Beet*).—Stem erect, branching, furrowed ; leaves alternate, nearly entire ; lower ones ovate, upper narrower ; flowers green, in dense, axillary, sessile clusters, arranged in spikes; root fleshy. Is cultivated. *August.*

3. **Spinacia.**—Flowers diœcious. Barren flowers—calyx 5-cleft; stamens 5. Fertile flowers—calyx 2-4-cleft. Styles 4, capillary. Utricle contained in the hardened and sometimes spiny calyx. ①

S. OLERACEA (*Spinage*).—Stem erect, branching; leaves hastate-lanceolate, tapering at base, on long petioles; flowers greenish; barren in a terminal panicle; fertile in dense, sessile racemes; fruit sessile, prickly, or unarmed. In kitchen gardens. Stem 1-2 feet high. *June-July.*

Order LXVII.—PHYTOLACCACEÆ.

Herbs, rarely shrubby. Leaves alternate, without stipules. Flowers perfect, in racemes. Sepals 4-5, petaloid. Stamens 4-5 and alternate with the sepals; or else 10 or more. Ovary 1 or several-celled. Styles and stigmas as many as the cells. Fruit dry or baccate, consisting of 1, or several 1-seeded carpels.

Phytolacca.—Calyx of 5 petaloid sepals. Stamens 5-30. Styles 5-12. Ovary composed of 5-12 united carpels, forming a 5-12-celled, half-globular berry. Cells 1-seeded.

P. DECANDRA (*Poke-weed*). — Smooth; stem tall, terete, branching, changing at length to deep purple; leaves large, ovate, acute at both ends, petiolate, entire; flowers greenish white, in long cylindrical racemes, at first terminal, but at length opposite the leaves; stamens 10; styles 10; fruit globose, depressed, dark purple, juicy. A tall, stout, poisonous plant, 6-8 feet high, rising from a very large, branching, poisonous root. *July-September.*

Order LXVIII.—POLYGONACEÆ (*Buckwheat Family*).

Herbs, or shrubs. Leaves alternate. Stipules ochreate, rarely none. Flowers usually perfect. Sepals 4-6, more or less united at base, often petaloid. Stamens definite, inserted on the base of the sepals. Ovary free, 1-celled, 1-ovuled. Styles or stigmas 2 or 3. Achenium usually triangular or oblong.

POLYGONACEÆ. { Sepals 4-6, embryo one side of albumen—*Polygonum.*
{ Sepals 6— { Sepals all alike—*Rheum.*
{ Sepals of 2 sorts—*Rumex.*

1. **Rheum.**—Calyx colored, of 6 sepals, persistent. Stamens 9. Styles 3. Stigma many-parted, reflexed. Achenia 3-angled; angles winged. ♃

R. RHAPONTICUM (*Garden Rhubarb*).—Stem erect, stout, fleshy, hollow; joints sheathed by large stipules; leaves ovate, cordate, obtuse, smooth;

petioles rounded beneath, channeled above ; flowers very numerous, greenish white, in fasciculate clusters. Cultivated for its large, acid petioles. 3–6 feet high. *May.*

2. **Polygonum.**—Calyx 4–6, mostly 5-parted ; segments often petaloid, persistent, and inclosing the achenium. Stamens 4–9, mostly 8. Styles 2–3, mostly 3, short, filiform. Achenium mostly triangular.

1. P. AVICULARE *(Knot-grass)*—Smooth ; stem procumbent, spreading ; branches ascending ; leaves oblong, rough on the margin, acute, sessile, mostly pale green ; stipules short, white, gashed ; flowers nearly sessile, 2–3 together in the axils of the leaves, greenish white, sometimes reddish white. In waste grounds, 4′–10′ long. *June–November.* ①

2. P. HYDROPIPER *(Water Pepper).*—Smooth ; stem erect or ascending, simple or branching ; leaves lanceolate, entire, acuminate, with pellucid dots, rough on the margin ; stipules inflated, fringed ; flowers pale greenish white, in loose, nodding spikes ; stamens 6–8 ; styles 2–3, united at base ; fruit lenticular or triangular. In low grounds, 1–2 feet high. *August–September.*

3. P. ORIENTALE *(Prince's Feather).*—Stem tall, erect, branching, somewhat hairy ; leaves ovate, upper stipules hairy, somewhat salver-form ; flowers large, open, rose-colored, in long, nodding, showy spikes ; calyx 5-parted ; stamens 7 ; style 2-cleft ; fruit lenticular. Cultivated, 4–8 feet high. *August.* ①

4. P. SAGITTATUM *(Arrow-leaved Bindweed).*—Stem weak, ascending or prostrate, square, with the angles bristly backward ; leaves sagittate, entire, acute ; stipules smooth ; flowers small, whitish, in small, terminal heads, on long, slender, smooth peduncles ; stamens mostly 8 ; styles 3 ; fruit acutely 3-angled. In wet grounds, 1–2 feet long. *June–August.* ①

5. P. CONVOLVULUS *(Black Bindweed).*—Stem angular, twining or prostrate, somewhat rough, naked at the joints ; leaves petiolate, hastate, cordate at base, acute, entire ; stipules nearly entire ; flowers greenish white, sometimes tinged with purple, in clusters of 3–4, nodding in fruit ; stamens 8 ; styles 3 ; fruit mostly smooth, triangular. In cultivated grounds, 1–5 feet long. *June–September.* ①

3. **Rumex.**—Calyx persistent ; sepals 6 ; stamens 6 ; styles 3. Stigmas forming a tuft. Achenium 3-angled, covered by the sepals.

1. R. OBTUSIFOLIUS *(Broad-leaved Dock).*—Stem erect, stout, somewhat rough, branching ; lower leaves ovate, obtuse, cordate at base ; upper ones oblong-lanceolate, acute ; flowers in loose, distant whorls, forming long, nearly naked racemes ; valves sharply toothed at base. Weed in cultivated grounds, 2–4 feet high. *July.*

2. R. CRISPUS *(Yellow Dock).*—Smooth ; stem erect, branching ; leaves lanceolate, acute, strongly waved on the margin ; flowers in numerous whorls, arranged in crowded racemes, leafless above, forming a large, terminal panicle. Root yellow, spindle-shaped. Weed in cultivated grounds, 2–3 feet high. *June–July.*

3. R. ACETOSELLA *(Sheep Sorrel).*—Stem erect, leafy, branching ; leaves

lanceolate-hastate.; upper ones lanceolate; all entire, petiolate, very acid to the taste; flowers small, tinged with a dull red, in slender, leafless racemes, anthers of the barren flowers yellow; valves ovate. Weed, in dry, sandy soils, 3'-8' high. *May–October.*

Order LXXI.—ARISTOLOCHIACEÆ (*Birthwort Family*).

Herbs, or shrubby plants; in the latter case often climbing. Leaves alternate or radical. Flowers perfect, solitary, of a dull brown or greenish color. Calyx-tube more or less adherent to the ovary; limb 3-cleft. Stamens 6 or 12, epigynous, or adherent to the base of the short and thick style. Ovary 3- or 6-celled. Stigmas radiate, as many as the cells of the ovary. Fruit a many-seeded capsule or berry.

Asarum.—Calyx campanulate; limb 3-cleft; tube adherent to the ovary. Stamens 12, inserted on the ovary. Anthers short. Style very short. Stigma 6-rayed. Fruit globular, fleshy, 6-celled. ♃

A. CANADENSE (*Wild Ginger*).—Pubescent, stemless; leaves 2, broad-reniform, large, on long, hairy petioles, soft-downy; flowers solitary, large, nodding, on a downy pedicel; calyx woolly; segments reflexed from the middle, brownish purple within. In rich woods. *July.*

Order LXX.—LAURACEÆ (*Laurel Family*).

Trees, or shrubs. Leaves alternate, usually punctate with pellucid dots, destitute of stipules. Flowers perfect, or diœciously polygamous. Sepals 4–6, more or less united at base, imbricated in 2 series. Stamens definite, usually more numerous than the sepals, inserted on their base. Anthers 2–4-celled, opening by recurved valves from base to apex. Ovary, style, and stigma single. Fruit a berry or drupe, usually with a thickened pedicel. Seed large, with a conspicuous embryo.

LAURACEÆ. { Fruit a drupe—*Sassafras.*
{ Fruit a berry—*Lindera.*

1. **Sassafras.**—Flowers diœcious. Calyx spreading, colored, 6-parted. Barren flowers with 9 stamens, in 3 rows. Anthers opening by 4 valves. Fertile flowers with 6 rudimentary stamens. Style filiform. Drupe obovoid, 1-seeded. Trees.

S. OFFICINALE (*Sassafras*).—Leaves ovate, entire, or 3-lobed, and tapering at base, alternate, petiolate, mucilaginous, as also the young shoots; flowers greenish yellow, in pedunculate clusters, appearing before the leaves; drupes dark blue, on a red stalk. Woodlands, 20–40 feet high. *April–May.*

2. Lindera.—Flowers diœciously polygamous.   Calyx 6-parted, open.   Sterile flowers with 9 stamens in 3 rows.   Anthers 2-celled, 2-valved.   Fertile flowers with 15–18 rudimentary stamens.   Berry obovoid, 1-seeded.

L. BENZOIN (*Spice Bush*).—Leaves oblong-obovate, entire, sessile, wedge-shaped at base, thin, paler beneath, nearly smooth ; flowers greenish yellow, in compound clusters ; pedicels scarcely as long as the flowers ; calyx-teeth oblong ; berries red.   In moist woods, 5–12 feet high.   *April.*

Order LXXI.—THYMELEACEÆ (*Mezereum Family*).

Shrubs with a tough bark.   Leaves alternate, or opposite, entire.   Flowers perfect.   Calyx petaloid, tubular, free from the ovary ; limb usually 4-cleft, the lobes imbricated in prefloration, sometimes entire.   Stamens definite, usually twice as many as the calyx-lobes, inserted in its throat.   Ovary with 1 ovule.   Style 1.   Stigma 1.   Fruit drupaceous.

THYMELEACEÆ. $\begin{cases} \text{Style short—}Daphne. \\ \text{Style long—}Dirca. \end{cases}$

1. Dirca. — Calyx petaloid, tubular, truncate ; margin waved. Stamens 8, long and slender, inserted in the calyx-tube.   Style filiform. Stigma capitate.   Berry oval, 1-seeded.

· D. PALUSTRIS (*Leather-wood*).—Shrubby ; stem very branching ; leaves oblong-obovate, entire ; flowers appearing before the leaves, pale yellow, rather small, funnel-form, 2–3 together ; berry small, reddish.   *April–May.*

2. Daphne.—Calyx 4-cleft, withering ; limb spreading.   Stamens 8, included in the calyx-tube.   Style 1.   Berry 1-seeded.   Shrubs.

D. MEZEREUM (*Mezereum*).—Leaves deciduous, lanceolate, entire, sessile, in terminal tufts ; flowers sessile in clusters of 3–4 ; calyx salver-form, with ovate, spreading segments ; stamens inserted in 2 rows near the top of the tube ; stigma sessile.   *March.*

Order LXXII.—SANTALACEÆ (*Sandalwood Family*).

Trees, shrubs, or herbs.   Leaves alternate, entire.   Flowers small, perfect, rarely diœcious, polygamous.   Calyx-tube adherent to the ovary ; limb 4–5-cleft.   Stamens as many as the lobes of the calyx, and inserted opposite them.   Ovary 1-celled, with 1–4 ovules.   Style 1.   Fruit indehiscent, crowned with the persistent calyx, often drupaceous.

Comandra.—Flowers perfect.   Calyx somewhat urceolate ; limb 4–5-parted, with an adherent, 5-lobed disk.   Stamens 4–5.   Fruit dry, 1-seeded, crowned with the persistent calyx-lobes.

C. UMBELLATA (*False Toad-flax*).—Very smooth ; stem erect, slender, branching above ; leaves oblong, entire, alternate ; flowers small, greenish white, in small clustered umbels of 3–5, each cluster with 4 deciduous bracts. In dry and rocky grounds, 6′–12′ high. *June.*

APETALOUS EXOGENS WITH IMPERFECT FLOWERS.

Order LXXIII.—EUPHORBIACEÆ (*Spurge Family*).

Herbs, shrubs, or even trees, often with a milky juice. Leaves opposite, alternate, or verticillate, usually simple, often stipulate. Flowers monœcious or diœcious. Staminate and pistillate flowers usually separate, but often combined and surrounded by a common, mostly petaloid involucre, the staminate being reduced to a single stamen, and the pistillate to a compound pistil, destitute of calyx, and supported on a conspicuous jointed pedicel. Calyx, when present, several-lobed. Petals sometimes present, and as many as the calyx-lobes. Anthers 2-celled. Ovary free from the calyx, when the latter is present, consisting of 2–9 more or less united carpels, attached to a prolongation of the axis. Styles as many as the carpels, distinct, often 2-cleft. Fruit a capsule separating into its component carpels.

EUPHORBIACEÆ. { Apparent flowers perfect—*Euphorbia.*

{ Flowers imperfect— { Evergreen shrubs—*Buxus.*
{ Tall annuals—*Ricinus.*
{ Low weeds—*Acalypha.*

1. **Euphorbia.**—Flowers monœcious, in a usually petaloid, 4–5-parted involucre. Sterile flowers numerous, included within the involucre, consisting of a single stamen on a jointed pedicel, and furnished with a bract at base. Anthers composed of 2 separate, globular cells. Fertile flower solitary, in the center, pedicellate, consisting of a 3-lobed, 3-celled ovary, destitute of a calyx. Styles 3, each 2-parted. Capsule consisting of 3 1-seeded carpels, opening each by 2 valves. Herbs with a milky juice.

1. E. LATHYRIS (*Caper Spurge*).—Smooth ; stem erect, stout, branching ; leaves linear-lanceolate, somewhat acute, entire, sessile ; leaves of the involucre oblong-ovate, cordate at base, acuminate ; fruit and seeds smooth. Stem 2–3 feet high. *July–September.*

2. E. HYPERICIFOLIA (*Spurge*).—Stem smooth, nearly erect, with spreading branches ; leaves opposite, oval-oblong, obliquely cordate at base, 3–5-

nerved beneath ; heads whitish, in axillary and terminal clusters, forming a sort of terminal corymb. In waste grounds, 8'–15' high. *July–August.* ①

3. E. MACULATA (*Spotted Spurge*).—Mostly hairy ; stem prostrate, diffusely branching ; leaves oval, sessile, smoothish above, pale and hairy beneath, often with large, purple spots above ; heads of flowers in axillary clusters, minute, whitish. A prostrate species, forming flat patches. Common in cultivated grounds. *June–September.*

2. Acalypha.—Flowers monœcious. Barren flowers very small ; calyx 4-parted ; stamens 8–16, united at base. Fertile flowers few ; calyx 3-parted. Styles 3, elongated, fringed.

A. VIRGINICA (*Three-seeded Mercury*).—More or less pubescent ; stem erect or ascending, branching ; leaves ovate or oblong-ovate, serrate ; barren flowers on short peduncles ; pistillate flowers 1–3 together in a large, leaf-like, broad, cordate-ovate, unequally lobed and toothed, acuminate bract, which is longer than the barren spike. In cultivated grounds, 6'–15' high. *August.*

3. Ricinus.—Flowers monœcious. Barren flowers—calyx 5-parted ; stamens numerous. Fertile flowers—calyx 3-parted ; styles 3, each 2-parted. Capsule prickly, 3-celled, 3-seeded. ①

R. COMMUNIS (*Castor-oil Plant*).—Herbaceous ; stem erect, branching, and mealy in appearance ; leaves peltate, palmate, with the lobes lanceolate, serrate, on long petioles ; fruit prickly. *July–August.*

4. Buxus.—Flowers monœcious. Barren flowers—calyx 3-leaved ; petals 2 ; stamens 4, with a rudimental ovary. Fertile flowers—calyx 4-leaved ; petals 3. Styles 3. Capsule 3-beaked, 3-celled, 2-seeded. Shrubs.

B. SEMPERVIRENS (*Box*).—Evergreen ; leaves opposite, ovate or obovate, entire, dark green, the petioles hairy on the margin ; anthers ovate-sagittate. The leaves are sometimes narrowly lanceolate. A dwarf variety is used for edgings.

Order LXXIV.—URTICACEÆ (*Nettle Family*).

Trees, or shrubs, usually with a milky or yellowish juice : or herbs with a watery juice. Leaves alternate or opposite, often rough or hispid, with stinging hairs, frequently stipulate. Flowers monœcious, diœcious, or polygamous, in panicles, aments, or fleshy heads. Calyx regular, persistent, rarely wanting, usually 3–5-parted. Stamens definite, distinct, inserted on the base of the calyx, usually as many as its lobes, and opposite them. Ovary free from the calyx, 1-ovuled. Style 1. Fruit a 1-seeded utricle, surrounded by the membranous or fleshy calyx.

URTICACEÆ.
- Trees—
  - Fruit dry— *Ulmus.*
  - Fruit a berry-like drupe—*Celtis.*
  - Fruit like a blackberry—*Morus.*
- Herbs—
  - Tall twiners—*Humulus.*
  - Style one—*Urtica.*
  - Styles two, long—*Cannabis.*

1. Morus.—Flowers monœcious or diœcious, the different kinds in separate spikes. Calyx 4-parted. Stamens 4. Styles 2. Achenia ovate, contained within the fleshy calyx, forming a juicy, berry-like fruit. Trees.

M. RUBRA (*Red Mulberry*).—Leaves rough-pubescent beneath, cordate or rounded at base, acuminate, entire, 3 to several-lobed ; flowers small, often diœcious ; fruit dark red, sweetish. *May.*

2. Cannabis.—Flowers diœcious. Barren flowers in axillary racemes or panicles ; sepals 5 ; stamens 5. Fertile flowers spicate, clustered ; calyx of 1 entire sepal inclosing the ovary. Herbs.

C. SATIVA (*Hemp*).—Stem erect, tall, branching, rough ; leaves opposite ; upper ones alternate ; all digitately parted ; leaflets lanceolate or linear-lanceolate, coarsely serrate, dark green above, paler beneath ; flowers green ; barren in terminal panicles, fertile in spikes. Weed-like plant, 4–6 feet high. *June.* ①

3. Humulus.—Flowers diœcious ; barren in axillary panicles ; sepals 5 ; stamens 5 ; fertile in axillary spikes or aments ; bracts leafy, imbricated. Achenia invested in the persistent, enlarged calyx, forming a strobile. ♃

H. LUPULUS (*Hop*).—Stem twining with the sun, rough backward with reflexed prickles ; leaves opposite, cordate, 3–5-lobed or undivided ; stipules ovate, persistent ; barren flowers very abundant, greenish, in axillary panicles ; fertile in large strobiles or cones. *July.*

4. Urtica.—Flowers monœcious or diœcious. Calyx mostly of 4 sepals. Stamens 4. Stigma sessile, globular. Achenium compressed, smooth, invested in the calyx. Stinging herbs.

U. DIOICA (*Nettle*).—Stem erect, branching, very hispid and stinging, obtusely 4-angled ; leaves opposite, ovate, cordate at base ; flowers monœcious or diœcious, in branching, panicled spikes. A stinging weed, 2–4 feet high. *July-August.*

5. Ulmus.—Flowers perfect, rarely polygamous. Calyx campanulate, 4–9-cleft. Stamens 4–9. Styles 2. Ovaries flat. Fruit a flat samara with a winged margin, by abortion 1-celled, 1-seeded. Trees.

1. U. AMERICANA (*Elm*).—Young branches nearly smooth ; leaves oblong-obovate, doubly serrate, smooth above, pubescent beneath ; flowers

small, purplish, pedicellate, in lateral clusters, appearing before the leaves ; fruit oval, fringed with dense down. *April.*

2. U. FULVA (*Slippery Elm*).—Young branches rough-pubescent ; leaves oblong-ovate, acute, doubly serrate ; buds covered with a rust-colored down ; flowers nearly sessile, in dense clusters ; calyx hairy ; fruit nearly orbicular, naked on the margin.

6. **Celtis.**—Flowers monœcious-polygamous. Calyx 5–6-parted, persistent. Stamens 5-6. Stigmas 2, long, recurved. Drupe globular, 1-seeded. Trees or shrubs.

C. OCCIDENTALIS (*Hackberry*).—Leaves ovate, entire, oblique at base, rough, often cordate at base ; flowers small, greenish white, axillary, pedunculate, appearing at the same time as the leaves ; fruit globular, with a thin, sweet flesh, small, dull red. *May.*

Order LXXV.—JUGLANDACEÆ (*Walnut Family*).

Trees. Leaves unequally pinnate. Stipules none. Flowers greenish, monœcious. Sterile ones in aments. Calyx membranous, irregular. Stamens indefinite. Fertile flowers usually in small clusters. Calyx-tube adherent to the ovary ; limb 3–5-parted. Petals sometimes present, and as many as the calyx-segments. Ovary 1-celled, partially 2–4-celled, 1-ovuled. Fruit drupaceous, the epicarp sometimes indehiscent, sometimes regularly dehiscent ; endocarp bony. Seeds single, oily, often edible.

JUGLANDACEÆ. { Husk 4-valved—*Carya.*
{ Husk not 4-valved—*Juglans.*

1. **Juglans.**—Barren flowers in long and simple aments ; stamens 8–40, with very short filaments. Fertile flowers solitary, or several together ; calyx 4-parted ; corolla 4-petaled ; stigmas 2. Fruit drupaceous, with a spongy, indehiscent epicarp, and irregularly-furrowed endocarp. Trees.

1. J. CINEREA (*Butternut*).—Leaves 15–19-foliate ; leaflets oblong-lanceolate, serrate, obtuse at base, acuminate, pubescent ; fruit oblong, about 2' in length, clothed with a clammy pubescence, tapering to an obtuse point ; nut rough, with sharp, ragged ridges. *April–May.*

2. J. NIGRA (*Black Walnut*).—Leaflets numerous, 15–21, ovate-lanceolate, long-acuminate, serrate, somewhat pubescent beneath ; fruit globose, covered with rough dots ; nuts marked with rough ridges. *May.*

2. **Carya.**—Barren flowers in slender aments ; calyx 3-parted ; stamens 3-8, nearly destitute of filaments. Fertile flowers 2–3 together ; calyx 4-parted ; corolla none ; stigma 4-lobed. Fruit globular, inclosed in a 4-valved epicarp, which opens. Nut smooth, 4–6-angled. Trees.

1. C. ALBA (*Shag-bark*).—Leaflets about 5, lanceolate-obovate or oblong-lanceolate, acuminate, serrate; fruit globular, depressed at apex; nut somewhat compressed, covered with a thick epicarp, tapering abruptly at the end, thin-shelled, with a large, oily, delicious kernel. *May.*

2. C. TOMENTOSA (*Thick-shelled Walnut*).—Leaflets 7–9, oblong-lanceolate or obovate-lanceolate, acuminate, aments hairy; fruit between ovoid and globose; epicarp thick and almost woody; nut marked with about 6 angles, with a well-flavored kernel. *May.*

3. C. PORCINA (*Pig-nut*).—Leaflets 5–7, lanceolate or ovate-lanceolate, serrate, acuminate; fruit obovate or pyriform, with a thin, dry epicarp, opening not more than half-way; nut small, extremely hard, with a thickish shell, and a small, bitterish kernel. *May.*

4. C. AMARA (*Bitter-nut*).—Leaflets 7–11, oblong-lanceolate, serrate, acuminate, smooth; fruit globular, with a very thin and soft husk, opening half-way down; nut with a very thin shell, capable of being crushed by the fingers; kernel very bitter. *May.*

Order LXXVI.—MYRICACEÆ (*Sweet-Gale Family*).

Shrubs. Leaves simple, aromatic, dotted with resinous glands. Flowers monœcious or diœcious. Sterile ones in aments, each in the axil of a bract. Stamens 2–6. Anthers 2–4-celled, opening lengthwise. Fertile flowers in aments or globose clusters. Ovary 1-celled, 1-ovuled, surrounded by several scales. Stigmas 2, subulate, or dilated and somewhat petaloid. Fruit a drupe-like, 1-seeded nut.

MYRICACEÆ. { Flowers diœcious—*Myrica.*
{ Flowers monœcious—*Comptonia.*

1. **Myrica.**—Flowers diœcious. Barren on oblong aments, each contained in a bract; stamens 2–8. Fertile flowers in ovoid aments. Ovary solitary, with 2 filiform stigmas. Fruit a globular nut, covered with resinous scales. Shrubs.

M. GALE (*Sweet Gale*).—Leaves lanceolate, cuneate at base, serrate near the apex, on very short petioles, after the flowers. Barren aments clustered; scales ovate, cordate; fruit in dense, oblong heads. A low shrub with a dark brownish bark. Shrub, 3–4 feet high. *April.*

2. **Comptonia.**—Flowers monœcious. Barren flowers in cylindrical aments; bracts reniform, cordate; stamens 3–6. Fertile flowers in globular aments; calyx-scales 5–6. Styles 2. Nut ovoid, smooth, 1-celled. Shrubs.

C. ASPLENIFOLIA (*Sweet Fern*).—Leaves linear or linear-lanceolate, pinnatifid with rounded segments, thin, dark green, numerous, on short petioles, fragrant; stipules semi-cordate, in pairs; barren aments erect, oblong; fertile, rounded burrs, situated beneath the barren ; nut ovate, brown. A low shrub, 1–3 feet high, in dry woods. *April–May.*

Order LXXVII.—Cupuliferæ (*Oak Family*).

Trees or shrubs. Leaves simple, alternate, with straight veins and deciduous stipules. Flowers usually monœcious. Sterile ones or both in aments. Calyx membranous and regular, or else scale-like. Stamens 1–4 times as many as the calyx-lobes. Ovary 2–6-celled, with 1 or more ovules in a cell. Fruit a 1-celled, 1-seeded nut or samara.

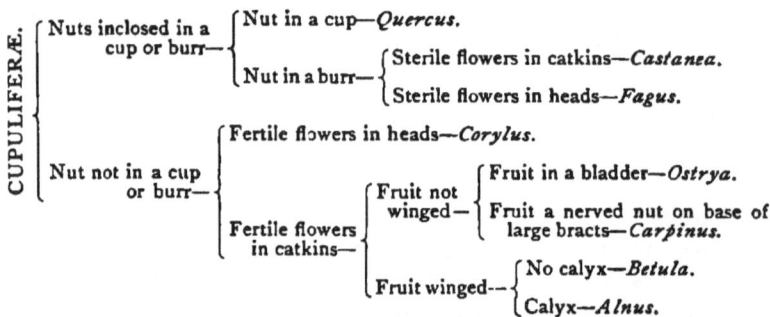

CUPULIFERÆ.

- Nuts inclosed in a cup or burr—
  - Nut in a cup—*Quercus.*
  - Nut in a burr—
    - Sterile flowers in catkins—*Castanea.*
    - Sterile flowers in heads—*Fagus.*
- Nut not in a cup or burr—
  - Fertile flowers in heads—*Corylus.*
  - Fertile flowers in catkins—
    - Fruit not winged—
      - Fruit in a bladder—*Ostrya.*
      - Fruit a nerved nut on base of large bracts—*Carpinus.*
    - Fruit winged—
      - No calyx—*Betula.*
      - Calyx—*Alnus.*

1. **Quercus.**—Barren flowers in loose, slender, nodding aments ; calyx 6–8-parted ; stamens 5–12. Fertile flowers solitary or clustered ; involucre or capsule cup-shaped, scaly ; ovary 3-celled, with 6 ovules ; stigma 3-lobed. Nut 1-celled, 1-seeded.

1. Q. ALBA (*White Oak*).—Leaves oblong or oblong-ovate, smooth, paler, and glaucous beneath, light green above, deeply and smoothly 5–7-lobed ; lobes oblong or oblong-linear, obtuse, nearly entire ; fruit pedunculate ; cup hemispherical, much shorter than the ovate acorn ; kernel sweetish, edible. A large forest-tree.

2. Q. BICOLOR (*Swamp White Oak*).—Leaves oblong-ovate, tapering and entire at base, white-downy underneath, coarsely and minutely 8–12-toothed, on short petioles ; teeth unequal, acutish ; fruit mostly in pairs, on long peduncles ; cup hemispherical, scarcely half as long as the oblong-ovate acorn ; kernel sweet. A tall tree, in swamps. *May.*

3. Q. PRINOIDES (*Dwarf Chestnut Oak*).—Shrubby ; leaves obovate, dentate, with coarse and nearly equal teeth, downy beneath, on short petioles ; fruit sessile, or on very short peduncles ; cup hemispherical ; acorn ovate ; kernel sweet. A dwarf species, in dry, sandy soils. *May.*

4. Q. RUBRA (*Red Oak*).—Leaves smooth, oblong, paler beneath, sinuately 7–11 lobed ; lobes spreading, entire or dentate, acute, with narrow sinuses between ; cup very flat and shallow, saucer-shaped, much shorter than the oblong-ovate acorn. A tall tree, in forests. *May.*

5. Q. COCCINEA (*Scarlet Oak*).—Leaves oblong or oval in outline,

smooth, deeply and sinuately 5-9-lobed, bright green on both sides, on long petioles ; lobes divaricate, sparingly toothed, with broad, open, deep sinuses, which extend two thirds to the mid-vein, or farther; cups very scaly. A large tree. *May.*

2. **Castanea.**—Barren flowers in separate clusters, in long aments ; calyx 5-6-parted ; perfect stamens 8-15. Fertile flowers 2-3 together, in a prickly, 4-lobed involucre ; calyx-border 5-6-lobed ; ovary 3-6-celled, with 10-15 ovules ; styles 3-6, capillary ; nuts 2-3 together, inclosed in the enlarged, thick, coriaceous involucre. Chiefly trees.

C. VESCA (*Chestnut*).—Leaves oblong-lanceolate, acuminate, marked with very prominent, straight veins ; sterile aments long, pendulous, fertile flowers yellowish white, very abundant, appearing after the leaves are full-grown ; nuts 2-3 together, of a peculiar, rich brown, hairy above, flattened on the sides, inclosed in a green, very prickly, 4-parted involucre. A large forest-tree. *July.*

3. **Fagus.**—Barren flowers in small heads, on nodding peduncles ; calyx 5-6-cleft ; stamens 5-12. Fertile flowers mostly 2 together, inclosed within a prickly involucre, bracted at base ; calyx with 4-5 subulate lobes ; ovary 3-celled ; cells 2-ovuled ; styles 3, filiform. Nuts 1-seeded, acutely triangular, 2 together in the prickly, 4-lobed involucre. Trees.

F. FERRUGINEA (*Beech*).—Leaves oblong-ovate, acuminate, with distinct teeth, light green, withering and mostly persistent in the winter ; flowers appearing with the leaves ; barren yellowish, soft-pubescent, in little, globular clusters, on slender peduncles, 2' long; nuts usually 2 together, dark brown, with an oily, sweet kernel. A forest-tree. *May.*

4. **Corylus.**—Barren flowers in long, drooping aments ; anthers 1-celled. Fertile flowers several together, in terminal and lateral heads ; calyx none ; ovaries several, 2-celled ; cells 1-ovuled ; stigmas 2, filiform. Nut bony, ovoid, surrounded by the enlarged, leafy coriaceous involucre. Shrubs.

C. AMERICANA (*Hazel-nut*).— Young branches glandular-pubescent ; leaves cordate, rounded, acuminate, coarsely serrate ; stigmas of the fertile buds red, forming a little tuft at the top of the bud ; involucre somewhat campanulate below, dilated. A shrub, 3-8 feet high. *April.*

5. **Ostrya.**—Barren flowers in drooping aments ; stamens 8-12, furnished with a roundish, ciliate bract, instead of a calyx. Fertile flowers numerous, with small, deciduous bracts in loose aments, each inclosed in an inflated, sac-like involucre. Ovary 2-celled, 2-ovuled.

O. VIRGINICA (*Iron-wood*).—Leaves ovate or oblong-ovate, acuminate, sheathing, and unequally serrate, petiolate, somewhat pubescent ; fertile aments resembling a cluster of hops ; involucre-scales bristly at base. A small, slender tree. *April-May.*

6. Carpinus.—Barren flowers in drooping, cylindrical aments; stamens 8–12, furnished with a roundish, ciliate bract instead of a calyx ; filaments very short ; anthers bearded at apex. Fertile flowers mostly in twos, each pair with a small, deciduous bract, contained in a large, 3-lobed involucre, each flower with a 2-celled, 2-ovuled ovary, terminating in 2 filiform stigmas. Nuts small, ovoid, furnished with an enlarged, open, and leaf-like scale. Trees.

C. AMERICANA (*Hornbeam*).—Leaves oval or oblong-ovate, acuminate, acutely and unequally serrate, petiolate ; fruiting aments drooping, long, loose, with the dark-brown nuts arranged by twos, each with an involucre. A small tree. *April–May.*

7. Betula.—Barren flowers in cylindric aments, each bract with 3 flowers, each flower consisting of 4 stamens. Fertile flowers in oblong-ovoid aments, 3 to each bract, with no calyx, each consisting of an ovary with 2 filiform stigmas. Fruit compressed. Trees.

1. B. LENTA (*Black Birch*).—Leaves ovate, serrate, cordate at base, acuminate, hairy on the veins beneath, as also the petioles ; fertile aments oval, erect, somewhat hairy ; lobes of the scales obtuse. A tree of rather large size, common in forests. *April–May.*

2. B. PAPYRACEA (*Paper Birch*).—Leaves ovate, acuminate, mostly cordate or obtuse at base, doubly serrate, dark green and smooth above, hairy on the veins beneath. A large tree. *April–May.*

3. B. ALBA, var. POPULIFOLIA (*White Birch*).—Leaves deltoid, with a very long acuminate point, truncate or hearted at base, smooth, and of a bright, shining green on both sides. A very slender and graceful tree, common in rocky woods and thickets. *April–May.*

8. Alnus.—Barren flowers in long, cylindrical, nodding aments ; scales 3-lobed, 3-flowered ; flowers with a 4-parted calyx and 4 stamens. Fertile flowers in ovoid aments ; bracts fleshy, 2-flowered, 3-lobed ; calyx-scales 4, minute. Shrubs.

1. A. INCANA (*Hoary Alder*).—Leaves broad-oval or ovate, somewhat cordate at base, sharply serrate, sometimes coarsely toothed, mostly white-downy underneath ; stipules oblong-lanceolate ; fertile aments oval. A shrub, 8–15 feet high. *March–April.*

2. A. SERRULATA (*Common Alder*).—Leaves obovate, acuminate, tapering at base, sharply and finely serrate, smooth and green on both sides, somewhat pubescent on the veins beneath ; fertile aments oblong-oval ; fruit ovate. A common shrub, 6–15 feet high. *March–April.*

Order LXXVIII.—SALICACEÆ (*Willow Family*).

Trees, or shrubs, rarely somewhat herbaceous. Leaves alternate, simple, with deciduous or persistent stipules. Flowers diœcious; both kinds in aments, achlamydeous, in the axils of 1-flowered bracts. Stamens 2–several, distinct or monadelphous. Ovary 1-celled, 2-valved. Seeds numerous, with a silky coma.

SALICACEÆ. { Catkin-scales entire, stigmas short—*Salix*.
{ Catkin-scales cleft, stigmas long—*Populus*.

1. **Salix.**—Aments cylindric; bracts entire, 1-flowered. Barren flowers—calyx none; stamens 2–8, rarely 1, accompanied by glands. Fertile flowers each with a gland at base; ovary simple; stigmas 2, short, mostly bifid. Shrubs and trees.

1. S. TRISTIS (*Sage Willow*).—Leaves nearly sessile, narrow-lanceolate, cuneate at base, acute or obtuse, woolly on both sides, at last nearly smooth; stipules minute, disappearing very early; aments small, nearly globular when young, loosely flowered; ovaries tapering to a long point, silky-pubescent; style short, stigmas bifid. A straggling shrub, 10′–18′ high. *April–May.*

2. S. HUMILIS (*Low Bush Willow*).—Leaves lanceolate or oblanceolate, abruptly acute or obtuse, petiolate, nearly smooth above; stipules usually present, varying from half ovate to lunate, entire or dentate; style long; stigma bifid. Stem 3–8 feet high. *April–May.*

3. S. VIMINALIS (*Osier, Basket Willow*).—Branchlets very long, straight and slender; leaves linear-lanceolate, very long-acuminate, nearly entire, white beneath, with silky pubescence; aments densely clothed with long hairs; ovary elongated; style filiform; stigmas linear. A large shrub, 10–15 feet high, in wet meadows. *May.*

4. S. CORDATA (*Heart-leaved Willow*).— Leaves lanceolate, or ovate-lanceolate, sharply serrate, acuminate at base, smooth, paler beneath; stipules reniform, dentate; aments appearing with the leaves, furnished with several leafy bracts at base; ovary pedicellate, smooth, lanceolate, acuminate. A variable shrub, 4–15 feet high, in wet grounds. *May.*

5. S. BABYLONICA (*Weeping Willow*).—Branchlets very long and slender, drooping; leaves lanceolate, acuminate, smooth, glaucous beneath; stipules roundish-oblique, acuminate; ovaries sessile, ovate, smooth. A beautiful tree, cultivated.

2. **Populus.**—Aments cylindrical. Bracts fringed and lobed at apex. Flowers on an oblique, turbinate disk. Stamens 8–30. Style very short. Stigma long, bifid. Aments drooping and preceding the leaves. Trees.

1. P. TREMULOIDES (*American Aspen, White Poplar*).—Leaves rounded-cordate, abruptly acuminate, dentate, smooth on both sides, pubescent on the margin, dark green, on flattened petioles; scales of the ament cut into several linear segments, fringed with hairs. A common tree in forests. *April.*

2. P. GRANDIDENTATA (*Large Poplar*).—Leaves roundish-ovate, acute, dentate, with large, unequal, sinuate teeth, white-downy beneath when

young; scales of the ament cut into several small, unequal segments, scarcely fringed with hairs. A large tree. *April.*

3. P. DILATATA (*Lombardy Poplar*).—Leaves deltoid, acuminate, smooth, serrate, as broad as long; trunk furrowed. A tall tree, of regular, pyramidal growth. *April.*

SUPERIOR ENDOGENS.

Order LXXIX.—ORCHIDACEÆ (*Orchid Family*).

Perennial, often acaulescent herbs. Leaves simple, entire, parallel-veined. Flowers very irregular. Perianth of 6 segments, all usually colored, and assuming various forms, especially the lowest of the 4 inner segments, or lip, which is often spurred. Stamens 3, consolidated with the style into a column, only the central one fertile; sometimes the two lateral ones fertile, and the central one abortive. Pollen sometimes granular and powdery, but more commonly cohering in wax-like masses, which are usually attached to a gland of the stigma. Ovary twisted, adherent to the tube of the perianth, 1-celled, many-ovuled. Fruit a 3-ribbed, 3-valved capsule.

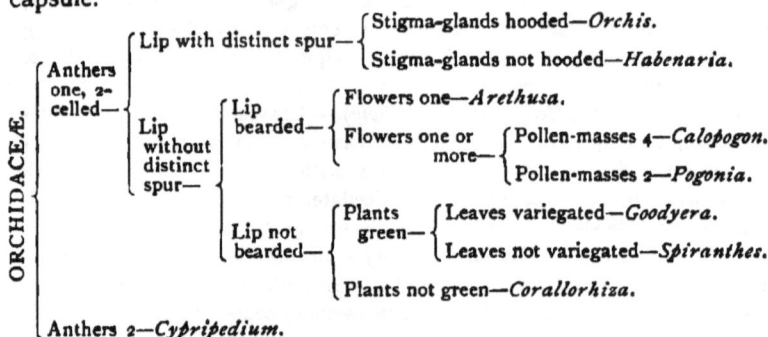

ORCHIDACEÆ—

Anthers one, 2-celled—
- Lip with distinct spur—
  - Stigma-glands hooded—*Orchis.*
  - Stigma-glands not hooded—*Habenaria.*
- Lip without distinct spur—
  - Lip bearded—
    - Flowers one—*Arethusa.*
    - Flowers one or more—
      - Pollen-masses 4—*Calopogon.*
      - Pollen-masses 2—*Pogonia.*
  - Lip not bearded—
    - Plants green—
      - Leaves variegated—*Goodyera.*
      - Leaves not variegated—*Spiranthes.*
    - Plants not green—*Corallorhiza.*

Anthers 2—*Cypripedium.*

1. **Corallorhiza.**—Sepals and petals nearly equal; lateral ones erect; upper vaulted. Lips spreading and recurved, usually produced behind into a short spur, which adheres to the top of the ovary. Pollen-masses 4, oblique to each other. Leafless.

C. MULTIFLORA (*Coral-root*).—Scape many-flowered; lip cuneate, oval, 3-lobed, spotted with bright purple; middle lobe recurved, much longer than the lateral, tooth-like lobes; spur conspicuous; capsule and ovary oblong. A leafless plant, 12′-18′ high, with a brownish-purple scape, probably parasitic on the roots of trees. *July–August.*

2. **Orchis.**—Flower ringent. Sepals and petals nearly equal, most of them converging and vaulted. Lip depressed, attached to the base

of the column, and terminating in a spur. Anthers all parallel, close together. Pollen in small grains combined in 2 large masses.

O. SPECTABILIS (*Showy Orchis*).—Acaulescent ; scape low, with about 5 acute angles ; leaves few, mostly 2, nearly as long as the scape, oblong-obovate, somewhat thick in texture, of a shining green ; flowers large, few, showy ; bracts lanceolate, acute, large and leaf-like ; petals and sepals arched, mostly pink, rarely white ; lip white, entire ; spur obtuse, white. In rich, rocky woods. *May.*

3. **Habenaria.**—Flower ringent. Sepals and petals nearly equal. Lip depressed, attached to the base of the column, terminating in a spur. Anther-cells parallel. Pollen-masses 2, pedicellate, attached to the 2 glands of the stigma.

1. H. TRIDENTATA (*Naked Orchis*).—Stem slender, bearing a large, oblong or oblong-ovate, obtuse leaf toward the base, and several smaller ones above ; flowers small, greenish, rather numerous, in a short, oblong spike ; lip truncate, 3-toothed ; spur slender, curved upward, clavate, longer than the ovary. *July.*

2. H. PSYCODES (*Small-fringed Orchis*).—Stem leafy, angled ; leaves oblong, upper ones lanceolate ; flowers small, purple, fragrant, in a dense raceme ; lower sepals roundish-elliptical, obtuse ; petals toothed at the extremity ; lip 3-parted, tapering at base ; spur longer than the ovary. Common in wet meadows. *July-August.*

4. **Arethusa.**—Flowers ringent. Sepals and petals lanceolate, nearly equal, cohering at base. Lip destitute of spur, spreading and recurved, bearded inside. Column dilated, attached at base to the lip. Anther-cells 2, approximate. Pollen-masses 4.

A. BULBOSA (*Arethusa*).—Mostly leafless in blossom ; scape low, sheathed, arising from a round bulb, and bearing a single, large, fragrant, purple flower. The sheaths contain a linear leaf, which sometimes appears with the flower, but generally follows it. In bogs, 4'-8' high. *May.*

5. **Pogonia.** — Flowers ringent. Sepals and petals somewhat spreading, not united at base. Lip crested or 3-lobed, bearded inside. Column clavate, not winged at apex. Anther pedicellate. Pollen-masses 2, 1 in each cell.

P. OPHIOGLOSSOIDES (*Adder's-tongue Arethusa*).—Stem slender, 1-leaved and 1-bracted ; leaf situated at the middle of the stem, oval or lanceolate, clasping ; flowers mostly solitary, rarely 2-3, large, pale purple or pink, somewhat nodding ; lip spatulate. In meadows and bogs. *June-July.*

6. **Calopogon.**—Sepals and petals nearly equal, distinct at base. Lip on the upper side of the flower, tapering to a claw or stalk, expanded at summit, bearded above. Column free, winged at apex. Pollen-masses 2, angular.

C. PULCHELLUS (*Grass Pink*).—Scape slender, 1-leaved, several-flowered, arising from a solid bulb; leaf linear, grass-like, veiny, sheathing the base of the scape; flowers large, bright reddish-purple, fragrant, 2–4 in number; bracts minute; lip concave, dilated at summit, bearded inside. In meadows and bogs, 8'–12' high. *July.*

7. **Spiranthes.**—Flowers somewhat ringent, in a mostly spiral spike. Upper sepal cohering with the petals. Lip oblong, concave, tapering at base, and furnished with 2 callous processes. Column arching, clavate, on a short, oblique pedicel. Stigma beaked, at length 2-cleft. Pollen-masses 2, composed of scarcely cohering grains.

1. S. GRACILIS (*Slender Ladies' Tresses*).—Scape very slender, smooth; leaves all radical, ovate, or oval-lanceolate, petiolate; flowers small, pearly white, arranged in a single row, which usually winds spirally round the axis; bracts ovate, acute; lip narrow-obovate, crisped at the summit. In dry fields, 6'–12' high. *July–August.*

2. S. CERNUA (*Ladies' Tresses*).—Stem rather stout, pubescent above; radical leaves linear-lanceolate; flowers whitish or cream-color, crowded in a dense spike, fragrant, somewhat pubescent; bracts ovate-lanceolate; lip oblong, dilated and crisped at apex. In wet ground, 7'–15' high. *August–October.*

8. **Goodyera.**—Flowers ringent. Upper sepal cohering with the petals, vaulted. Lower sepals situated beneath the lip, which is sessile. Column straight. Pollen-masses 2, composed of angular grains.

G. PUBESCENS (*Rattlesnake Plantain*).—Scape pubescent; leaves all radical, in tufts, ovate, petiolate, dark green, veined and blotched with white; flowers small, greenish white, like the scape glandular-pubescent, arranged in a dense spike; the roundish, inflated lip ovate, ending in an abrupt point. In rich woods, 6'–12' high. *July–August.*

9. **Cypripedium.**—Sepals spreading; the 2 lower or forward ones united into 1 below the lip, rarely distinct. Petals alike. Lip a large, inflated, obtuse sac. Column 3-lobed; lateral lobes anther-bearing.

1. C. PARVIFLORUM (*Yellow Lady's Slipper*).—Stem leafy, pubescent; leaves oval, acuminate, strongly parallel-veined, pubescent; flowers large, mostly solitary; sepals ovate, or ovate-lanceolate; petals elongated, greenish, striped and spotted with brownish purple; lip large, inflated, yellow, spotted inside. In moist, rich woods, 10'–20' high. *May–June.*

2. C. ACAULE (*Lady's Slipper*).—Acaulescent; stem naked, 2-leaved at base, 1-flowered; leaves oval or oblong, acute, strongly veined, pubescent; flower large, bracted; sepals oblong-lanceolate; petals linear; lip dependent, purple, reticulate, longer than the petals, opening by a fissure on the upper side; middle lobe of the column rhomboidal, acuminate, deflexed. In moist, and especially in evergreen woods, 6'–12' high. *May–June.*

Order LXXX.—IRIDACEÆ (*Iris Family*).

Perennial herbs, arising from rhizomas, bulbs, or corms, rarely with fibrous roots. Leaves equitant. Flowers often showy, usually arising from a spathe. Perianth-tube adherent to the ovary; limb colored, 6-parted; the divisions usually in 2 obvious, often unequal series. Stamens 3, distinct or monadelphous, with extrorse, 2-celled anthers. Ovary 3-celled. Style 1. Stigmas 3, dilated or petaloid. Capsule 3-celled, 3-valved, with loculicidal dehiscence.

IRIDACEÆ.
{ Perianth of unequal parts, stamens 3—*Iris.*
{ Perianth of equal parts— { Leaves grass-like—*Sisyrinchium.*
{ Leaves not grass-like—*Crocus.*

1. **Iris.**—Perianth 6-cleft, 3 outer segments reflexed, larger than the 3 inner ones. Stamens 3, distinct, placed before the 3 inner segments of the perianth. Stigmas 3, petaloid. Capsule 3-6-angled. Rhizoma creeping or tuberous.

1. I. VERSICOLOR (*Blue Flag*).—Stem stout and thick, acute on one side, simple or branching; leaves sheathing at base; flowers beardless, large, blue, showy; outer segments of the perianth variegated with green, yellow, and white, and veined with purple; ovary obtusely triangular. Common in wet meadows, 1-2 feet high. *June.*

2. I. GERMANICA (*Flower-de-Luce*).—Stem often branching, many-flowered; leaves shorter than the stem; spathe membranous at apex; flowers large, beardless, light blue or bluish white; segments of the perianth emarginate; stigmas acute, serrate. In gardens, 1-3 feet high. *May-June.*

2. **Sisyrinchium.**—Perianth arising from a 2-parted spathe, 6-parted; segments similar, spreading. Stamens monadelphous. Stigmas 3, filiform. Capsule globose, somewhat 3-angled.

S. BERMUDIANA (*Blue-eyed Grass*).—Scape simple or branching, compressed, winged, naked or 1-2-leaved; leaves narrow-linear, grass-like; spathe with 2-5 blue flowers, opening in succession; pedicels filiform; segments of the perianth obovate. In moist meadows and grassy fields, 6'-12' high. *June-July.*

3. **Crocus.**—Perianth funnel-form, with a very long, slender tube, arising from a radical spathe. Stigma 3-cleft, crested; segments convolute.

C. VERNUS (*Crocus*).—Flowers large, with a very long, slender tube, sessile on the bulb; anthers sagittate, yellow; stigma included in the perianth, with short segments; scape triangular, rising after flowering, and bearing the ovary; leaves narrow-linear, following the flowers. Garden plant. *March-April.*

Order LXXXI.—Amaryllideæ (*Amaryllis Family*).

Perennial herbs, arising from bulbs, rarely with fibrous roots. Leaves parallel-veined. Flowers showy, mostly on scapes, and arising from spathes. Perianth regular. Stamens 6, with introrse anthers, arising from the segments of the perianth. Ovary 3-celled. Style 1. Stigma 3-lobed. Fruit a capsule or berry.

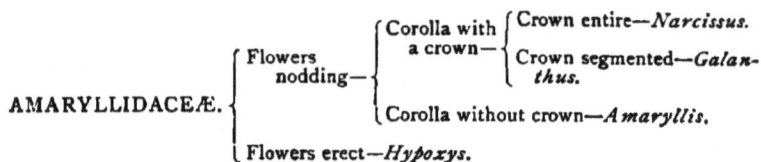

AMARYLLIDACEÆ.
— Flowers nodding —
— Corolla with a crown —
— Crown entire—*Narcissus.*
— Crown segmented—*Galanthus.*
— Corolla without crown—*Amaryllis.*
— Flowers erect—*Hypoxys.*

1. **Amaryllis.**—Perianth 6-parted, nodding, somewhat funnel-shaped ; segments petaloid, nearly similar, somewhat unequal. Stamens 6, inserted at the throat. Capsule 3-parted. Seeds flat, numerous.

A. FORMOSISSIMA (*Jacobæan Lily*).—Acaulescent ; scape 1-flowered ; leaves linear, or narrow-oblong, thick ; flower large, nodding, of a brilliant dark red ; segments very much spreading ; tube fringed ; spathe red. Cultivated. *June–August.*

2. **Narcissus.**—Perianth of 6 equal divisions, furnished with a crown at the throat, consisting of a whorl of sterile stamens completely united in a tube. Fertile stamens 6, inserted within the crown. Root bulbous.

1. N. JONQUILLA (*Jonquil*).—Scape slender, 1-2-flowered ; leaves narrow ; flowers large, rich yellow, very fragrant, rising from a long spathe ; segments of the perianth spatulate, reflexed ; crown shorter than the segments, flat, shallow, and somewhat like a saucer, spreading, crenate. Cultivated. *May-June.*

2. N. POETICUS (*Narcissus*).—Scape 1-flowered ; leaves linear, as long as the scape ; flower large, mostly white, fragrant ; segments imbricated at base, reflexed ; crown expanded, flat, white variegated with circles or rings of crimson and yellow. Cultivated. *June.*

3. N. PSEUDO-NARCISSUS (*Daffodil*).—Scape erect, 2-edged, striate, 1-flowered ; leaves linear, striate ; flower very large, sulphur-yellow ; crown very long, cup-shaped, serrate on the margin. A garden plant. *April-May.*

3. **Galanthus.**—Flowers arising from a spathe. Perianth 6-parted, outer segments concave ; tube furnished with a crown of 3 small, emarginate segments.

G. NIVALIS (*Snow-drop*).—Stem usually 2-leaved near the summit, 1-flowered ; leaves linear, acute ; flower large, nodding, snow-white. Common in gardens, 3'-8' high. *April.*

4. Hypoxys.—Perianth persistent, 6-parted, spreading. Stamens 6. Capsule crowned with the withered perianth, narrowed at base. Seeds roundish.

H. ERECTA (*Star-grass*). — Hairy, acaulescent ; scape 1-4-flowered ; leaves linear, grassy, longer than the scape ; flowers in a sort of umbel, each with a minute bract at base ; segments hairy, yellow within, greenish without. In meadows and grassy fields, 3'-6' high. *June-July.*

INFERIOR ENDOGENS.

Order LXXXII.—ALISMACEÆ (*Water-Plantain Family*).

Marsh herbs, acaulescent. Leaves parallel-veined, but often with reticulated veinlets, sometimes linear and fleshy. Flowers regular, perfect, or monœcious, usually in racemes or panicles. Perianth of 6 sepals, in 2 series. Sepals herbaceous. Petals often colored, sometimes similar to the calyx. Stamens definite or indefinite. Carpels 3 or more, 1-celled, 1-seeded. Styles and stigmas as many as the carpels.

ALISMACEÆ. { Leaves rush-like—*Triglochin.*
 Leaves with blades— { Flowers perfect—*Alisma.*
 Flowers imperfect—*Sagittaria.*

1. Alisma.—Flowers perfect. Stamens 6. Ovaries numerous, arranged in a circle. Achenia compressed, coriaceous.

A. PLANTAGO (*Water Plantain*).—Leaves oval or ovate, varying to oblong, or even lanceolate, acuminate, on long petioles, 3-9-nerved ; flowers in a loose panicle ; petals small, deciduous, white, with a purplish tinge, longer than the green sepals. In wet grounds, 1-2 feet high. *July-August.*

2. Sagittaria.—Flowers monœcious, rarely diœcious. Stamens numerous. Ovaries many, aggregated in a spherical head, and forming in fruit a globose head of compressed, winged achenia. ♃

S. VARIABILIS (*Arrowhead*).—Scape simple, sheathed at base by the petioles, acaulescent ; flowers in verticils of 3 ; sterile ones at the summit of the scape, fertile ones below ; petals white ; leaves extremely variable, generally triangular with sagittate lobes, varying from very broad to very narrow. In other forms, the leaves are oval or oblong, with thick, spongy petioles. Again, the leaves are linear, and the scape and petioles very slender. Common in wet grounds. *June-July.*

3. Triglochin.—Sepals and petals concave, greenish, deciduous. Stamens 6. Filaments very short. Anthers large, oval. Pistils a compound ovary of 3–6 cells. Stigmas sessile. Capsule dividing at length into 3–6 1-seeded carpels.

T. MARITIMUM (*Arrow-grass*).—Scape naked, fleshy, angled; leaves very narrowly linear, fleshy, semi-cylindric, shorter than the scape; flowers small, green, pedicellate, arranged in a long, loose raceme, destitute of bracts; fruit ovate, composed of 6 carpels, rounded at base. In salt marshes, 8′–15′ high. *August.* ♉

Order LXXXIII.—NAIADACEÆ (*Pondweed Family*).

Aquatic plants, with cellular leaves. Flowers inconspicuous, perfect, monœcious, or diœcious. Perianth of 4 distinct sepals, rarely monosepalous. Stamens definite, 4, 2, or 1. Ovaries 1; or else 2–4, free from the perianth, distinct. Stigma simple, often sessile. Fruit 1-celled, 1-seeded, indehiscent.

NAIADACEÆ. { Flowers perfect—*Potamogeton.*
{ Flowers imperfect—*Naias.*

1. **Naias.**—Flowers monœcious or diœcious, axillary, solitary. Barren flowers—stamen 1, in a little spathe. Fertile—style 1, short; stigmas 2–4; ovary 1. Achenium 1-seeded, in a loose, membranous sheath.

N. FLEXILIS (*Water Nymph*).—Aquatic; stem very slender, dichotomously branched; leaves opposite or whorled, very narrowly linear, sheathing at base; flowers minute, axillary. In ponds and slow waters. Stem 5′–20′ long. *July–September.*

2. **Potamogeton.**—Flowers perfect, spicate. Sepals 4. Stamens 4. Anthers 2-celled. Ovaries 4. Achenia 4, sessile, flattened on the inner side. Floating aquatics. ♉

1. P. NATANS (*Broad-leaved Pondweed*).—Stem nearly or quite simple; upper leaves ovate or broadly elliptical, varying to oblong-lanceolate, rounded or cordate at base; immersed ones lanceolate, linear or capillary, all on long petioles; spike of purplish flowers raised above the water. Ponds and slow waters. *July–September.*

2. P. GRAMINEUS (*Various-leaved Pondweed*).—Stems slender, mostly branched below; upper leaves oval, on long petioles; immersed ones lanceolate, varying to narrow-linear, or even capillary; lower ones sessile; spikes cylindrical, somewhat loose, on long peduncles thicker than the stem. Common in shallow, stagnant, and slow waters. *July–August.*

3. P. PERFOLIATUS (*Clasping Pondweed*).—Stem branching dichotomously; leaves alternate, ovate, sometimes broad-ovate, obtuse, cordate and clasping at base; spike purplish, loosely-flowered, on a short peduncle. *July–August.*

Order LXXXIV.—Typhaceæ (*Cat-tail Family*).

Herbs of marshes and ditches. Stems without joints. Leaves rigid. Flowers monœcious, arranged in a spadix, which is destitute of a spathe, or in globose heads. Perianth of 3 sepals, or none. Stamens 3-6, with long and slender filaments, and cuneiform anthers. Ovary free from the perianth, 1-celled, 1-seeded. Stigmas 1-2. Fruit a utricle.

TYPHACEÆ. { Flowers in heads—*Sparganium.*
{ Flowers in spikes—*Typha.*

1. **Typha.**—Flowers in a long, terminal, cylindrical spike, upper part staminate. Stamens intermingled with hairs, with 3 anthers on a common filament. Ovaries below, surrounded by numerous bristles. ♃

T. LATIFOLIA (*Cat-tail*).—Stem erect, simple ; rhizoma creeping ; leaves ensiform, nearly flat ; barren and fertile spikes contiguous, mostly forming one long, compact, cylindrical spike, turning brownish in fruit. Common in swamps, 3-5 feet high. *July.*

2. **Sparganium.**—Flowers in separate, globose heads, with leafy bracts ; upper ones barren scales ; lower fertile ; ovaries surrounded by 3-6 calyx-like scales. Fruit turbinate, 1-2-celled, 1-2-seeded. ♃

S. EURYCARPUM (*Burr-reed*).—Stem erect, branching above ; leaves linear, triangular at base, with concave sides ; flowers in globular clusters, resembling burrs, of a whitish green, lowest mostly pedicellate ; stigma linear, longer than the style, often 2. Around ponds and in ditches. *July-August.*

Order LXXXV.—Araceæ (*Arum Family*).

Herbs, or tropical shrubs, with a fleshy rhizoma or corm. Leaves sheathing at base, simple or compound, sometimes with more or less reticulated veins. Flowers usually sessile in a terminal or lateral spadix, sometimes monœcious ; sometimes perfect with a perianth of 4-6 sepals. Stamens definite in the perfect flowers, 4-6, usually indefinite in the monœcious flowers. Ovary free from the perianth, 1-several-celled. Seeds solitary, or several. Fruit usually a proper berry, sometimes dry.

ARACEÆ. { Leaves with expanded blade— { Flowers without calyx and corolla— { Leaves compound—*Arisæma.*
{ Leaves simple— { Leaves sagittate—*Peltandra.*
{ Leaves cordate—*Calla.*
{ Flowers with perianth—*Symplocarpus.*
{ Leaves linear—*Acorus.*

1. **Arum.**—Flowers mostly monœcious, situated at the base of a spadix, which is naked above. Fertile flowers below. Barren above, inclosed in a cucullate spathe. Perianth none. Berries distinct, 1-celled, several-seeded. ♃

A. TRIPHYLLUM (*Indian Turnip*).—Acaulescent; leaves mostly 2, on long petioles, sheathing at base, ternate; leaflets ovate or oval, acuminate, sessile; spadix shorter than, and included within the ovate, acuminate spathe, which is flattened and bent over the top of the spadix, and is frequently marked with dark purple or whitish spots or stripes, otherwise of a dark, shining green. Common in rich, rocky woods. *May.*

2. **Peltandra.**—Flowers monœcious, covering the long spadix. Spathe elongated, convolute. Perianth none. Anthers on the upper part of the spadix, sessile, peltate. Ovaries at the base of the spadix. Berries distinct, 1-celled, 1–3-seeded. ♃

P. VIRGINICA (*Arrow Arum*).—Acaulescent; leaves oblong-sagittate, acute at apex, with obtuse lobes, on long petioles, dark, shining green, of large size; spathe elongated, curved at apex, dark green, enveloping the slender spadix; ovaries sessile, becoming a bunch of green berries. A water-plant, 10′-18′ high. *June–July.*

3. **Calla.**—Spathe ovate, spreading, persistent. Spadix covered with flowers, the lower perfect, the upper often entirely staminate. Perianth none. Berries distinct, several-seeded.

C. PALUSTRIS (*Water Arum*).—Acaulescent; leaves cordate, on long, sheathing petioles, shining green, large and smooth. Herb in cold bogs with creeping root-stock.

4. **Acorus.**—Spathe none; spadix lateral, sessile, densely flowered. Flowers perfect. Sepals 6. Stamens 6. Anthers reniform. Stigmas sessile, minute. Ovaries 2–3-celled, becoming dry and few-seeded. ♃

A. CALAMUS (*Sweet Flag*).— Rhizoma creeping; leaves long, light green; scape long, resembling the leaves, bearing the sessile spadix on its edge, just above the middle; spadix covered with yellowish-green flowers. Common in wet grounds. *June–July.*

5. **Symplocarpus.**—Spathe with an incurved point, fleshy. Spadix pedunculate, oval, entirely covered with the perfect flowers. Sepals 4, persistent. Stamens 4. Style 4-angled. Stigma minute. Seeds large, globular, imbedded in the enlarged, spongy spadix.

S. FŒTIDUS (*Skunk Cabbage*).—Acaulescent; leaves ovate, cordate at base, acute, on short petioles, at length very large; spadix preceding the leaves, enveloped in a spathe, striped with purplish brown; flowers crowded on the spadix, dull-purple. Common in swamps and wet meadows, with offensive odors, resembling that of a skunk. *March–April.*

Order LXXXVI.—Liliaceæ (*Lily Family*).

Herbs, arising from bulbs or tubers, rarely with fibrous or fas-cicled roots. Leaves simple, sheathing or clasping at base. Flowers regular, perfect, often showy. Perianth with 6, rarely 4, equal, usually colored segments, free from the ovary. Stamens 6, rarely 4, inserted on the segments of the perianth. Anthers introrse. Ovary 3-celled. Styles united into 1. Stigma simple, or 3-lobed. Fruit capsular, with several or many seeds in each cell.

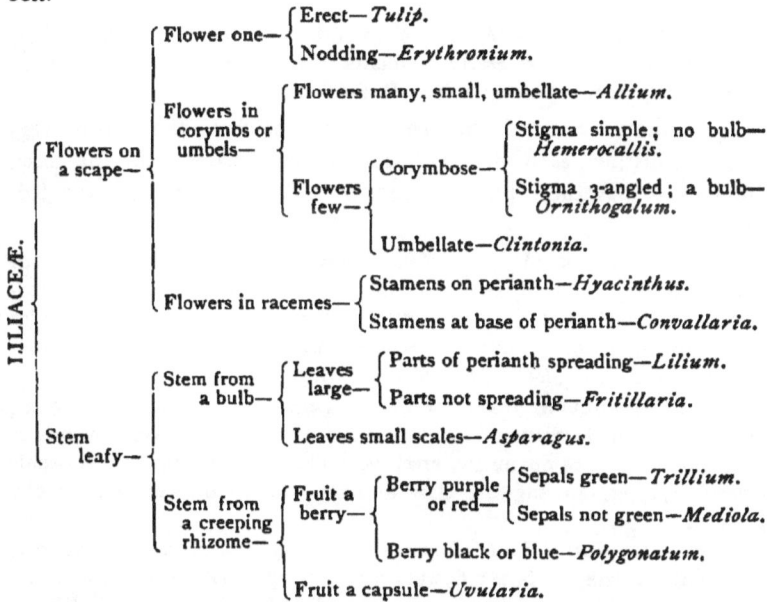

LILIACEÆ.

```
                    ┌ Flower one—  ┌ Erect—Tulip.
                    │              └ Nodding—Erythronium.
                    │
                    │              ┌ Flowers many, small, umbellate—Allium.
  ┌ Flowers on      │ Flowers in   │                ┌ Stigma simple; no bulb—
  │   a scape—      │ corymbs or   │                │   Hemerocallis.
  │                 │ umbels—      │ Flowers ┌ Corymbose— 
  │                 │              │ few—    │        └ Stigma 3-angled; a bulb—
  │                 │              │         │            Ornithogalum.
  │                 │              └         └ Umbellate—Clintonia.
  │                 │
  │                 └ Flowers in racemes— ┌ Stamens on perianth—Hyacinthus.
  │                                       └ Stamens at base of perianth—Convallaria.
  │
  │                 ┌ Stem from   ┌ Leaves  ┌ Parts of perianth spreading—Lilium.
  │                 │   a bulb—   │ large—  └ Parts not spreading—Fritillaria.
  │ Stem            │             └ Leaves small scales—Asparagus.
  └ leafy—          │
                    │             ┌ Fruit a  ┌ Berry purple ┌ Sepals green—Trillium.
                    │ Stem from   │ berry—   │   or red—    └ Sepals not green—Mediola.
                    │ a creeping  │          └ Berry black or blue—Polygonatum.
                    └ rhizome—    └ Fruit a capsule—Uvularia.
```

1. **Trillium.**—Sepals 3, green, persistent. Petals 3, colored, at length withering. Stamens 6. Anthers linear, with short filaments. Stigmas persistent. Berry 3-celled ; cells several-seeded. Stem simple, 1-flowered, whorl of 3 leaves.

1. T. CERNUUM (*Nodding Trillium*).—Leaves broad-rhomboidal, abruptly acuminate, nearly sessile ; flower nodding beneath the leaves, on a recurved peduncle, white ; petals oblong-ovate, acute, recurved, scarcely longer than the sepals. In wet woods, 8'–15' high. *May-June.*

2. T. ERECTUM (*Purple Trillium*).—Leaves broad-rhomboidal, abruptly and sharply acuminate, sessile ; peduncle nearly erect, soon reclining ; petals ovate, flat, spreading, scarcely longer, but much broader than the sepals, dull purple ; ovary brownish purple. In low, rich woods, 10'–15' high. *May.*

2. Medeola.—Perianth revolute, consisting of 6 petaloid, similar, oblong, deciduous segments. Stamens 6, with filiform filaments. Stigmas 3, long and recurved. Berry globose, 3-celled, several-seeded.

M. VIRGINICA (*Cucumber Root*).—Stem erect, slender, simple, covered with soft locks of wool; leaves in 2 whorls; lower one near the middle of the stem, consisting of 5–8 obovate-lanceolate, acuminate leaves; upper one of 3 ovate, acuminate, smaller leaves; flowers appearing in succession, yellowish green; styles dark red. In rich, damp woods. *June–July.*

3. Tulipa.—Perianth campanulate; segments 6. Stamens 6, short, subulate. Anthers 4-angled. Stigmas thick. Capsule oblong, 3-angled.

T. GESNERIANA (*Tulip*)—Scape smooth, 1-flowered; leaves radical, ovate-lanceolate; flowers erect; segments of the perianth obtuse, smooth. A universally admired exotic bulb. *May–June.*

4. Lilium.—Perianth campanulate or somewhat funnel-form; segments 6, distinct, each with a honey-bearing furrow near the base. Stamens 6. Anthers linear. Capsule oblong, somewhat 3-angled. Seeds flat.

1. L. CANADENSE (*Yellow Lily*).—Leaves in several remote whorls of 3–6, lanceolate, 3-nerved, rough on the margins and nerves; flowers nodding, campanulate, few, yellow, often tinged with scarlet, spotted with purple inside, on long peduncles; sepals sessile, revolute from the middle. In wet meadows, 2–3 feet high. *June–July.*

2. L. PHILADELPHICUM (*Red Lily*).—Leaves linear-lanceolate, acute; lower ones usually scattered; upper ones verticillate in several whorls of 5–7; flowers 1–4, campanulate, erect, vermilion-red, spotted inside; sepals lanceolate, erect, tapering to a claw at base. In dry thickets and shrubby pastures.

3. L. CANDIDUM (*White Lily*).—Stem erect, thick; leaves scattered, lanceolate, tapering at base; flowers large, campanulate, snow-white, in a terminal umbel, very fragrant, smooth inside. In gardens, 3–4 feet high. *July.*

4. L. TIGRINUM (*Tiger Lily*).—Stem tall, bulb-bearing; leaves scattered, 3-veined, lanceolate; upper ones ovate, cordate at base; flowers large, dark orange, spotted with brownish purple, in a pyramidal raceme; segments of the perianth revolute, covered with glandular projections on the inside. In gardens, 5–6 feet high. *July–August.*

5. Fritillaria. — Perianth campanulate; segments 6, broad at base, with a honey-bearing cavity just above the claw. Stamens 6, as long as the petals.

F. IMPERIALIS (*Crown-Imperial*). — Stem thick, leafy below, naked above; leaves mostly linear-lanceolate, long and narrow, entire; flowers large, nodding, pedicellate, in a terminal cluster; pedicels each furnished with a pair of small, narrow leaves, which, together, form a sort of terminal crown, beneath which the flowers hang. In gardens, 2–3 feet high. *April–May.*

6. Erythronium.— Perianth campanulate ; segments 6, distinct, recurved, deciduous, the 3 inner usually with a groove at base. Filaments 6, subulate. Style elongated. Capsule obovate, 3-valved, Seeds ovate.

E. AMERICANUM (*Dog-tooth Violet*).—Nearly stemless ; scape about 2-leaved near the base, 1-flowered ; leaves oval-lanceolate or lanceolate, green, spotted with brownish purple, nearly equal in length ; flower nodding, pale yellow, spotted at base inside ; style clavate ; stigma undivided. Common on rich hill-sides, 3′–6′ high. *May.*

7. Hemerocallis. — Perianth funnel-form ; tube short ; limb spreading, 6-parted. Stamens 6, inserted at the throat. Filaments long and filiform. Stigma simple.

H. FULVA (*Day Lily*).—Scape erect, smooth, corymbosely branching above ; leaves long-linear ; flowers large, erect, bracted, of a tawny red on the inside, in a corymb ; outer sepals with branching veins, inner wavy, obtuse. In gardens, 2–4 feet high. *July.*

8. Allium.—Flowers in an umbel, with a spathe at base. Perianth of 6 distinct sepals. Stamens 6. Capsule 3-lobed. Seed black.

1. A. TRICOCCUM (*Wild Leek*).—Scape naked, leaves oval-lanceolate, flat, thin, smooth, tapering to a petiole, withering before the appearance of the flowers ; umbel not bulb-bearing, many-flowered, globose ; flowers white ; filaments undivided ; pod deeply 3-lobed. In damp, rich woods, 8′–15′ high. *June–July.*

2. A. CEPA (*Onion*).—Scape stout, hollow, swelling below the middle, glaucous ; leaves round, hollow, swelling below the middle, glaucous, shorter than the scape ; umbel globose, many-flowered ; flowers greenish white. 3–4 feet high. *July.*

3. A. SATIVUM (*Garlic*).—Bulb compound, consisting of several smaller ones united, and included in one covering membrane ; stem leafy, bulbiferous ; leaves linear ; flowers small, white ; stamens 3-cleft. Scape 2-feet high. *July.*

9. Ornithogalum.—Perianth leafy, 6-parted ; segments spreading above the middle. Filaments 6, dilated at base. Stigma 3-angled. Capsule roundish, angled. Seeds roundish.

O. UMBELLATUM (*Star of Bethlehem*).—Scape naked ; leaves narrow-linear, channeled, as long as the scape ; flowers few, loosely corymbose, pedicellate, bracted ; sepals white, marked with a green stripe on the outside. Cultivated, 5′–8′ high. *May.*

10. Hyacinthus.—Perianth varying from funnel-form to campanulate, subglobose ; segments 6, similar. Stamens 6, inserted near the middle of the segments. Ovary with 3 honey-bearing pores.

H. ORIENTALIS (*Hyacinth*).—Scape naked ; leaves linear-lanceolate, half as long as the scape ; flowers in a dense, terminal, and somewhat thyrsoid raceme ; perianth funnel-form, cleft to the middle, swelling at base. Cultivated and admired, 6'–12' high. *April–May.*

11. **Convallaria.**—Flowers racemed. Perianth campanulate, 6-parted ; segments spreading, united at base. Stamen 6, at the base of the segments. Berry globose, 2-celled.

C. MAJALIS (*Lily of the Valley*).—Scape smooth, naked, semi-cylindric ; leaves usually 2, situated near the base of the scape, ovate or elliptic-ovate ; flowers white, fragrant, in a simple, loose, 1-sided raceme. Cultivated in gardens, 5'–6' high. *May.*

12. **Clintonia.**—Flowers umbellate. Perianth campanulate. Sepals 6, distinct, deciduous. Stamens 6, inserted at the base. Style long, filiform, columnar. Ovary 2-celled.

C. BOREALIS (*Wild Lily of the Valley*).—Rhizoma slender, creeping ; scape naked, 2–4-leaved at base ; leaves large, oval or oblong, petiolate, smooth and shining ; flowers few, rarely single, greenish yellow, nodding, in a terminal umbel ; berries blue. In damp woods, 8'–12' high. *June.*

13. **Polygonatum.**—Perianth tubular, 6-cleft at summit. Stamens 6, inserted at or above the middle of the tube, and inclosed in it. Ovary 3-celled. Berry globular. Cells 1-seeded.

1. P. PUBESCENS (*Solomon's Seal*).—Stem recurved at summit, round, rarely marked with a single furrow ; leaves.oval-lanceolate, glaucous and very slightly pubescent beneath, 3–5-veined, sessile ; peduncles axillary, smooth, nodding, 1–2-flowered ; flowers greenish. In woods and thickets, 1–2 feet high. *June.*

2. P. CANALICULATUM (*Large Solomon's Seal*).—Stem tall and stout, channeled, recurved ; leaves oblong-ovate, somewhat clasping at base, marked with numerous prominent veins ; peduncles nodding, smooth, 2–5-flowered ; flowers greenish. In rich, moist thickets, 2–5 feet high. *June.*

14. **Uvularia.**—Perianth nearly campanulate, deeply 6-parted. Stamens 6, short, adherent to the base of the segments. Anthers long. Style deeply 3-cleft. Capsule 3-angled, 3-celled, opening by 3 valves at top. Seeds few in a cell, arilled. Rhizoma creeping.

U. GRANDIFLORA (*Large Bellwort*).—Stems branching above, recurved ; leaves perfoliate, oblong ; flowers large, greenish yellow, lily-like, nodding, terminating the branches ; sepals smooth within. In rich woods, 8'–18' high. *May–June.*

15. **Asparagus.**—Perianth erect, 6-parted ; segments spreading above, with the 6 stamens at base. Style short. Stigma 3-lobed. Berry globular, 3-celled. Cells 2-seeded.

A. OFFICINALIS (*Asparagus*).—Stem erect, very branching, herbaceous ; leaves in clusters, pale green ; flowers small, axillary, yellowish green ; berries red. Cultivated for its young shoots, 2-4 feet high. *June.*

Order LXXXVII.—SMILACEÆ (*Smilax Family*).

Herbs, or shrubs, often climbing. Leaves reticulately-veined. Flowers diœcious. Perianth free from the ovary, 6-parted, regular. Stamens 6, inserted at the base of the segments. Ovary 3-celled. Fruit a globular, few or many-seeded berry.

Smilax.—Flowers diœcious or polygamous. Perianth campanulate, with 6 equal, spreading, deciduous segments. Stamens 6, attached at base, with short filaments. Anthers linear, attached by the base. Stigmas 3, nearly sessile. Berry globose, 1-3-celled, 1-seeded.

1. S. ROTUNDIFOLIA (*Greenbrier*).—Stem climbing, prickly, woody ; branches round or somewhat 4-angled ; leaves roundish-ovate, somewhat cordate at base, 5-nerved, abruptly acuminate, on short petioles, pale beneath ; berries round, bluish, with a glaucous bloom. *June.*

2. S. HERBACEA (*Carrionflower*).—Stem herbaceous, unarmed, angular, erect, recurved or climbing by tendrils ; leaves ovate, often roundish, 7-9-veined, mucronate or acuminate, usually cordate at base, smooth, paler beneath ; flowers yellowish green, in dense umbels of 20-40, on long peduncles, extremely fœtid ; berries dark blue, covered with a bloom. In moist thickets, with a disgusting, carrion-like odor to the flowers. *June.*

Order LXXXVIII.—JUNCACEÆ (*Rush Family*).

Herbaceous plants, generally coarse and grass-like, often leafless. Flowers usually greenish, small, dry, glumaceous, in cymose clusters. Leaves fistular, or else flat, often channeled, sometimes none. Perianth regular, in 2 series of 3 segments in each. Stamens 6, or 3. Ovary 3-celled, or 1-celled, because the placentæ do not reach the axis. Styles united into 1. Stigmas 3. Capsule 3-valved.

JUNCACEÆ. $\begin{cases} \text{Capsule 1-celled—}Luzula. \\ \text{Capsule 3-celled—}Juncus. \end{cases}$

1. **Luzula.**—Perianth persistent. Stamens 6 Stigmas 3. Capsule 1-celled, 3-seeded. Leaves grass-like.

L. CAMPESTRIS (*Field Rush*).—Leaves linear, hairy, especially on the margin ; flowers in little spikes, the central one being nearly sessile ; sepals acuminate, awned, longer than the obtuse capsule. In fields and open woods, 3'-10' high. *May.*

2. **Juncus.**—Perianth persistent, 6-parted. Stamens 6, rarely 3. Stigmas 3. Capsule 3-celled, loculicidal, many-seeded.

1. J. EFFUSUS (*Bog Rush*).—Scape erect, soft and flexible, striate, sheathed at base ; flowers in a sessile, very branching panicle, small, numerous, greenish ; sepals acute ; stamens mostly 3 ; capsule obtuse. In wet grounds, 2–3 feet high. *June–July.*

2. J. ACUMINATUS (*Bog Rush*).—Stem erect, round ; leaves few, round or nearly so ; flowers in many or few brownish, few-flowered heads, arranged in a panicle ; sepals linear-lanceolate, very acute, much shorter than the acute, triangular capsule. In bogs and along ponds, 10′–18′ high. *August.*

3. J. TENUIS (*Slender Rush*).—Stem very slender, erect, leafless, except at base ; leaves linear, setaceous, shorter than the stem ; flowers separate, rarely sessile, in a loose, somewhat umbelled, cymose panicle, with unequal branches ; sepals lanceolate, acuminate, longer than the ovoid-globose capsule.

4. J. BUFONIUS (*Annual Rush*).—Stem slender, leafy, often branching at base ; leaves channeled, very narrowly linear ; flowers greenish, remote, sessile, forming a spreading dichotomously branching panicle ; sepals lanceolate, much longer than the obtuse capsule. Along road-sides, 3′–6′ high. *June–August.*

Order LXXXIX. — Pontederiace.e (*Pickerel-weed Family.*)

Aquatic herbs. Flowers solitary, or spicate, arising from a spathe, or from a fissure in the petiole. Perianth tubular, 6-cleft, persistent and withering, colored, often irregular. Stamens 3, inserted on the throat of the perianth ; or 6, and variously attached to the perianth. Ovary 1–3-celled. Style 1. Stigma 1. Capsule 3 valved, 1 or many-seeded.

Pontederia.—Perianth funnel-form, bilabiate, upper lip 3-parted ; lower lip of 3 spreading divisions. Stamens 6, 3 inserted near the summit of the tube, and exsert ; 3 near the base (often imperfect) with very short filaments. Ovary 3-celled, 1-ovuled. ♃

P. CORDATA (*Pickerel-weed*).—Stem thick and stout, erect ; leaves mostly radical, cordate-sagittate, smooth and glossy green, petiolate ; flowers blue, in a dense spike, with a bract-like spathe ; anthers blue. In shallow water, 10′–18′ above water. *July.*

Order XC.—Cyperace.e (*Sedge Family*).

Herbs, usually perennial, coarse, grassy, cæspitose plants. Culms usually solid, without joints or nodes, mostly triangular. Leaves with entire sheaths, sometimes wanting. Flowers solitary, each in the axil of a glume-like bract. Perianth wanting, or else reduced to mere bristles. Stamens usually 3, sometimes 2, or 1. Styles 2–3, more or less united. Fruit an achenium.

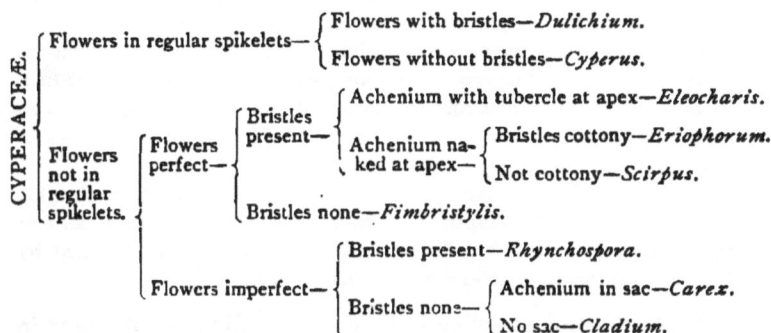

```
CYPERACEÆ.
 ┌ Flowers in regular spikelets— ┌ Flowers with bristles—Dulichium.
 │                               └ Flowers without bristles—Cyperus.
 │
 │ Flowers   ┌ Flowers   ┌ Bristles  ┌ Achenium with tubercle at apex—Eleocharis.
 │ not in    │ perfect—  │ present—  │ Achenium na-  ┌ Bristles cottony—Eriophorum.
 │ regular   │           │           └ ked at apex—  └ Not cottony—Scirpus.
 │ spikelets.│           └ Bristles none—Fimbristylis.
 │           │
 └           └ Flowers imperfect— ┌ Bristles present—Rhynchospora.
                                  │ Bristles none—  ┌ Achenium in sac—Carex.
                                  └                 └ No sac—Cladium.
```

1. Cyperus.—Spikelets few, many-flowered, in loose or dense clusters. Glumes arranged in 2 rows, decurrent at base. Stamens 1–3, usually three. Style 2–3-cleft. Achenium lenticular or triangular. Culm triangular, with 1–3 leaves at summit, forming an involucre to the umbel.

1. C. DIANDRUS (*Brown Sedge*).—Culm slender, usually decumbent; spikelets flat, oblong-lanceolate, acutish, 14–20-flowered, more or less in fascicles forming an umbel with 2–4 very short, sometimes unequal rays; glumes oblong, brown on the margin; stamens 2. Stem 6'–10' long. *August–September.*

2. C. STRIGOSUS (*Bulbous Sedge*).—Culm erect, leafy, tuberous at base; leaves broad-linear; umbel simple or decompound; spikelets narrow-linear, flat and few-flowered, at length reflexed, on spikes forming an umbel; scales oblong-lanceolate, yellowish; stamens 3; achenium narrow-oblong. In wet grounds, 1–2 feet high. *August.*

2. Dulichium.—Spikelets linear, compressed, arranged in 2 rows, on solitary, axillary peduncles. Glumes arranged in 2 rows. Perianth reduced to bristles. Stamens 3. Style 2-cleft above, the lower portion persistent, forming a beak to the compressed achenium.

D. SPATHACEUM (*Sheathed Sedge*).—Culm erect, simple, leafy, sheathed below; leaves alternate, linear, flat, short, arranged on the stem in 3 rows; spikes proceeding from the sheaths, consisting of 8–12 linear-lanceolate spikelets, in 2 rows; spikelets 5–9-flowered, rather long. Along rivers and borders of ponds, 1–2 feet high. *August.*

3. Eleocharis.—Spikes single, terminal. Glumes imbricated all round without much order. Perianth reduced to 3–12, mostly 6, rigid, persistent bristles. Stamens 3. Style 2-3-cleft, bulbous and persistent at base. Culms leafless, simple.

15

1. E. OBTUSA (*Spike Rush*).—Culms nearly terete in tufts ; spike globose, at length becoming somewhat cylindrical, obtuse, densely many-flowered ; glumes very obtuse, light brown, whitish on the margin ; achenium obovate, of a shining brown, surrounded by 6 bristles. In shallow water and muddy grounds, 6'–12' high. *July–August.*

2. E. PALUSTRIS (*Round Rush*).—Culm nearly terete ; spike oblong-lanceolate, acute, many-flowered, often obliquely attached ; glumes reddish-brown, achenium obovate, surrounded by about four bristles longer than itself, and crowned with a small, ovate, flattened tubercle. In shallow water and low grounds, 6'–20' high. *June–July.*

3. E. TENUIS (*Slender Rush*).—Culm very slender and wiry, 4-angled with concave sides ; spike elliptical, 20–30-flowered ; glumes ovate, obtuse ; achenium obovate, with a small, depressed tubercle, and surrounded by bristles. In wet meadows, 6'–12' high. *June–July.*

4. **Scirpus.**—Spikes cylindrical, clustered. Glumes imbricated in no order. Perianth of 3–6 bristles. Stamens 3. Style 2–3-cleft. Culms sheathed at base.

1. S. VALIDUS (*Bulrush*).—Culm tall, cylindric, filled with spongy pith, tapering above, dark green ; spikes oblong-ovate, numerous, arranged in a compound panicle ; glumes ovate, achenium obovate, mucronate, surrounded by 4–5 bristles. In wet grounds, 4–8 feet high. *July.*

2. S. PUNGENS (*Acute Club Rush*).—Culm acutely angled with concave sides ; leaves few below, channeled above, often 6'–8' long ; spikes ovoid, sessile, 1–5 in a cluster ; glumes ovate, 2-cleft, mucronate ; style 2-cleft ; bristles 2–6, shorter than the obovate, mucronate achenium. In ponds and streams. *July–August.*

3. S. ATROVIRENS (*Umbelled Club Rush*).—Culm obtusely triangular, rigidly erect, leafy ; leaves broad-linear, rough on the margin, flat ; spike ovoid, crowded in dense, globular, dark, dull-green heads, containing 10–20 spikes. *July.*

5. **Eriophorum.**—Spikes many-flowered. Scales imbricated all round without order. Perianth consisting of numerous, rarely 6, woolly, persistent bristles. Stamens mostly 3. Style 3-cleft.

1. E. POLYSTACHYON (*Cotton Grass*).—Culm rigidly erect, obscurely 3-angled ; leaves linear, flat, terminating in a triangular point ; involucre 2–3-leaved ; spikes about 10, on slender, nodding peduncles. Wool straight, nearly an inch long. Bogs and meadows, 1–2 feet high. *June.*

2. E. VIRGINICUM (*Brown Cotton Grass*).—Culm rather stout, rigid, nearly terete, leafy ; leaves long, flat, narrowly-linear, rough on the margin ; involucre 2–4-leaved ; spikes erect, crowded in a dense head ; wool dense, of a rusty-brown color, 3 or 4 times as long as the scale ; stamen 1. *July–August.*

6. **Fimbristylis.**—Spike several or many-flowered. Glumes imbricated in regular rows. Perianth of bristles none. Stamens 1–3.

Style 2–3-cleft, with a bulbous base, which is deciduous or persistent.

1. F. AUTUMNALIS (*Autumn Club Rush*).—Culms low, slender, compressed, tufted ; leaves narrow linear, flat, acute, shorter than the stem ; involucre 2-leaved ; spikes oblong, acute ; glumes ovate-lanceolate, brownish, mucronate ; stamens 2–3 ; style 3-cleft, entirely deciduous. In muddy grounds, 3′–8′ high. *August–October*.

2. F. CAPILLARIS (*Annual Club Rush*).—Culms nearly naked, capillary in dense little tufts ; leaves setaceous, shorter than the culms, with sheaths, hairy at the throat ; involucre 2–3-leaved ; spikes ovoid ; glumes oblong, brownish ; stamens 2 ; style 3-cleft, the bulbous base persistent. In sandy fields, 3′–6′ high. *August*. ①

7. **Rhyncospora.**—Flowers in ovate, several-flowered, loose spikes. Lower glumes usually empty. Perianth of 6 bristles. Stamens usually 3, Style 2-cleft. Achenia coherent with the bulbous, persistent base of the style.

1. R. ALBA (*White Beak Rush*)—Culm slender, 3-angled above, leafy ; leaves linear, very narrow, spikes lanceolate, white, in corymbose, axillary and terminal fascicles, on slender peduncles ; glumes lanceolate ; achenium ovoid, shorter than the bristles, with a slender beak or tubercle nearly as long as itself. *July–August*.

2. R. GLOMERATA (*Beak Rush*).—Stem slender, triangular, leafy ; leaves linear, flat, rough on the edge ; spikes oblong-ovate, in dense, very distant clusters, on long peduncles, sometimes in pairs from the same axil ; achenium obovoid ; bristles rough backwards. In wet grounds, 10′–20′ high. *July–August*.

8. **Cladium.**—Flowers polygamous, in a loose spike ; lower glumes empty ; terminal ones bearing a perfect or fertile flower. Perianth of bristles none. Style 2-3-cleft, deciduous. Achenium hard and corky, without a tubercle.

C. MARISCOIDES (*Twig Rush*).—Culm leafy, obscurely triangular, erect ; leaves narrow-linear, channeled ; spikes in heads or clusters of 5–8, arranged in small, compound cymes or umbels, on very long peduncles ; glumes light tawny-brown ; styles 3-cleft ; achenium ovoid-globose, with a short beak. In meadows and low grounds. 12′–21′ high. *July–August*.

9. **Carex.**—Flowers monœcious ; the two kinds are either combined in the same spike, or else arranged in different spikes ; rarely diœcious. Glumes single, 1-flowered, imbricated without order. Stamens 3, rarely 2. Stigmas 2–3. Achenium inclosed in a perigynium, or inflated persistent sac, contracted and closed at apex, and crowned with more or less of the persistent base of the style.

1. C. BROMOIDES (*Slender Swamp Sedge*).—Stem slender, leafy ; spikes several, approximate, oblong-lanceolate, alternate, lower ones barren, or all

often so; perigynia erect, lanceolate, acuminate, bifid, longer than the lanceolate glume. In swamps and meadows. 10′-20′ high.

2. C. CEPHALOPHORA (*Pasture Sedge*).—Stem rather stout; spikes 4–6, closely aggregated in an ovoid, bracteate head; the lower ones sometimes a little remote; perigynium compressed, broad ovate, green when mature. In dry fields and woods, 6′-12′ high.

3. C. SPARGANIOIDES (*Pale Sedge*).—Spikes 7–10, ovoid; upper ones more or less aggregated; lower usually distinct, and more or less remote; perigynium broad ovate, not nerved, rough on the margin, compressed, margined, diverging, hispid, green when mature; style short, swelling at base.

4. C. VULPINOIDEA (*Fox Sedge*).—Spikes numerous, very dense, generally branching, closely aggregate, forming an oblong, dense, compound spike; perigynium ovate, broad at base, small, compressed, margined, nerved with a short, bifid, abrupt beak, yellowish when mature.

5. C. STIPATA (*Three-cornered Sedge*).—Culm thick, sharply 3-angled, with concave sides; spikes 6–12, aggregated, lower ones often distinct; perigynium lanceolate, round and truncate at base, destitute of a margin, on a short stalk, nerved, tapering to a long, bifid beak. In wet grounds, 10′-18′ high.

6. C. ROSEA (*Rose Sedge*).—Culm low, slender; spikes several-flowered, 3–5, two uppermost usually approximate, the rest distinct, more or less remote; perigynia oblong, margined, rough on the margin, compressed with a bifid beak, diverging and stellate when mature, twice as long as the broad-ovate, obtuse glume, green at maturity. In low grounds, 8′-15′ high.

7. C. SCOPARIA (*Brown Sedge*).—Spikes 5–10, somewhat clavate when young, at length ovate, approximate, sometimes aggregated in a dense head; perigynia narrow lanceolate, nerved, margined, longer than the lanceolate, acuminate glume. In low grounds and meadows, 1–2 feet high.

8. C. STRAMINEA (*Winged Sedge*).—Spikes 3–6, roundish-ovoid, alternate, approximate; perigynia roundish-ovate, much compressed, broadly winged, with a short, abrupt, bifid beak, somewhat longer than the lanceolate glume. A common species in fields along woods, distinguished by its broad, broadly-winged perigynia.

9. C. POLYTRICHOIDES (*Dwarf Sedge*).—Culms very low, setaceous; spike linear, staminate above; perigynia few, alternate, oblong, somewhat triangular, obtuse, smooth, emarginate, twice as long as the ovate, mostly obtuse glume. In cold swamps and bogs, 2′-4′ high.

10. C. TENTACULATA (*Burr Sedge*).—Fertile spikes 2-3, ovoid-cylindrical, densely flowered, approximate, upper one sessile, the others on short peduncles; bracts leafy, much longer than the culm; perigynia much inflated, spreading, smooth, with a long, bifid beak, twice longer than the lanceolate, awned glume.

11. C. STRICTA (*Rigid Sedge*).—Staminate spikes 1-3, cylindric; fertile 2-4, long-cylindric, usually barren above, sessile; lower one often on a short peduncle; bracts rarely longer than the culm, auricled at base; perigynia ovate-acuminate, or elliptical, nerveless, not beaked. Tufts in wet grounds, 2–3 feet high.

Order XCI.—GRAMINEÆ (*Grass Family*).

Perennial herbs with fibrous roots, rarely arising from bulbs, sometimes annual or biennial. Culms cylindrical, usually fistular, closed at the nodes, sometimes solid. Leaves entire, usually narrow, alternate, with the sheath split from one node down to the next, usually with a membranous ligule between the base of the leaf and sheath. Flowers in spikes, racemes, or panicles, usually perfect, in 1 to many-flowered spikelets, composed of glume-like bracts in 2 rows. Outer bracts (*glumes*) 2, rarely 1, often unequal; the inner 2 immediately inclosing each flower (*paleæ*), alternate. Perianth none, or consisting of very small, membranous scales (*squamulæ*). Stamens 1-6, commonly 3, Anthers versatile. Ovary 1-celled. Styles 2. Stigmas 2, feathery. Fruit a caryopsis.

1. **Spikelets 1-flowered, with 2 glumes and 2 paleæ, or less.**

* GLUMES ABSENT, OR BOTH VERY MINUTE. PALEÆ 2. *Leersia.*

* * SPIKELETS IN PANICLES, OFTEN CONTRACTED, AND APPARENTLY RACEMOSE.

† *Paleæ awnless or with inconspicuous awns.*

Glumes equal, or the lower somewhat longer. Paleæ thin, not coriaceous, obtuse, upper smaller; one often minute or wanting. *Agrostis.*

Lower glume smaller, often minute, usually shorter. Paleæ surrounded by short hairs at base; lower 3-nerved, usually mucronate. Stamens 3. *Muhlenbergia.*

Spikelets with a rudimentary, plumose pedicel. Glumes mostly nearly equal, longer than the flower. Paleæ surrounded at base by a tuft of white bristles; lower paleæ awnless, or with inconspicuous awn attached to the back. Stamens 3. *Calamagrostis.*

Spikelets with 2 rudiments of abortive flowers, 1 on each side of the perfect flower. Paleæ shorter than the equal glumes. *Phalaris.*

†† *Paleæ with a conspicuous awn.*

Lower glume smaller than the upper, sometimes minute. Paleæ 2, herbaceous. Awn of the lower palea single, not jointed on its apex. *Muhlenbergia.*

Glumes not equal, often mucronate. Lower palea with 3 awns at tip, much larger than the upper. *Aristida.*

* * * FLOWERS ARRANGED IN SIMPLE OR NEARLY SIMPLE SPIKES.

† *Spike terminal.*

Spike simple, dense, cylindrical. Glumes mucronate or awned. Paleæ both present, awnless. *Phleum.*

Spikelets in threes at each joint of the rachis. Glumes side by side, mucronate, awned. Lower palea awned at apex. *Hordeum.*

†† *Spikes more than one.*

Spikelets loosely arranged in 2-4 rows, on one side of the flattened rachis. Spikes few. *Paspalum.*

Flowers oblong, crowded in clusters of 2-3, in 1-sided, approximate, slender spikes. *Panicum.*

2. **Spikelets 2-flowered, the upper perfect, the lower staminate or neutral, and in the latter case usually reduced to a single palea.**

Spikelets single, with the terminal flower perfect, not surrounded by bristles, arranged in racemes, panicles, or compound spikes. *Panicum.*

Spikelets surrounded by several or many bristles, and arranged in a cylindrical more or less compound spike. *Setaria.*

Spikelets in pairs, one pedicellate and sterile ; the other with the terminal flower perfect and awned, all arranged in spikes or racemes. *Andropogon.*

Spikelets in clusters of 2–3, 1 only with the terminal flower perfect and awned, in panicles. *Sorghum.*

Flowers monœcious ; barren flowers in terminal, numerous spikes ; fertile in solitary, lateral spikes, inclosed in a leafy involucre of numerous bracts. Styles long and slender. *Zea.*

3. **Spikelets 3-flowered, 2 of them imperfect or abortive, mostly reduced to single paleæ.**

Spikelets in contracted panicles ; lower flower abortive or obsolete ; middle flower perfect ; upper flower staminate. *Holcus.*

Spikelets in nearly simple, cylindrical spikes ; the two lateral flowers neutral, reduced to awned paleæ ; middle flower perfect, with awnless palea. *Anthoxanthum.*

4. **Spikelets more than 2 flowers, or if only 2 both are perfect.**

* SPIKELETS IN PANICLES OFTEN MUCH CONTRACTED AND SPIKE-LIKE.

† *Lower palea with a conspicuous awn.*

Spikelets 3–6-flowered, in dense, 1-sided clusters, forming a crowded panicle. Glumes awned, somewhat rough and ciliate on the back, as also the lower palea. *Dactylis.*

Spikelets 3–10-flowered, in open panicles, or contracted, spicate ones. Glumes unequal. Paleæ rounded on the back, awned at the apex. ·*Festuca.*

Spikelets 5–12-flowered, in loose, at length drooping panicles. Glumes unequal. Lower palea mostly 2-cleft, with a straight awn below the tip. *Bromus.*

Spikelets 7-flowered, in a loose, racemose panicle. Lower palea 2-toothed at apex, with an awn composed of the three twisted nerves rising from between the teeth. *Danthonia.*

Spikelets 3–6-flowered ; uppermost imperfect. Glumes unequal. Lower palea rounded on the back, with a twisted awn, consisting only of the middle nerve, and arising on the back below the 2-cleft tip. *Avena.*

†† *Lower paleæ awnless.*

Spikelets compressed, 2–7-flowered, in open panicles, clothed with more or less of a web-like down, but not bearded at base. Upper palea 2-toothed, deciduous, together with the larger, 5-nerved, lower palea. *Poa.*

Spikelets 3–10-flowered ; flowers not webbed nor bearded at base. Paleæ rounded on the back ; upper one adhering to the inclosed grain, which is somewhat downy at apex. *Festuca.*

Spikelets 3–7-flowered ; flowers with a copious silky beard at base ; lower flower with 1 stamen ; the others with 3 stamens. Glumes very unequal. *Phragmites.*

* * SPIKELETS IN SIMPLE, TERMINAL SPIKES.

Spikelets 3–10-flowered, attached singly to the joints of the rachis, with the side against it. Glumes 2, on opposite sides of the spikelet. *Triticum.*

Spikelets solitary at each joint of the rachis, 2–3-flowered. Glumes 2, opposite, subulate, shorter than the flowers. *Secale.*

Spikelets 2–7-flowered, in clusters of 2–4 at each joint of the rachis. Glumes side by side on the front of the spikelets, rarely none. *Elymus.*

1. **Leersia.**—Spikelets 1-flowered, compressed, perfect, in secund racemes, arranged in panicles. Paleæ 2, compressed, awnless, nearly equal ; lower broader. Stamens 1–6. Stigmas plumose. ♃

L. ORYZOIDES (*Cut-Grass*).—Culm rough backward, with hooked prickles, as also the lanceolate leaves; panicle sheathed at base, with numerous diffuse branches ; stamens 3 ; paleæ whitish, ciliate on the keel. In wet grounds, 1–2 feet high. *August.*

2. **Phleum.**—Glumes 2, much longer than the paleæ, mucronate or awned. Paleæ 2, unequal, truncate, included in the glumes. Stamens 3. ♃

P. PRATENSE (*Timothy. Herd's Grass*).—Culm erect, simple, smooth, sometimes bulbous at base ; leaves flat, glaucous ; flowers in a long, dense, simple, terminal, cylindrical spike ; glumes ciliate on the back, truncate, tipped with a very short awn ; anthers purplish. *June–July.*

3. **Agrostis.**—Spikelets 1-flowered, paniculate. Glumes 2, sub-equal, or the lower one larger, mostly longer than the paleæ. Paleæ 1–2 ; lower one larger, often awned ; upper one often wanting, or minute. Stamens 3. Caryopsis free. ♃

A. VULGARIS (*Red-top. Herd's Grass in Penn.*).—Culm mostly erect, slender ; leaves linear, with smooth sheaths ; ligule short, truncate ; panicle spreading, with slender, purplish branches ; lower palea 3-veined, twice as long as the upper, equaling the glumes, rarely awned. Culm 1–2 feet high. *July.*

4. **Muhlenbergia.**—Spikelets 1-flowered, in more or less contracted, often spicate panicles. Glumes acute, mucronate, persistent ; lower one smaller, often minute. Paleæ 2, usually bearded at base, inclosing the grain. Stamens 3. ♃

M. MEXICANA (*Mexican Drop-seed*).—Culm erect, ascending, very branching, and very leafy above ; leaves linear, short ; panicles numerous, contracted, densely flowered ; branches somewhat spicate ; glumes acute, unequal ; upper glume as long as the very acute, lower palea. In damp grounds, 1–3 feet high. *August.*

5. **Calamagrostis.**—Spikelets 1-flowered, often with a rudimentary flower, in a loose panicle. Glumes 2, nearly equal in length, surrounded by a tuft of white bristles. Lower palea awned on the back, below the tip, or awnless, mostly longer than the upper one. Stamens 3.

C. CANADENSIS (*Blue-joint*).—Culm simple, tall, stout, rigid ; leaves with smooth sheaths, linear-lanceolate ; panicle oblong, loose ; lower palea 3–5-nerved, equaling the tuft of hairs surrounding it, and with a very fine awn on the back below the tip, scarcely exceeding it. A reedy grass, 2–5 feet high. *July–August.*

6. Aristida.—Spikelets 1-flowered. Glumes 2, unequal, often mucronate ; paleæ pedicellate ; lower tipped with 3 awns ; upper palea much smaller, minute. Stamens 3. Stigmas plumose.

A. DICHOTOMA (*Poverty Grass*).—Culms tufted, dichotomously and very branching ; leaves very narrow, more or less revolute ; spikelets on clavate pedicels, in short, appressed racemes ; lateral ones minute, middle one as long as or longer than the palea, bent or twisted. In sandy fields, 6'–12' high. *August–September.*

7. Dactylis.—Spikelets 2-3-flowered, aggregated in dense clusters, forming a dense, branching, 1-sided panicle. Glumes unequal, carinate, mucronate. Stamens 3. ♃

D. GLOMERATA (*Orchard Grass*).—Culm erect, somewhat rough ; leaves linear-lanceolate, rough, somewhat glaucous ; panicle dense, with remote branches ; glumes very unequal ; anthers large, yellow. In shaded fields, especially orchards. 2–3 feet high. *June.*

8. Poa.—Spikelets compressed, ovate or oblong, few-flowered, in loose, open panicles. Glumes usually shorter than the flowers ; the lower one smaller. Lower palea 5-nerved, with a soft, web-like down. Upper palea smaller, 2-toothed, deciduous with the rest of the flower. Stamens 2-3.

1. P. ANNUA (*Low Meadow Grass*).—Culms low, mostly decumbent or spreading, somewhat compressed ; leaves short, smooth, with smooth sheaths ; panicle nearly as long as broad, with mostly solitary, at length horizontal branches ; spikelets 3-6-flowered, on very short pedicels, much crowded ; flowers slightly downy near the apex. Common everywhere, 3'–6' high. *May–November.*

2. P. PRATENSIS (*Meadow Grass*).—Culm erect, round, smooth ; leaves with smooth sheaths ; ligules short, truncate ; panicle pyramidal, diffuse ; branches spreading, in half-whorls of 4-5 ; spikelets 3-5-flowered, nearly sessile, densely crowded on the branches. *May–July.*

9. Festuca.—Spikelets 3-10-flowered, in open or racemose panicles. Flowers not webbed at base. Glumes unequal. Paleæ rounded on the back, entire, acute, and often awned at apex. Upper palea usually adhering to the ripe caryopsis. Stamens mostly 3.

1. F. TENELLA (*Slender Fescue Grass*).—Culm very slender, wiry, mostly simple ; leaves very narrow ; panicle simple, with racemose branches, contracted so as to resemble a spike ; spikelets 6-8-flowered ; palea tipped with an awn shorter than itself. In dry soils, 6'–12' high. *June–July.*

2. F. ELATIOR (*Tall Fescue Grass*).—Culm smooth, erect ; leaves broad-linear ; panicle open, loosely branching, with spreading, drooping branches ; spikelets crowded, 4-6-flowered ; lower palea awnless. In meadows, 2-4 feet high. *June.*

10. **Bromus.**—Spikelets 5–15-flowered, panicled. Glumes unequal, shorter than the flower, nerved. Lower palea convex, with an awn proceeding from below the tip. Upper palea convolute, adhering to the caryopsis at the groove. Stamens 3.

B. SECALINUS (*Chess*).—Culm smooth, erect; leaves flat, rough above, with margins and sheaths smooth; panicle spreading, with branches nearly simple, drooping; spikelets ovate, cylindrical, tumid, smooth, about 10-flowered; awn of the lower palea usually shorter than the spikelet. Troublesome in wheat-fields, 2–3 feet high. *June.*

11. **Pragmites.**—Spikelets 3–8-flowered. Flowers with tufts of white, silky hairs at base. Lower flower either neutral, or with a single stamen; the others perfect with 3 stamens. Glumes 2, shorter than the flowers, very unequal. Lower palea twice as long as the upper. ♃

P. COMMUNIS (*Water Reed*).—Culm erect, smooth, very stout; leaves very long, broad-lanceolate, glaucous; panicle very large, loosely branched, at length diffuse; branches in half whorls, erect, spreading, slender; spikelets 3–5-flowered, erect. On the borders of ponds and streams, 6–10 feet high. *July-August.*

12. **Triticum.**—Spikelets 3–8-flowered, spicate, each attached to a separate joint of the rachis. Glumes 2, nearly equal, opposite, ovate. Paleæ 2; lower one awned or mucronate; upper compressed. Stamens 3.

1. T. VULGARE (*Wheat*).—Culm erect, terete, smooth; leaves linear, somewhat rough above; spike somewhat 4-sided; spikelets crowded, about 4-flowered; glumes ventricose; awns longer than the flower. The most valuable species of grain cultivated. *June.*

2. T. REPENS (*Couch Grass*).—Culm erect, arising from a creeping rhizoma; leaves linear-lanceolate, somewhat rough or hairy above; spike compressed; spikelets remote, alternate, 4–8-flowered; glumes 5–7-veined, lanceolate. A weed in gardens. *June-August.*

13. **Secale.**—Spikelets 2–3-flowered, spicate, each attached to a single point of the rachis; lower flowers sessile and opposite; upper often abortive. Glumes opposite, shorter than the flowers. Lower palea ciliate on the keel and margin, tipped with a very long awn.

S. CEREALE (*Rye*).—Culm tall, erect, slender, hairy below the spike; leaves rough above and on the margin, glaucous; spike long, compressed; awns long and straight. A valuable grain, 3–6 feet high. *June.*

14. **Elymus.**—Spikelets 2–6-flowered, attached, 2–4 together at each joint of the rachis. Glumes 2, side by side, and not opposite, nearly equal, subulate. Lower palea convex on the back, mostly awned at apex. ♃

E. VIRGINICUS (*Wild Rye*).—Culm erect, stout, smooth; leaves broadlinear, rough, deep green; spike erect, rigid, thick, on a short peduncle;

spikelets mostly in pairs, 2–3-flowered, smooth ; glumes rough, lanceolate, tipped with a short awn ; lower palea with a short awn. In damp thickets, 2–4 feet high. *August.*

15. Hordeum.—Spikelets 1-flowered, with an awn-like rudiment at base on the inner side, 3 at each joint of the rachis ; lateral ones often abortive. Glumes 2, side by side, and not opposite, subulate. Lower palea long-awned at apex.

H. VULGARE (*Barley*).—Culm erect, smooth ; leaves broad-linear, nearly smooth ; spike erect, thick ; spikelets all fertile ; glumes shorter than the flowers ; lower palea very long-awned ; fruit arranged in 4–6 rows. Frequently cultivated, 2–3 feet high. *May.* ①

16. Danthonia.—Spikelets 3–8-flowered, in a spicate panicle. Glumes 2, longer than the flowers. Lower palea nerved, 2-toothed at apex, with a twisted awn arising from between the teeth ; upper palea obtuse, entire. ♃

D. SPICATA (*Oat Grass*).—Culms slender, erect, ascending, tufted ; leaves mostly radical, somewhat involute ; cauline leaves much shorter than the radical ones, erect, with sheaths hairy at the throat ; panicle slender, bearing a few, appressed, 7-flowered spikelets ; lower palea hairy. In dry fields, 1–2 feet high. *June–July.*

17. Avena.—Spikelets 2–7-flowered in panicles ; uppermost flower imperfect. Glumes 2, large and somewhat unequal. Lower palea rounded on the back, nerved, 2-toothed at apex, with a twisted or bent awn on the back.

A. SATIVA (*Oat*).—Culm erect, smooth ; leaves broad-linear, rough above ; panicle loose, with slender, drooping branchlets ; spikelets 2–4-flowered, on slender, drooping peduncles ; lower flower awned, rarely both awnless ; upper palea closely investing the grain. Valuable grain, 2–3 feet high. *June.*

18. Holcus. — Spikelets 2–3-flowered, in a contracted panicle. Flowers pedicellate, shorter than the glume ; lowest flower neutral, often wanting ; middle flower perfect, awnless ; upper staminate, with its lower palea awned on the back. ♃

H. LANATUS (*Velvet Grass*).—Culm and broad-linear leaves pale green, covered with soft, velvety down ; panicle oblong, dense-flowered, whitish, tinged with purple ; staminate upper flower with a recurved, short awn. In wet meadows, 10′–20′ high. *June–July.*

19. Anthoxanthum.—Spikelets 3-flowered, in a spicate panicle ; lateral flowers neutral, consisting of 1 hairy palea, awned on the back ; central flower perfect. Glumes 2, very unequal, the upper one larger and equaling the flowers. ♃

A. ODORATUM (*Sweet Vernal Grass*).—Culm erect, slender ; leaves short, pale green ; panicle spicate, oblong, with short, nearly simple branches ;

spikelets pubescent, green ; paleæ of the lateral flowers ciliate on the margin, one with a bent awn ; the other with a short, straight awn below the tip. An early grass in fields, fragrant when drying. *May–June.*

20. Phalaris.—Spikelets in dense panicles, with 1 perfect flower, and 2 neutral, abortive rudiments at base. Glumes 2, equal, longer than the 2 coriaceous, awnless paleæ which inclose the compressed grain. ♃

P. ARUNDINACEA (*Canary Grass*).—Culm tall, erect, simple or branching ; leaves lanceolate, rough on the margin ; panicle contracted, oblong, branching ; branches somewhat spicate, densely flowered, at length somewhat spreading ; rudimentary flowers hairy, much shorter than the perfect ones. In wet grounds, 2–5 feet high. *July.*

21. Paspalum.—Spikelets roundish, flat on the under side, convex above, in several rows, on one side of a flattened rachis, apparently 1-flowered, with only a single glume. Paleæ 2. Stigma plumose, colored. ♃

P. SETACEUM (*Wild Millet*).—Culm slender, decumbent or ascending, simple or branching at base ; leaves linear-lanceolate, flat, ciliate, and with the sheaths softly hairy ; spikes mostly solitary, very slender, terminal one on a long, very slender peduncle ; lateral ones often with included peduncles ; spikelets orbicular, in 2 rows. Weed in sandy fields, 1–2 feet long. *August.*

22. Panicum.—Spikelets panicled, racemed, or somewhat spicate, 1 neutral or staminate flower, and 1 perfect flower. Glumes 2, usually minute, sometimes wanting. Lower flower with a single palea usually awnless, sometimes awned, rarely with 2 paleæ ; upper flower perfect, with 2 awnless paleæ which inclose the free grain. Stigmas plumose, mostly colored.

1. P. SANGUINALE (*Crab Grass*).—Culm decumbent at base, rooting at the joints, then erect ; leaves linear-lanceolate, and with the sheath somewhat hairy ; spikes 3–12, in digitate, terminal clusters ; spikelets mostly in pairs, oblong ; upper glume shorter than the flower. Troublesome weed in cultivated grounds, 1–2 feet high.

2. P. CAPILLARE (*Hair Panic Grass*).—Culm erect, simple above, often branched at base ; leaves broad-linear, hairy as well as the sheaths ; panicle very large, pyramidal, very loose, with numerous capillary branches ; spikelets small, on long, slender pedicels ; neutral flower consisting of 1 palea, much longer than the perfect flower. In cultivated grounds, 1–2 feet high. *August.*

3. P. LATIFOLIUM (*Broad-leaved Panic Grass*).—Culm erect, smooth, nearly simple, usually bearded with soft hairs at the nodes ; leaves oval-lanceolate, cordate and clasping at base, nearly or quite smooth, except at the usually bearded throat ; panicle loose, short, on an exserted peduncle, with short, nearly simple, spreading branches. In thickets and damp woods, 1–2 feet high. *June–July.*

4. P. DICHOTOMUM (*Forked Panic Grass*).—Culm slender, erect, or somewhat decumbent, generally very branching, and somewhat dichotomous; leaves linear-lanceolate, flat; radical ones usually much shorter and broader, dark green, as also the sheaths, sometimes pale green, and more or less hairy; panicles compound, terminal and lateral, with spreading branches; lateral panicles often short and simple; lower flower neutral, with a single palea. In moist situations, 4'-20' high. *June-September.*

5. P. CRUS-GALLI (*Barnyard Grass*).—Culm stout, branching at base; leaves lanceolate, flat, rough on the margin, otherwise smooth, as also the sheaths; spikes alternate, compound, forming a dense panicle; lower flower neutral, with 2 paleæ; lower palea awned; awn rough, usually long. In rich, waste grounds. *August-September.*

23. Setaria.—Spikelets in compound, cylindrical spikes, furnished with 1 or more bristles, resembling awns; otherwise as in the Panicums proper. ①

S. GLAUCA (*Bottle Grass*).—Culm erect; leaves lanceolate, rough, hairy at base; spike dense, cylindric, of a dull-yellowish color when mature; bristles in clusters of 6-10, much longer than the spikelets. Weed in waste places, 1-2 feet high. *July-August.*

24. Andropogon.—Spikelets in pairs at the joints of the rachis, spicate or racemed, one of them pedicellate, barren, often rudimentary; the other with the lower flower neutral; the upper perfect. Paleæ thin and delicate; lower one awned at the tip. Stamens 1-3. ♃

A. FURCATUS (*Broom Grass*).—Culms erect, branching, nearly smooth; leaves linear-lanceolate, nearly smooth; radical ones very long; spikes straight, 3-6, hairy, digitate or clustered at the top of the culm, usually purple; spikelets approximate, hairy, appressed; sterile spikelets staminate, awnless; stamens 3. In dry soils, 4-6 feet high. *September.*

25. Sorghum.—Spikelets pedicellate, in clusters of 2-3, forming an open panicle; lateral ones barren; middle spikelet fertile. Glumes coriaceous; 2 lower flowers neutral; upper flower with 2 paleæ, the highest awned at the tip.

S. VULGARE (*Broom Corn*).—Culm tall, erect, solid; leaves lanceolate, pubescent at base; panicle very large, diffuse; branches long, slender, whorled, at length drooping; perfect spikelets with hairy, persistent glumes. Extensively cultivated, 6-12 feet high. ①

26. Zea.—Flowers monœcious. Barren flowers in terminal, clustered racemes. Spikelets 2-flowered; glumes 2, obtuse, nearly equal; paleæ obtuse, awnless. Fertile flowers lateral and axillary. Spikelets 2-flowered, 1 flower abortive. Glumes 2, obtuse. Paleæ awnless. Style 1, very long, filiform. Grains in 3-12, usually regular rows, compressed. ①

Z. MAYS (*Indian Corn*).—Culm erect, leafy, branching only at base; leaves very long, channeled, recurved, entire, 2-4 feet long; barren spikes 6-12, in terminal, nearly digitate clusters; fertile spikes, 1-4, nearly sessile, 6'-15' long, and even longer. Cultivated, 5-20 feet high. *July.*

GYMNOSPERMS.

Order XCII.—CONIFERÆ (*Pine Family*).

Trees or shrubs; the wood abounding in a resinous juice. Leaves scale-like, almost always evergreen. Flowers monœcious, or diœcious, destitute of calyx and corolla. Stamens 1, or more, often monadelphous, forming a sort of loose ament. Fertile flowers usually in aments, consisting of open carpellary scales, sometimes solitary and destitute of any form of carpel. Ovary, style and stigma wanting. Ovules naked, 1, 2, or more, erect, or sometimes turned downward. Fruit a strobile, or cone, sometimes drupaceous, or a solitary drupaceous seed. Embryo with 2, or frequently more cotyledons.

CONIFERÆ.
- Fertile flowers many; with bracts; cone large—
  - Leaves in clusters—*Pinus.*
  - Leaves single—*Abies.*
- Fertile flowers many; no bracts—
  - Fruit a cone—
    - Scales shield-shaped—*Cupressus.*
    - Scales not shield-shaped—*Thuya.*
  - Fruit a berry—*Juniperus.*

1. **Pinus.**—Flowers monœcious. Barren aments in spikes; stamens numerous, with very short filaments; anthers 2-celled; pollen consisting of 3 united grains. Fertile aments terminating the branches, consisting of imbricated scales, each with a deciduous bract outside, and a pair of ovules inside, attached to the base. Fruit a cone. Seeds sunk in hollows at the base of the scales. Cotyledons 3-12. Trees with acerose leaves.

1. P. RESINOSA (*Red Pine*).—Leaves in pairs, half cylindrical, elongated, arising from long sheaths; cones ovoid-conical, solitary or several together, half as long as the leaves; scales dilated in the middle, unarmed. A tall species. *May.*

2. P. RIGIDA (*Pitch Pine*).—Leaves in threes, with short sheaths; cones ovoid-conical, mostly in clusters; scales ending in short, recurved spines. Common in sandy soils. *May.*

3. P. STROBUS (*White Pine*).—Leaves in fives, slender, with very short, deciduous sheaths; cones cylindrical, pendulous, solitary, loose, somewhat curved. A tall tree, common in rich woods. *May.*

2. **Abies.**—Barren aments scattered, or clustered near the ends of the branchlets. Strobile small, round oblong, with thin, flat scales, not thickened at apex. Seeds winged. Cotyledons 3–9.

A. BALSAMEA (*Balsam Fir*). — Leaves narrow-linear, flat, obtuse, bright green above, silvery white beneath ; cones erect, large, cylindrical, violet-colored; bracts obovate, appressed. A slender and beautiful fir, in cold woods and swamps. *May.*

3. **Thuya.**— Flowers monœcious, the two kinds on separate branches. Barren flowers in small, ovoid aments; stamens attached by a scale-like filament, with 4 anther-cells. Fertile flowers consisting of a few loose scales, each bearing 2 erect ovules at the base inside. Seed winged. Cotyledons 2.

T. OCCIDENTALIS (*Arbor-vitæ*).—Leaves rhomboid-ovate, appressed, imbricate in 4 rows on the 2-edged, flat branchlets; cones nodding, oblong ; scales without joints, 1-seeded ; seeds with broad wings. An evergreen tree, common in swamps. *May.*

4. **Cupressus.**—Flowers monœcious, the two kinds on separate branches. Barren flowers in ovoid aments. Fertile flowers in globular aments ; scales peltate, in 4 rows, bearing several erect ovules at base. Cone globular, not opening till mature. Seeds compressed, winged. Cotyledons 2–3.

C. THYOIDES (*White Cedar*).—Leaves ovate, with a gland on the back, imbricated in 4 rows on the compressed branchlets, minute, dull green. A moderately large tree in swamps.

5. **Juniperus.**—Flowers diœcious, sometimes monœcious, in very small aments. Barren aments—scales peltate, bearing 3–6 anther-cells on their lower margin. Fertile aments globose ; scales few, fleshy, fruit a scaly berry containing 2–3 bony seeds. Cotyledons 2.

1. J. COMMUNIS (*Juniper*).—Leaves linear, subulate, spreading, mucronate, bright green below, glaucous above ; barren flowers in small, axillary aments ; fertile flowers axillary, sessile, in small, globular cones ; berries round-cylindrical, dark blue, and of a sweetish turpentine taste. A shrub with rigid branches, prostrate on the ground.

2. J. VIRGINIANA (*Red Cedar*).—Leaves in 4 rows, crowded in pairs or threes, on young or rapidly growing shoots, subulate, in other cases very small, triangular-ovate; closely imbricated ; barren flowers in small, oblong aments ; berries small, blue, with a glaucous bloom. A very small tree, of irregular growth. *April–May.*

GLOSSARY.

Abortion. Failure to grow. Imperfect development.

Acaules'cent. With no proper stem. See caulescent.

Acces'sory. Something superadded.

Accessory fruits. Those formed by the union of many flowers. 114.

Accum'bent. Lying against. Cotyledons are accumbent when the radicle is folded against their edges. 123.

Ac'erose. Needle-shaped. 9, 162.

Ache'nium. A small, indehiscent, seed-like fruit. 111.

Achlamyd'eous. Without floral envelopes. 45.

Acotyled'onous. Without cotyledons.

Ac'rogens. Growing by the terminal buds only. End-growers.

Acu'minate. Ending in a tapering point. 7.

Acute'. Ending in a sharp angle. 7.

Adhe'sion. The growing to each other of different floral whorls. 62.

Adnate'. United to. Adnate anther is one grown by its whole length to the connective. 16, 85.

Æstiva'tion. The arrangement of the parts of a flower in the bud. 132.

Affin'ity. Near relationship among plants shown by resemblances of character. 72–76.

Ag'gregate. Crowded densely together without cohering. 114.

Albu'men (of seeds). The tissue in which the embryo is imbedded and by which it is nourished. 118.

Al'gæ. Sea-weeds and other cryptogamous water-plants. 183.

Al'veolate. Honey-comb like.

Am'ent. A scaly, deciduous spike. 31.

Anat'ropous or anat'ropal. Reversed, so as to bring the micropyle to the hilum. 123.

Andrœ'cium. The stamens of a flower taken collectively. 90.

An'giosperm. A plant having its seeds within a pericarp. 184.

An'nual. Living one year only. 21.

An'nular. In rings.

An'ther. The thickened pollen-bearing head of a stamen. 81–87.

Antherid'ia. Organs of cryptogamous plants that correspond to the anthers of flowering plants. 180.

Apet'alous. Without petals.

Apical. Relating to the apex. 7.

Apocar'pous. Having separate carpels. 52.

Append'age. Any superadded part. 89.

Appendic'ular. Furnished with an appendage. 89.

.Appressed'. Pressed closely together but not adherent.

Aquat'ic. Living in water.

Arach'noid. Like cobweb.

Archego'nium or archegone. The same as pistillidia. 180.

Aril. A false or extra coat of a seed.

Armed. Bearing prickles, stings, etc.

Ascending ovules. Rising upward obliquely. 103.

Atten'uate. Becoming slender.

Auric'ulate. With ear-like lobes. 11.

Awn. Any bristle-like appendage. 174.

Axial. Situated in or belonging to the axis.

Axil (leaf). The space just above the point where the leaf is joined to the stem. 21.

Ax'illary. Starting from the axil of a leaf. 21.

Baccate. Berry-like ; pulpy.

Banner. The upper petal of a papilionaceous corolla. 39.

Barb. A hooked or appendaged bristle.

Ba'sal. Belonging to the base. 93.

Bas'sifixed. Attached by the base.

Beak. A prolonged tip.

Bearded. Having long, stiff hairs. 174.

Berry. A thin-skinned, indehiscent, juicy fruit, having its seeds imbedded in a pulpy mass. 109.

Biden'tate or bicus'pid. Two-toothed. 5.

Bien'nial. Living two years. 21.

Bi'fid. Two-cleft to about the middle.

Bila'biate. Two-lipped. See Labiate. 42.

Biloc'ular. Two-celled.

Bi'nary. Arranged in twos. 44.

Bisect'ed. Completely divided into two parts.

Biter'nate. Twice ternate.

Bladdery. Thin and inflated.

Blade. The flattened green part of a leaf. 1.

Bloom. The powdery, waxy covering of many fruits and leaves.

Botanical name. 175.

Bract. A small leaf or scale of a flower-cluster. 29.

Bulb. A leaf-bud with fleshy scales: usually underground. 28.

Bulblets. Small bulbs borne above-ground.

Cadu'cous. Falling off very early ; or, when the flower opens.

Calyp'tra. The cap or hood of a sporange. 181.

Ca'lyx. The outer floral whorl. 32, 33.

Cam'bium. A sappy layer between the wood and bark.

Campan'ulate. Bell-shaped. 41.

Campylot'ropous (ovule). Having the apex bent over close to the base. 105.

Cap'illary. Hair-like.

Cap'itate. Head-shaped. Having a globular apex.

Cap'sule. The pod of a compound pistil. The dry dehiscent fruit of syncarpous pistils. 113.

Car'pel. A simple pistil ; one of the parts of a compound pistil. 35.

Car'pophore. The stalk of a pistil within the flower. 154.

Caryop'sis. A one-celled, one-seeded fruit, with membranous pericarp united to the seed. 111.

Cat'kin. A scaly, deciduous spike ; an ament. 31.

Caules'cent. Having a stem. 25.

Cau'line. Belonging to the stem. 28.

Chaff. Small membranous scales.

Chala'za. The place where the nucleus and coats of an ovule grow together. 104.

Characters of plants. Their permanent features. 72.

Chlo'rophyl. The green coloring-matter of plants.

Cil'iate. Having a fringe of hairs on the margin.

Circumscis'sile. Divided by a circular line round the sides.

Cla'vate. Club-shaped.

Claw. The narrowed base or stalk of a petal. 38.

Cleft. Cut about half-way down.

Coher'ent or cohesion. The growing together of like parts. 52.

Col'umn. Results from the union of the filaments of a stamen with each other, or with the style and stigma. 52.

Colum'nar. Shaped like a column.

Coma. A tuft of hairs upon a seed.

Com'missure. The face by which two carpels cohere. 153.

Complete. Having all the parts. 44.

Compound pistil. Consists of several united carpels. 93.

Compressed. Flattened on two opposite sides.

Cone. An inflorescence formed mostly of scales, as the fruit of the pine. 164.

Con'ical. Round and decreasing to a point. 19.

Con'nate. When opposite leaves are grown together from the first, they are said to be connate. 25.

Connect'ive. The prolongation of a filament uniting the lobes of an anther. 81.

Conniv'ent. Converging. 91.

Contorted. Twisted together. 133

Con'volute. Rolled up lengthwise. 133.

Cor'date. Heart-shaped. 9.

Coria'ceous. Like leather in texture.

Corm. A solid bulb. 28.

Cor'neous. Like horn.

Corol'la. The inner whorl of floral leaves. 32.

Coro'na or **crown.** An appendage at the top of a claw. 43.

Cortical layer. A layer of the bark.

Cor'ymb. A kind of flat flower-cluster. 31.

Cotyle'dons. The first leaves of the embryo. 119.

Cremocarp. The fruit of umbelliferæ. 153, 111.

Crested. Having a crest-like appendage.

Cru'ciform. Cross-shaped.

Cryp'togams. Flowerless plants.

Cucul'late. Hooded.

Culm. The stem of grasses and sedges. 174.

Cu'neate. Same as cuneiform.

Cu'neiform. Wedge-shaped. 9.

Cu'pule. A little cup. The involucre of a nut.

Cus'pidate. Tipped with a sharp, stiff point. 7.

Cyme. A loose, irregular, definite inflorescence. 135.

Cyp'sela. An achenium with an adherent calyx-tube. 111.

Decan'drous. With ten stamens. 49.

Decid'uous leaves fall off in autumn. Flower-leaves that fall before the fruit develops are said to be deciduous.

Dec'linate. Turned downward or sidewise.

Decompound. Redivided many times.

Decum'bent. Prostrate, but tending to rise at the summit.

Decur'rent leaves are prolonged below their insertion on the stem. 25.

Decus'sate. Crossed in pairs.

Definite. Not exceeding twelve. 49, 115.

Dehis'cent. Bursting open, as of a pod or anther. 112.

Deflexed. Bent downward.

Depressed. Flattened.

Diadel'phous. Having the filaments grown together in two bundles. 51.

Dian'drous. Having two stamens. 49.

Dichlamyd'eous. Having both calyx and corolla. 45.

Dichot'omous. Regularly dividing by pairs. 121.

Dic'linous. With stamens in one flower and pistils in another. 45.

Dicotyle'dons. Plants in which the embryo has two cotyledons. 121.

Didyn'amous. Having two long and two short stamens. 50.

Diffuse. Spreading widely and irregularly.

Dig'itate. Palmately compound. 13.

Dimid'iate. Where one lobe or side is undeveloped.

Diœ'cious. With staminate flowers on one plant and pistillate on another. 46.

Discoid flower-heads are destitute of rays. 145.

Disk. A fleshy rim round the base of the pistil. The central part of a head of flowers. 69, 140.

Disk-florets. The inner florets of a flower-head. 141.

Dissected. Cut deeply into many divisions.

Dissep'iments. Partitions in an ovary. 94.

Dis'tichous. In two rows.

Distinct. Not coherent. 62.

Divar'icate. Widely divergent.

Dodecan'drous. Having twelve stamens. 49.

Dorsal. Pertaining to the back of an organ.

Downy. Covered with soft, short hairs.

Drupe. A stone-fruit. 110.

Dura'men. Heart-wood.

Dwarf. Diminutive.

Ech'inate. Having prickles.

Ellip'tical. Oblong with rounded ends. 9.

Emar'ginate. Notched at the summit. 7.

Embryo. The plant-germ in a seed. 118.

En'docarp. The inner coat of a pericarp. 108.

En'dogens. Inside-growing plants. 71.

En'siform. Sword-shaped.

Entire. Having a smooth margin. 4.

Ep'icarp. The outer covering of a fruit. 108.

Epider'mis. The skin of a plant.

Epig'ynous. Having the stamens inserted upon the ovary. 62.

Epipet'alous. When the stamens are inserted upon the corolla. 62.

Eq'uitant. Riding straddle. Fig. 96.

Erect' (ovules). Rising upright from the base of the cell. 103.

Essential organs. The stamens and pistil. 45.

Exalbu'minous seeds. Have no albumen. 118.

Excen'tric embryo is on one side of the center of the albumen. 123.

Ex'ogens. Outside-growers. 71.

Exot'ic. Of foreign origin.

Exsert'ed (stamens). Extending beyond the corolla. 90.

Ex'tine. The outer coat of a pollen-grain. 88.

Extrorse'. Facing outward. 85.

Fal'cate. Scythe-shaped.

Farina'ceous. Mealy in texture.

Fas'cicle. A close cluster. 25.

Feather-veined. Where the veins all arise from the midrib. 4.

Fenes'trated. Having chinks or slits.

Ferru'ginous. Resembling iron-rust.

Fertile. Bearing fruit. 45.

Fil'ament. The stem-like part of a stamen. 35.

Fil'iform. Thread-like. 87.

Fim'briate. Fringed.

Fis'tular. Hollow and cylindrical.

Flo'rets. The flowers of a flower-head. 141.

Folia'ceous. Belonging to or resembling foliage leaves.

Fol'licle. A simple carpel opening down the inner suture. 112.

Free. With no adhesion. 62.

Frond. The leaf of a fern. 177.

Fruit. The mature ovary. 106.

Frutes'cent. Shrub-like.

Fuga'ceous. Soon falling off.

Funic'ulus. The stem of an ovule or seed. 104.

Fur'cate. Forked.

Furrowed. Having longitudinal grooves or furrows.

Fu'siform. Spindle-shaped. 20.

Gamopet'alous. With the petals united. 37.

Gamophyl'lous. Of united floral leaves. 37.

Gamosep'alous. Formed of united sepals. 36.

Germination. The early development of a plant from the embryo. 119.

Gibbous. Tumid in places.

Glands. Small cellular parts, mostly on the surface, that secrete oil or other matters.

Glau'cous. Covered with bloom.

Glom'erule. A dense cluster.

Glume. The floral covering of grasses and sedges. 174.

Gon'ophore. A stem supporting stamens and pistil. 69.

Gyn'obase. A dilated base, or receptacle, bearing a many-celled ovary.

Gyn'ophore. The stem that raises the pistil above the stamens.

Gynosper'mous. Naked-seeded. 184.

Has'tate. With a spreading lobe each side of the base. 11.

Heptan'drous. Having seven stamens. 49.

Herba'ceous. Like herbage. Not woody.

Hermaph'rodite. Having both stamens and pistil.

Hesperid'ium. A fruit like the orange. 109.

Heterog'amous. Bearing flowers of different kinds as to stamens and pistil. 145.

Hexan'drous. With six stamens. 49.

Hi'lum. The scar left after separation of seed from placenta. 105.

Hirsute'. Hairy.

His'pid. Beset with stiff hairs.

Hoary. Grayish white.

Homog'amous. Bearing like flowers as to stamens and pistil. 145.

Homo'tropous. Curved with the seed. Fig. 382.

Hybrid. Cross-breed.

Hypog'ynous. Inserted under the pistil. 60.

Im'bricate. Overlapping.

Imperfect (flowers). Wanting either stamens or pistil. 45.

Incised. Cut deeply and irregularly.

Included. Inclosed within. 90.

Incum'bent (cotyledons). With the radicle folded back upon one of the cotyledons. 123.

Indefinite. More than twelve. 49.

Indehis'cent. Not bursting open. 109.

Indig'enous. Native to the country.

Inferior. Growing below some other organ. 59.

Inflated. Turgid and bladdery.

Inflected. Bent inward.

Inflores'cence. The arrangement of flowers on the stem. 29.

Insertion. The place and mode of attachment of a part to its support. 23.

In'ternode. The space between two nodes. 21.

In'tine. The inner coat of a pollen-grain. 88.

Introrse'. Facing inward. 85.

Involu'cre. The outer green circle of bracts of a flower-head. 140.

In'volute. Rolled inward. 22.

Irreg'ular dehis'cence. When seeds escape through chinks or pores or other irregular openings. 102.

Keel. A projecting ridge on a surface. 39.

La'biate. Two-lipped. 42.

Lacin'iated. Cut into deep narrow lobes.

Lactes'cent. Having a milky juice.

Lam'ina. The blade of a leaf.

La'nate. Woolly.

Lan'ceolate. Lance-shaped. 11.

Leg'ume. A pod which has two valves, and its seeds attached to one suture. 113.

Lentic'ular. Lens-shaped.

Li'ber. The inner bark next the wood.

Li'chen. A class of low plants ; known as rock-moss. 183.

Limb. The blade of a leaf or petal.

Lin'ear. Narrow with parallel margins.

Lip. A lobe of a labiate corolla.

Lobe. A certain marked-off portion of an organ. 6.

Loculic'idal dehis'cence. When seeds escape by the dorsal suture.

Lo'ment. A long pod with transverse divisions. 113.

Ly'rate. Lyre-shaped.

Male (flowers). Having stamens but no pistil. 45.

Marces'cent (floral whorls). Persisting in a dry and withered state.

Medul'lary rays. Rays of cellular tissue reaching from pith to bark.

Medul'lary sheath. A vascular thin layer around the pith.

Membrana'ceous. Of the texture of membrane.

Mer'icarp. Half the fruit of an umbellifer. 154.

Me'socarp. The middle layer of a pericarp. 108.

Mi'cropyle. The passage through the coats of an ovule or seed. 122.

Midrib. The main rib of a leaf. 2.

Monadel'phous stamens are united by their filaments into one set. 51.

Monan'drous. Having one stamen. 49.

Monochlamyd'eous. Having a calyx and no corolla. 45.

Monœ'cious. With staminate and pistillate flowers on the same plant. 46.

Mu'cronate. Tipped with a short, abrupt point. 8.

Multiple (pistil). Composed of several distinct carpels. 94.

Myce'lium. The vegetative part of a fungus. 182.

Nec'tary. A gland that secretes a sugary liquid. 71.

Nerves. The ribs and veins of leaves.

Neutral (flowers). Having neither stamens nor pistil.

Nu'cleus. The center of an ovule where the embryo is formed. 104.

Nut. A hard, one-seeded, indehiscent fruit. 111.

Ob. When prefixed to words, inverts their meaning ; as—

Obcor'date. The reverse of cordate, the notch being at the apex. 10.

Oblong. Longer than wide, with rounded ends. 9.

Ob'ovate. The inverse of ovate. 9.

Obsolete. Not apparent. Rudimental.

O'chreate. With sheathing stipules.

Octan'drous. With eight stamens. 49.

Oper'culum. The lid of a sporange. 181.

Organog'raphy. A description of the organs of plants.

Orthot'ropous (ovule). Straight. See Fig. 329. 105.

Os'mose. The tendency of fluids to intermix.

O'vary. The lowest part of the pistil, containing the seeds. 35.

O'vate. Shaped like an egg. 9.

O'vule. A rudimentary seed. 103.

Pa'lea. Chaff. A bract-like scale among florets. 175.

Pal'mate. A form of venation. 4, 13.

Pan'icle. A kind of flower-cluster. 31.

Pap'pus. A downy calyx. 147.

Paraph'yses. Filaments along with antheridia. 180.

Pari'etal placentation. Having the placenta attached to the walls of the ovary. 99.

Parted. Deeply divided into parts.

Ped'ate. Shaped like a bird's foot. 16.

Ped'icel. The flower-stalks in a cluster. 31.

Pedun'cle. The flower-stem. 29.

Pel'tate. Shield-shaped. 11.

Pentam'erous. Arranged in fives.

Pentan'drous. Having five stamens. 49.

Pentas'tichous. In five rows. 129.

Perfect flowers. Having stamens and pistil.

Per'ianth. Floral leaves. 33.

Per'icarp. The wall of the ovary. 107–109.

Per'igone. See page 180.

Per'istome. A fringe of teeth around the mouth of a sporange. 181.

Persist'ent. Remaining long in place.

Per'sonate. A labiate corolla with a projection in the throat. 45.

Peryg'inous. Having the stamens inserted upon the calyx. 62.

Pet'al. A leaf of the corolla. 33.

Pet'iole. Leaf-stalk. 1.

Pet'iolule. The stalk of a leaflet. 12.

Phyllotax′is. Leaf-arrangement. 126.

Pi′leus. A cap. The head of a fungus. 183.

Pin′na. A leaflet of a pinnate leaf. 13.

Pin′nule. One of the subdivisions of a pinna.

Pis′tillate. Having a pistil but no stamens. 45.

Pistillid′ia. The pistil-like organ of cryptogamia. 180.

Placen′ta. That part of the ovary which bears ovules. 95.

Plu′mule. The first bud of a growing plant. 119.

Pollen. The powdery contents of anthers. 88.

Pollin′ia. Pollen-grains cohering in masses. 89.

Polyadel′phous. Having the filaments grown together in three or more bundles. 51.

Polyan′drous. Having more than twelve stamens. 49.

Polyg′amous. Having pistillate, staminate, and hermaphrodite flowers on the same plant. 47.

Polysep′alous. With distinct sepals. 48.

Pome. A fleshy, indehiscent, many-celled fruit. 110.

Pri′mine. The outer sac of an ovule. 104.

Protecting organs (of flowers). The calyx and corolla. 45.

Pubes′cent. Having fine short hairs or down.

Pyxis. A pod which dehisces by the falling off of a sort of lid. 113.

Raceme′. A flower-cluster of one-flowered pedicels along a rachis. 31.

Rad′icle. Belonging to or coming from the root. 119.

Raphe. The connection between the base of the nucleus and of the ovule. 104.

Ray florets. The outer petal-like florets of a flower-head. 141.

Receptacle. The top of the peduncle supporting the flower or flower-head. 67.

Reflexed. Bent outward or backward.

Regular. Having all the like parts of the same size and form. 38.

Reniform. Kidney-shaped. 10.

Retic′ulated. Resembling net-work.

Retuse′. Blunt and slightly indented. 7.

Rev′olute. Rolled backward. 22.

Rhizo′ma. An underground stem. 26.

Rib. A part of the framework of a leaf. 2.

Rin′gent. Gaping open.

Root-stock. Root-like parts of stems on or under the ground.

Rosa′ceous. Having petals like a rose.

Ro′tate. Wheel-shaped.

Rugose′. Wrinkled.

Run'cinate. When the leaf-lobes point backward. 7.

Runner. A slender branch growing along the ground and rooting at the joints.

Sag'ittate. Shaped like an arrow-head. 10.

Salver-form. A form of corolla. 41.

Sama'ra. A dry, indehiscent, winged fruit. 112.

Scape. A flower-stem apparently arising from the ground.

Scorpioid. Curved like the scorpion's tail.

Sec'undine. The inner sac of an ovule. 104.

Segment. One of the parts of any cleft or divided body.

Se'pal. A leaf of the calyx. 33.

Septici'dal dehiscence. Opening through the partitions. 101.

Septif'ragal dehiscence. By the falling away of the valves. 102.

Ser'rate. With forward-pointing teeth. 4.

Ses'sile. Without a stalk. 2.

Seta. The stalk of a sporange. 181.

Seta'ceous. Bristle-like.

Sig'moid. Shaped like the letter S.

Sil'icle. A short, broad silique. 113.

Sil'ique. The peculiar pod of cruciferous plants. 113.

Simple pistil. Consisting of only one carpel. 93.

Si'nus. The space between two lobes. 6.

Soro'sis. A kind of multiple fruit. 114.

So'rus. A cluster of fruit-dots on the frond of ferns. 179.

Spat'ulate. Shaped like a spatula. 9.

Spike. An elongated flower-cluster with sessile flowers. 29.

Spike'let. A small spike.

Spi'nous. Thorny.

Spiral. Winding like the thread of a screw.

Spore-case or **sporange.** Cells containing spores. 179.

Spores. The powdery grain in cryptogamous plants corresponding to seeds. 171.

Spur. Any projecting appendage of flower-leaves. 43.

Squa'mulæ or **lodicules.** Minute scales at base of ovary in grasses. 174.

Stam'inate. Having stamens but no pistil. 45.

Stand'ard. Same as banner.

Ster'ile. Not producing seed. Barren or imperfect. 45.

Stipe. The stem of a mushroom. 183.

Stip'itate. Having a stipe. 69.

Stip'ule. One of the parts of a complete leaf. 2.

Stom'ata. Breathing-pores of leaves and other green parts.

Stri'ated. Grooved or channeled.

Style. The stem of the pistil next above the ovary. 35.

Su'bulate. Tapering like an awl. 9.

Suspended ovules. Hanging from the summit of the cell. 103.

Su'ture. The seam formed by the union of two margins. 94.

Symmet'rical. Having an equal number of parts in each set. 43.

Syncar'pous. United carpels. 52.

Syngene'sious. United anthers. 50.

Tendril. Any thread-shaped body used for climbing.

Terete'. Same as cylindrical.

Ter'nary. Arranged in threes.

Tetradyn'amous. Having four long and two short stamens. 49.

Thal'amus. The receptacle of the flower.

Theca. A spore-case. 179.

Thyrse. A compact panicle.

To'mentose. Covered with woolly hairs.

To'rus. Same as thalamus.

Trian'drous. Having three stamens. 49.

Tri'fid. Three-cleft.

Trime'rous. Arranged in threes.

Tris'tichous. In three ranks. 129.

Triter'nate. Thrice ternate.

Trun'cate. Abrupt ending of leaf-blade. 7.

Tu'ber. A thickened portion of an underground stem, having eyes or buds on its surface. 28.

Tuber'culated. Having a pimpled surface. 20.

Um'bel. A flower-cluster umbrella-like in form. 31.

Unguic'ulate. Having a claw.

Urce'olate. Urn-shaped.

U'tricle. An achenium with thin membranous pericarp. 111.

Vag'inule. The collar at the base of a sporange-stalk. 181.

Valv'ate. Opening by valves.

Valve. One of the parts of a pericarp or anther. 83, 95.

Ven'tral suture. The suture that looks toward the axis of the flower. 94.

Ven'tricose. Swelled out on one side.

Ver'satile. Attached by one point. 85.

Whorl. Leaves placed in a circle round the stem.

Zo'ospore. A spore of certain water-plants which moves by means of vibratile cilia.

INDEX.

THE END.

* BOTANY BY OBSERVATION. *

SCIENCE PRIMER OF BOTANY.

By J. D. HOOKER, C. B., P. R. S.

FULLY ILLUSTRATED.

18mo. Flexible cloth.

A very interesting and valuable little work, designed to supply an elementary knowledge of the principal facts of plant-life, together with the means of training beginners in the way to observe plants methodically and accurately.

FIRST BOOK OF BOTANY.

By ELIZA A. YOUMANS.

Designed to Cultivate the Observing and Reasoning Powers of Children.

The true objective method applied to elementary science-teaching. Plants themselves are the objects of study. The pupil is told very little, and from the beginning throughout he is sent to the plant to get his knowledge of the plant.

DESCRIPTIVE BOTANY.

By ELIZA A. YOUMANS.

A Practical Guide to the Classification of Plants, with a Popular Flora.

Introduces the pupil to the study of Botany by the direct observation of vegetable forms.

This book takes the place of the author's "Second Book of Botany," but provides a complete course in itself, no other book being necessary.

PHYSIOLOGICAL BOTANY.

By ROBERT BENTLEY, F. L. S., Prof. of Botany in King's College, Lond.

Prepared as a Sequel to "Descriptive Botany," by Eliza A. Youmans.

Designed to give an elementary account of Structural and Physiological Botany, or of the inner and minute mechanism and activities of plants.

It treats of what the parts of a plant are built up, and what functions they perform in its history as a living being.

HENSLOW'S BOTANICAL CHARTS.

Thoroughly Modified and Adapted for Use in the United States, by Eliza A. Youmans.

Six Charts mounted on rollers, containing nearly five hundred figures colored to the life, which represent twenty-four orders and more than forty species of plants. An invaluable aid in making the study of Botany interesting and attractive.

New York: D. APPLETON & CO., 1, 3, & 5 Bond Street.

HENSLOW'S BOTANICAL CHARTS.

Modified and adapted for Use in the United States,

By ELIZA A. YOUMANS.

---

*One of the most attractive, interesting, and instructive accessories
for the school-room ever published.*

A diagram system of illustration, as an aid in teaching a subject like botany, is so valuable that the publishers have been induced, at the risk of heavy expense, to issue this series of charts for school use.

In the plan of illustration adopted, the plant is first represented in its natural size and colors; then a magnified section of its flowers is given, showing the relations of the parts to each other, and also magnified views of the different floral organs. The charts contain nearly five hundred figures colored to the life, and which represent twenty-four orders and more than forty species of plants, showing a great variety of forms and structures of leaf, stem, root, flower, fruit, and seed.

The Charts are not designed to supersede the study of plants, but only to facilitate it. Their office is the same as the illustrations of the book; but they are more perfect, and bring the pupil a step nearer to the objects themselves. Many plant characters are so minute that they are difficult to find, and much is gained by referring first to the enlarged and colored representations. Besides this special assistance in object-study, they will be of great value in bringing into a narrow compass a complete view of the structures and relations of the leading types of the vegetable kingdom. While individual characters are distinctly shown, they are brought collectively before the eye, so as to be readily compared and contrasted with each other. As the natural objects are imaged with such fullness and truthfulness of detail, they will be found equally valuable to the beginner, the intermediate pupil, and the advanced student—equally useful in helping to a knowledge of the rudiments, and in guiding to the higher classifications of the science.

They can be used with any botanical text-books, and during the season of plants they should be upon the walls of every school-room where botany is studied.

D. APPLETON & CO., PUBLISHERS, 1, 3, & 5 BOND STREET, N. Y.

FOR ELEMENTARY SCIENCE STUDY.

HOW WE LIVE;

OR, THE HUMAN BODY, AND HOW TO TAKE CARE OF IT. An Elementary Course in Anatomy, Physiology, and Hygiene. By JAMES JOHONNOT, EUGENE BOUTON, Ph. D., and HENRY D. DIDAMA, M. D.

A text-book thoroughly adapted to elementary instruction in the public schools, giving special attention to the laws of Hygiene (including the effects of alcohol), with a special chapter on Alcohol and Narcotics by Dr. Didama.

INTRODUCTION PRICE, 40 cents.

FIRST BOOK OF CHEMISTRY.

By MARY SHAW-BREWSTER.

A course of simple experiments for beginners, giving great prominence to practical work by the pupil. The experiments are of the most elementary character, and the simplest apparatus is employed.

INTRODUCTION PRICE, 66 cents.

FIRST BOOK OF BOTANY.

Designed to Cultivate the Observing Powers of Children. By ELIZA A. YOUMANS. Revised edition. 12mo. 158 pages.

In this book the true objective method is applied to elementary science-teaching. Plants themselves are the objects of study, and the knowledge thus gained becomes at once accurate and of practical value as a preparation for study in other departments of science.

INTRODUCTION PRICE, 64 cents.

FIRST BOOK OF ZOÖLOGY.

By EDWARD S. MORSE, Ph. D., formerly Professor of Comparative Anatomy and Zoölogy in Bowdoin College. 12mo. 190 pages.

Professor Morse has adapted this First Book of Zoölogy to the pupils of the United States. The examples presented for study are such as are common and familiar to every school-boy—as snails, insects, spiders, worms, mollusks, etc.

INTRODUCTION PRICE, 87 cents.

Sample copies will be mailed, post-paid, to teachers or school-officers, for examination, at the introduction price. Send for full descriptive circulars, catalogue, "Educational Notes," etc.

D. APPLETON & CO., Publishers,

New York, Boston, Chicago, Atlanta, San Francisco.

THE ELEMENTS

OF

POLITICAL ECONOMY.

By J. LAURENCE LAUGHLIN, Ph. D.,

ASSISTANT PROFESSOR OF POLITICAL ECONOMY IN HARVARD UNIVERSITY.

NATIONAL prosperity depends, in no slight degree, upon the diffusion of satisfactory economic and political education; and, if properly presented, the average student in the high-school or academy can intelligently cope with public questions and the economic principles which underlie them.

In this book the elementary principles of political economy are presented in a plain and simple form. The main topics are treated; the fundamental principles are emphasized; but no effort is made to produce a detailed and exhaustive treatise; Socialism, Taxation, the National Debt, Free Trade and Protection, Bi-metallism, United States Notes, Banking, the National Banking System, the Labor Question, Co-operation, and other leading questions of the day are treated in a brief and simple manner.

12mo, 363 pages. Introduction price, $1.20.

Specimen copy, for examination, will be mailed, post-paid, to any teacher on receipt of the introductory price. Send for full descriptive circulars of the series of Science Text-Books.

D. APPLETON & CO., Publishers,

NEW YORK, BOSTON, CHICAGO, ATLANTA, SAN FRANCISCO.

THE NEW PHYSICS.

A Manual of Experimental Study for High Schools and Preparatory Schools for College.

By JOHN TROWBRIDGE,

PROFESSOR OF PHYSICS, HARVARD UNIVERSITY.

With Illustrations - - - - - *12mo. Cloth, $1.50.*

Prepared with special reference to the present advanced scientific requirements for admission to the leading colleges.

THE NEW PHYSICS is intended as a class manual of experimental
study in Physics for colleges and advanced preparatory schools.
It involves the use of simple trigonometrical formulas in experimental
demonstrations and in the discussions and mathematical computations of
various forms of energy.

In THE NEW PHYSICS, Professor Trowbridge has so presented the
subjects treated, theoretically and practically, as to furnish to the student
the means of rigid and thorough mental discipline, and at the same time
of acquiring that practical knowledge of the subject which will properly
prepare him for subsequent and deeper study in the sciences. The
modern tendency of physical science is carefully noted and clearly shown
by means of the illustrations employed and their mutual relations.

Professor Trowbridge's NEW PHYSICS is a successful and complete
refutation of the fallacy which has long prevailed among those who adhere
exclusively to the classics for purposes of mental discipline. Its text
shows that the mastery of certain definite and proportionate requirements
in the sciences, as requisites for college admission, calling for definite
attainment before entrance upon a collegiate course of study, will furnish,
in due proportion, that mental training and development which are a
necessary preparation for the broader training of the college curriculum.

THE NEW PHYSICS is also adapted to the use of colleges and special
training-schools, and will be found a convenient and practical text-book
for such institutions.

For sale by all booksellers; or sent by mail, post-paid, on receipt of price.

New York: D. APPLETON & CO., Publishers, 1, 3, & 5 Bond Street.